Alice Hamilton

Alice Hamilton at twenty-four, the year she graduated from medical school (1893).

ALICE HAMILTON
A Life in Letters

Barbara Sicherman

A COMMONWEALTH FUND BOOK
Harvard University Press
Cambridge, Massachusetts
and London, England
1984

LIBRARY OF CONGRESS CATALOGING IN PUBLICATION DATA

Sicherman, Barbara.
Alice Hamilton, a life in letters.

"A Commonwealth Fund book."
Includes index.
1.   Hamilton, Alice—1869–1970.
2.   Physicians—United States—Biography.
3.   Toxicologists—United States—Biography.
I.   Hamilton, Alice, 1869–1970.   II.   Title.
R154.H238S53   1984   616.9'803'0924 [B]   83-26521
ISBN 0-674-01553-3 (alk. paper)

This volume is published as part of a long-standing
cooperative program between Harvard University Press
and the Commonwealth Fund, a philanthropic foundation,
to encourage the publication of significant and
scholarly books in medicine and health.

To the memory of my mother and father

PREFACE

THIS VOLUME ORIGINATED nearly a decade ago when I discovered the letters of Alice Hamilton, the legendary pioneer of industrial medicine. It was late in the afternoon of a long day spent in pursuit of my subject when I arrived at the home of W. Rush G. Hamilton in Deep River, Connecticut. A telephone conversation two days earlier had led me to believe that Mr. Hamilton, Alice's cousin once removed and also her heir, had something in the way of family papers. But nothing had prepared me for the profusion of letters, photographs, and memorabilia that littered the dining room and spilled over into the rest of the house. It was the realization of every historian's dreams. It was also the first of several lodes that eventually yielded records of five generations of Hamiltons.

As I read over this historical treasure in the next days and weeks, it came to me that the best way to bring Alice Hamilton to public view was through her own words. Her letters provided an altogether more personal view than she had permitted herself in her autobiography, *Exploring the Dangerous Trades* (1943), written in her early seventies. A Victorian and an intensely private person, she did not believe in exposing family secrets or her inner self to public view. A reader of the manuscript for the Atlantic Monthly Press concluded from this reticence: "She has little of the devil in her certainly." But to friends and family she showed a different side, delighting them with her keen perceptions, intensity of feeling, and a wit that was sometimes wicked. These qualities were all apparent in her letters. In them she emerged as an engaging personality, at once earnest and funny, self-doubting and self-preserving, passionate and irreverent. In spare prose she captured the essence of scenes and people, while also conveying her complex reactions to them. And the early letters, written to her cousin and confidante Agnes Hamilton, recorded her efforts—never mentioned in her autobiography—to find a true vocation and to balance her personal ambition with her obligations to her family.

I originally intended to let the letters stand on their own, supplemented by an introduction and explanatory notes. But as I neared what then seemed to be the end of the project I reconsidered. Since Alice Hamilton was neither a literary figure nor a household word and had not been the subject of scholarly attention, there was no ready body of knowledge for readers to draw on. Moreover, letters are by their nature episodic, descriptive of fleeting moments, some less important than events left unrecorded or for which the record has been lost. How then were readers to interpret the letters—and the silences— of a woman who did not easily give herself away? Having intended to follow the letters with a full-scale biography, I had visited scores of archives, interviewed relatives and friends, and sought out additional information on Hamilton and the large cast of characters that appear in her letters. Contemplating this material, I decided that however powerful Alice Hamilton's letters were, understanding of them— and therefore of her—would be enhanced by an integrated work of biography and letters.

The result is a work in two voices: Alice Hamilton's and mine. Her voice is preserved in 131 letters written over a period of 76 years. Ranging widely in subject and mood, they resemble a series of snapshots that record the experiences and impressions of many isolated moments in a long life. But despite their fragmentary nature, the letters collectively create a lifeprint as characteristic of the individual as are fingerprints or an artist's brushstrokes. They also have a freshness and immediacy that autobiography, an act of reinterpretation (often many years after the event), cannot have. My part of the text is intended to provide both a life in brief and the context in which to read and interpret the letters. It fills in some of the inevitable gaps in a work as selective as this one, supplementing, amplifying, and in some instances modifying the self-portrait conveyed by the letters. As a biographer and historian, I have also sought to interpret the essential meaning of Alice Hamilton's life and career.

In selecting a manageable number of letters from the rich store I eventually collected (more than 1300 when I stopped counting), three considerations were paramount: that the letter be of biographical importance, that is, especially revealing of Alice Hamilton's character, modes of thinking, or way of life; that it be of high quality, that is, of literary merit, historical importance, or both; and that, taken as a whole, the letters reflect the range of her personal, professional, and political interests. I attempted no exact balancing of periods or subjects: there are many more letters concerning her medical training, peace activities during World War I, and political views during the McCarthy era than about her laboratory work or her early involvement in industrial medicine.

Specialists in her field may wish for less correspondence about Alice Hamilton's life and more about her work, the principal reason, after all, for her historical importance. But, as the letters reveal, her work, though central, was but one facet of a rich and varied life. Moreover, Hamilton's most important professional work in the early years took the form of government investigations, which are recorded in official reports and articles; the surviving correspondence is often sparse or perfunctory. The most revealing letters concern her efforts, at a time when the federal government had no powers of enforcement in the field, to use her personal influence to secure better working conditions in factories. A representative selection of the genre, mainly letters to businessmen, is included here. In my own text I have tried to assess the nature and significance of her contribution to industrial toxicology.

Since my goal was a readable volume rather than a definitive textual edition, I have silently corrected typos and spelling errors (few of the latter, since Hamilton was a good speller), removed inconsistencies of punctuation, and standardized the format of the place names and dates that appear at the beginning of each letter. At the same time, in order to preserve the flavor of the original, I have kept the British orthography that Alice Hamilton favored as a young woman and adhered to her custom of placing book titles in quotation marks rather than italics, as well as to her sometimes inconsistent practices in the matter of capitalization. Hamilton left many letters undated and sometimes misdated a letter by as much as two years; in such instances I have supplied the correct date in brackets and indicated the incorrect date in a note. With two exceptions—one a letter partly obscured by an inkblot, the other a long travelogue intended as a diary—the selected letters are published in their entirety.

The greatest challenge was Alice Hamilton's handwriting. (Fortunately she learned to type in her middle years.) I became adept at deciphering her scrawl and at interpreting her intention from the context, but even after considerable effort, some uncertainties remained, most notably in differentiating such sometimes interchangeable word pairs as "there" and "then." Rather than litter the pages with question marks, I made educated guesses in most cases. In some instances I have resorted to bracketed question marks, most often in the case of proper names, and, just once I believe, to "illegible," the editor's admission of defeat.

Alice Hamilton has been a large part of my life for the past decade, and it is with some regret that I loosen the ties that have bound me so closely to her for so long. But any sense of loss I feel is more than balanced by my eagerness to share with other readers a subject who has provided me with so much interest, puzzlement, wonder, and also laughter. On occasion I have been given pause by the realization that almost certainly Alice Hamilton would not have approved of my project. She was too private a person to have wanted her letters published. And she would have denied that they were of any value, except momentarily to those to whom they were written. But she was aware that scholars might some day wish to use the old family letters (presumably those of her paternal grandfather, founder of the American line of Hamiltons). And if she did not see herself as a person of historical importance, she might nevertheless have respected the interest of later generations in understanding the past of which she was a part. In any event, my scruples did not go very deep. I have too much of the historian-biographer's fascination with deciphering the lives of those who have gone before and attempting to discern patterns of meaning for our own age. I am also too much a product of that age to withhold material that so stunningly captures a memorable life.

CONTENTS

Introduction   1

I. The Hamiltons of Fort Wayne   11

II. Medical Training, 1890–1894   33

III. "I Shall Know, Being Old": Career and   88
Family, 1895–1897

IV. Hull House, 1897–1907   111

V. Exploring the Dangerous Trades, 1908–1914   153

VI. The War Years, 1915–1919   184

VII. The Harvard Years, 1919–1927   237

VIII. Elder Stateswoman, 1928–1935   311

IX. Semiretirement, 1935–1949   356

X. "Old-Old Age," 1950–1970   380

Abbreviations   418
Notes   419
Sources   441
Acknowledgments   448
Index   451

ILLUSTRATIONS

Frontispiece

Alice Hamilton at twenty-four, the year she graduated from medical school (1893)
>   Courtesy of Madeleine P. Grant and Michigan Historical Collections, Bentley Historical Library, University of Michigan

Following page 200

Gertrude Pond Hamilton
>   Courtesy of the Schlesinger Library, Radcliffe College

Montgomery Hamilton
>   Courtesy of the Schlesinger Library, Radcliffe College

"The Three A's"
>   Courtesy of Russell Williams

The Hamilton sisters
>   Courtesy of the Schlesinger Library, Radcliffe College

The Hamilton cousins
>   Courtesy of the Schlesinger Library, Radcliffe College

Medical clinic, University of Michigan
>   Courtesy of Madeleine P. Grant and Michigan Historical Collections, Bentley Historical Library, University of Michigan

U.S. delegates to the First International Congress of Women, The Hague, 1915
>   Courtesy of Swarthmore College Peace Collection

Alice Hamilton at fifty (1919)
    Courtesy of Madeleine P. Grant

Alice Hamilton around the time of her retirement from Harvard (1935)
    Courtesy of Madeleine P. Grant

Margaret and Alice Hamilton in Hadlyme, with portraits of female ancestors, about 1959
    Courtesy of the Schlesinger Library, Radcliffe College

Alice Hamilton at ninety (1959)

Alice Hamilton

HAMILTON FAMILY

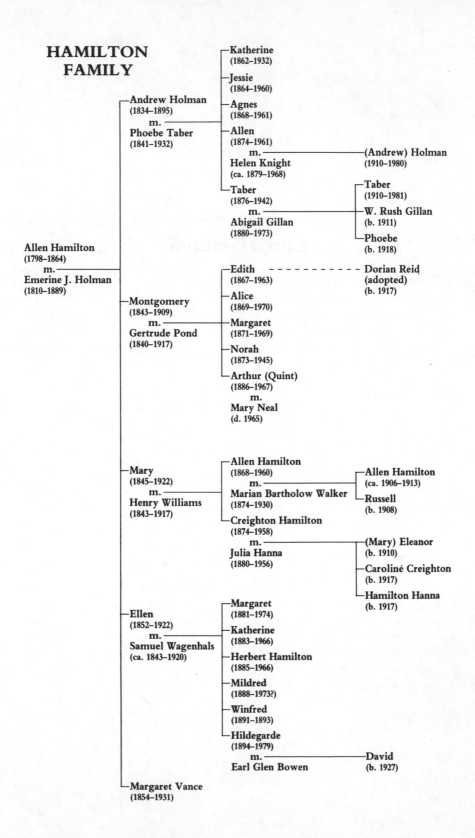

Allen Hamilton
(1798–1864)
m.
Emerine J. Holman
(1810–1889)

Andrew Holman
(1834–1895)
m.
Phoebe Taber
(1841–1932)

Katherine
(1862–1932)

Jessie
(1864–1960)

Agnes
(1868–1961)

Allen
(1874–1961)
m.
Helen Knight
(ca. 1879–1968)

(Andrew) Holman
(1910–1980)

Taber
(1876–1942)
m.
Abigail Gillan
(1880–1973)

Taber
(1910–1981)

W. Rush Gillan
(b. 1911)

Phoebe
(b. 1918)

Montgomery
(1843–1909)
m.
Gertrude Pond
(1840–1917)

Edith
(1867–1963) - - - - - - - - **Dorian Reid**
(adopted)
(b. 1917)

Alice
(1869–1970)

Margaret
(1871–1969)

Norah
(1873–1945)

Arthur (Quint)
(1886–1967)
m.
Mary Neal
(d. 1965)

Mary
(1845–1922)
m.
Henry Williams
(1843–1917)

Allen Hamilton
(1868–1960)
m.
Marian Bartholow Walker
(1874–1930)

Allen Hamilton
(ca. 1906–1913)

Russell
(b. 1908)

Creighton Hamilton
(1874–1958)
m.
Julia Hanna
(1880–1956)

(Mary) Eleanor
(b. 1910)

Caroline Creighton
(b. 1917)

Hamilton Hanna
(b. 1917)

Ellen
(1852–1922)
m.
Samuel Wagenhals
(ca. 1843–1920)

Margaret
(1881–1974)

Katherine
(1883–1966)

Herbert Hamilton
(1885–1966)

Mildred
(1888–1973?)

Winfred
(1891–1893)

Hildegarde
(1894–1979)
m.
Earl Glen Bowen

David
(b. 1927)

Margaret Vance
(1854–1931)

INTRODUCTION

NO ONE DID MORE during the first half of the twentieth century to alert Americans to the danger of industrial diseases than Alice Hamilton. A physician and reformer, she specialized in industrial toxicology, a branch of public health that investigates the hazards to workers of poisonous substances used in manufacturing. If she did not single-handedly found the field, as has sometimes been claimed, she was its foremost practitioner in the early years of this century. Trained in pathology and bacteriology, she became an expert on lead, TNT, and benzene poisoning and conversant as well with the intricacies of the storage battery, pottery, and lead smelting industries, among others. No narrow specialist, she combined exacting technical knowledge with an empathic concern for the sick and disabled, for anyone, in fact, less privileged than she. An unlikely blend of crusader and scientist, C.-E. A. Winslow, Yale's noted public health specialist, called her.[1]

Alice Hamilton belonged to the extraordinary generation of women who pioneered in the professions and social reform between 1890 and 1920. Collectively designated the first generation of college women, they left a record of substantial achievement in many professions, including such traditionally male preserves as medicine. Those identified with the social settlements were among the first to devise humane solutions for the problems of industrial society, and during the Progressive era at the beginning of this century they attained national as well as local influence. Jane Addams, often considered the exemplar of the group, provided a rationale for social service for a generation of men as well as women. She also drew to Hull House, the most

1

famous settlement, a strikingly talented group of women who together learned investigative and political skills that transformed them from well-meaning dilettantes into effective agents of social change. Besides Alice Hamilton, they included Florence Kelley, the crusading head of the National Consumers' League, Julia Lathrop and Grace Abbott, the first two chiefs of the Children's Bureau, and Edith Abbott and Sophonisba Breckinridge, prominent social work educators.[2]

Most women of this generation engaged in work that was sex-segregated, sex-typed, or both. Hamilton's colleagues in reform concentrated for the most part on activities, such as child labor and working conditions for women, that fell into the sphere generally acknowledged by both sexes to belong to women. Even among scientists, a large proportion found employment in the women's colleges, while those who worked in universities specialized in fields like child welfare and home economics that did not challenge the conventional division of labor.

Rather than concentrating on the problems of women and children, Alice Hamilton went straight to the center of the male world: the laboratory, the factory, and Harvard. Her work exposed her to adventures at odds with nineteenth-century definitions of womanhood. In her search for poisonous dusts, she jumped on table tops, interviewed workers in saloons, climbed dangerous catwalks, and descended deep into mine shafts, clad in workers' overalls. She also slept in unlocked mining shacks, was mistaken for a prostitute, and at least once wangled her way in through the back door of a large manufacturing company that feared her presence. "I used to have to do those things," she explained matter-of-factly to an interviewer late in life.[3]

This was unconventional work for a woman, especially one who struck others as "almost fragile" and of whom the young Walter Lippmann said: "In a platonic world she will represent the idea of feminism no amendments required. She has the most satisfying taste of all personalities I've ever met—wine and silver and homespun." She enjoyed her adventures and retained a fresh approach to politics even in her nineties. But she also held fast to the standards of her late-Victorian childhood. Modest about her achievements, she was addicted to the comforts of home and hearth. Scorning modernism in literature and art, she found the narrator in Camus' *The Fall* "shallow" and admired Pasternak's treatment of "passionate love" in *Doctor Zhivago*: "It is poetical, lovely, I only wish Faulkner, Hemingway and a lot of them would read it."[4]

Like most of her generation, she thought women were different from men, wiser, more empathic, more generous in sympathy. Admiring those of her sex who followed traditionally "masculine" pursuits in their own, more "feminine" ways, she disapproved the single-minded feminists of the 1920s not only because she feared they would undermine the protective labor legislation that she and her associates had worked so hard to attain but also because they assumed that women should in all respects emulate men. Alice Hamilton never fully resolved a question that has often confronted professional women: the degree to which women are and should strive to be like their male colleagues. While maintaining that women had their own distinctive contributions to make to society and adopting a sometimes apologetic and self-blaming tone about their public record to date, she insisted that women should have the same professional opportunities men did and recognized that "so far men have never given us a fair field." Indeed, despite her own path-breaking career, she remained pessimistic about the possibility of attaining fundamental changes in relations between the sexes. She thought it would be a long time before men would take care of children or accept women as equals, and in her sixties she observed: "My idea is that the American man gives over to the woman all the things he is profoundly disinterested in, and keeps business and politics to himself."[5]

Contrary to her own expectations, Alice Hamilton found that her sex did not stand in the way of her industrial work. Rather, she believed that factory owners, steeped in a culture that proclaimed women the guardians of the nation's morals, more readily tolerated a woman investigator. Certainly few could resist the combined force of her meticulous research and her uncommon powers of persuasion. She could extract hidden information from hostile and even rude managers, who grudgingly had to acknowledge that she was both fearless and scrupulously honest. There was also the dramatic incongruity that a woman of breeding and exquisite appearance should engage in such work, one that undoubtedly served her purposes well. After presenting Alice Hamilton with the National Achievement Award of the Chi Omega Sorority in 1936, Eleanor Roosevelt observed: "When she rose and said that in her field she had encountered very little opposition, your instinctive reaction was that no one could help wanting to be of service to her."[6]

Alice Hamilton became interested in industrial diseases at Hull House, where she went to live in 1897 while beginning her career as professor of pathology at the Woman's Medical School of Northwestern University. The settlement was an ideal place from which to observe the connections between environment and disease, and while establishing herself as a research pathologist and bacteriologist she also involved herself in neighborhood activities broadly related to health: she started a well-baby clinic, investigated the cocaine traffic and the causes of a typhoid epidemic, and took part in several city-wide ventures to improve the health of the poor. Then in 1908 Governor Charles S. Deneen appointed her to the Illinois Commission on Occupational Diseases. After preliminary investigation revealed the need for a full-scale study, she resigned from the commission to direct the survey.

This was the first state study of its kind, and her work on lead poisoning proved to be its outstanding feature. Pioneering in what is today known as shoe-leather epidemiology, she tracked down dispensary and union death records and persuaded workers and their families to cooperate—often a difficult task, since admission of illness meant almost certain loss of an already precarious livelihood. Her careful work established beyond a doubt which Illinois industries used lead and demonstrated shockingly high morbidity and mortality rates. She went on to become a special investigator for the federal Bureau (later Department) of Labor, a position she held from 1911 to 1920. From the start, her work was characterized by painstaking accuracy, scrupulous honesty—she would not state as a fact what she did not know to be true—and a disinclination for publicity. She preferred personal persuasion to the techniques of exposure perfected by contemporary muckrakers.

Her work was so exceptional that in 1919 Harvard broke tradition by appointing her assistant professor of industrial medicine. The medical school was inaugurating a degree program in industrial hygiene— the first of its kind. Had there been a qualified man, she would not have been hired. ("I was really about the only candidate available" was her own characteristically modest way of stating this.) But it was a new field, one of low standing in the larger profession, and consequently of little interest to most ambitious male physicians. To Alice Hamilton, by contrast, the work provided the opportunity for "definite achievement" for which she had been waiting. She brought to it unusual dedication as well as skill.[7]

The appointment attracted considerable attention. Not only was she Harvard's first woman professor in any field, but the university did not then admit women to its medical school—and did not until 1945.

4

Harvard's president A. Lawrence Lowell raised no objections, but, mindful of a recent appeal by women leaders (among them Alice Hamilton) to open the medical school to women, he maintained that the appointment "seems to me quite independent of co-education in the school." To reassure anxious members of the Harvard Corporation, Alice Hamilton agreed to forgo such perquisites of faculty standing as access to the Harvard Club and tickets to football games. She was such an anomaly that each year her printed invitation to commencement carried a handwritten warning: "Under no circumstances may a woman sit on the platform."[8]

Less interested in the trappings than the substance of her work, Alice Hamilton actually went to Harvard on her own terms. She rejected a suggestion that she study department stores and held out for an appointment that left her half of every year free to do her own work and to return to Hull House, which she considered home. If the arrangement kept her perpetually in search of supplemental income, it satisfied her instinct for independence and variety. Her unusual situation may also account in part—though only in part—for the fact that she was still an assistant professor when she retired in 1935.

During her Harvard years, Alice Hamilton became prominent in reform as well as medical circles. She had arrived at Hull House with her political views largely unformed but leaning toward the classical liberalism of her patrician family. Her new environment plunged her into a world of poverty, trade unions, radical talk, and social action far removed from the circumstances of her protected childhood in Fort Wayne, Indiana. Although she initially felt most comfortable on the political fence, she had a strong capacity for empathy and was deeply moved by the daily hardships of her poor and mainly immigrant neighbors as well as by recurring episodes of intolerance and police brutality.

World War I completed Alice Hamilton's transformation into an effective public person and a passionate advocate of her beliefs, some of which outraged superpatriots. Like Jane Addams, though less outspokenly, she opposed American entry into the war; her commitment to pacifism, she observed, finally enabled her "to leave muddled thinking behind." In 1919 she and Jane Addams visited Germany to observe the effects of starvation there. It was one of the most shattering experiences of her life. In the past she had responded to such feelings with a sense of her own helplessness. This time, on her return to the United States, she appealed for public support of the Quaker relief effort. When an influential Harvard colleague tried to silence her because a backer of the medical school considered her stand "pro-German," Alice Hamilton put the cause and her personal integrity ahead of financial security and professional standing.[9]

In later years, her forthright stands on such controversial subjects as birth control, recognition of the Soviet Union, and civil liberties sometimes embroiled her in public controversy, earning her a reputation for radicalism and a place on various lists of "subversives" from the 1920s through the 1950s. (The FBI was still keeping tabs on her in the 1960s, when she was in her nineties.) But although she was drawn to radicals of various persuasions, she was too pragmatic as well as too skeptical of authority to be a true believer. She could be devastatingly critical of government, business, the courts, but she could not join her radical associates in their wholesale condemnation of American institutions and values. Preferring persuasion to denunciation and tangible gains to ultimate (though far-off) victory, she was a true progressive in her conviction that the system could be improved by the exposure of evils and the efforts of forward-looking men and women to correct if not to eliminate them. But if she was unable to give wholehearted support to many radical causes, she stood by controversial individuals like Elizabeth Gurley Flynn when their own political rights came under attack and contributed to their defense from her meager resources. Nothing stirred her more than authoritarianism or injustice wherever she found it—in the Sacco-Vanzetti case, in Nazi Germany or Soviet Russia, in America during the McCarthy era—and it was in defense of the martyrs to individual conscience that she reached her highest eloquence.

During a long and graceful old age—she died in 1970 at the age of 101—she continued to reassess her positions in the light of new evidence. Her abhorrence of Hitler led her to support the American effort during World War II. And in 1952 she reversed her long-standing opposition to the Equal Rights Amendment after she was convinced that it would not endanger protective legislation for working women. Though inclined to believe that the pendulum had swung rather far in the direction of labor unions, she remained committed to the aspirations of working-class men and women and supported legislation on their behalf. In her nineties, she deplored the renewed attacks on "subversives" and protested the Vietnam war.

Those who knew Alice Hamilton in her later years were impressed by her intelligence and vitality, as well as by her passionate commitment to her beliefs. Two comments suggest something of the reverence in which she was held. A former student ranked her, along with Harvey Cushing and Walter B. Cannon, as one of his three most brilliant professors at Harvard. And Felix Frankfurter, a friend for half a century and the target of many of her reproofs, considered her "the finest combination of exquisiteness and expertness" he knew. She was calm and gracious but also forceful on behalf of her beliefs (though never for herself). With her aristocratic and mellow voice, and her sure sense of the ridiculous, she could gently deflate even the most seasoned opponent. Her self-sufficiency made her seem a little remote to some; others admired her serenity. An admirer of the "disinterested" motives and wisdom in which many of her fellow progressives also believed, she probably cultivated this detachment.[10]

This then was the public Alice Hamilton. But spirited and open as she was in most respects, she had a deep reticence about herself that intrigued the discerning observer. Elizabeth Glendower Evans, a Boston reformer and friend, found her in maturity "a perfectly harmonious woman," but wondered: "Had she no storms in her youth, no warring loyalties, no cravings for gaieties, for husband and children? When her classmates went home . . . and used their lives as other women had used theirs since the world began, why did she enter a medical school as if it were a fate ordained? One asks these questions; but one gets no answer."[11]

The discovery of a large body of personal letters makes it possible, if not to answer these questions entirely, at least to perceive a more complex person than the public biography conveys. The letters reveal the storms and warring loyalties, if not the craving for a husband, suspected by her colleague. Written mainly to her cousin and lifelong friend Agnes Hamilton, the early letters chart Alice Hamilton's evolution from a spirited but highly uncertain young girl into the principled, risk-taking woman of maturity. So striking was this transformation that Harriet Hardy, who collaborated with Alice Hamilton in bringing out a new edition of her textbook *Industrial Toxicology* in 1949 and saw her frequently in later years, did not recognize in the early letters the Alice Hamilton she had known.[12]

The letters document how difficult it was for even an unusually privileged and gifted woman to attain a sure sense of personal and professional autonomy. Despite high purpose, a family that for the most part supported her aspirations, and the encouragement of her male professors, it took Alice Hamilton many years to overcome the constraints, in her case more internal than external, to full realization of her goals.

The search for a meaningful way of life and work is a human rather than an exclusively female problem. But to educated women of the late nineteenth century it proved particularly problematic. Members of an elite group—less than 3 percent of the population, male or female, attended college in 1890—many escaped the pressures to conform to the social conventions of the middle class. Daughters in families like the Hamiltons often strove to live up to a "family culture" that fostered learning and moral goodness rather than conventional social success. But it was easier to know what to avoid than what to become. The first generation of women to seek careers outside the home, they had to create their own models. Their sense of being different, while sometimes disturbing, helped them to persist against external and internal obstacles. The latter, including pressure to sacrifice oneself to "the family claim" or the temptation to fall back on it when things went badly, the burden of overwork, ill health, and depression, and profound self-doubt, often proved most difficult to overcome.[13]

Alice Hamilton's letters richly convey the sense of specialness that characterized so many of these women, the sense of being uniquely privileged and thus obliged to prove oneself worthy—to society, family, above all to self. Like other professional women of her generation, she insisted that a woman must choose between career and marriage. A good Victorian, she did not cease to glorify the family claim for those who chose it: she simply decided it was not for her. Having thus declared herself, she must have been profoundly disturbed to find that her chosen career brought few rewards. She had early rejected clinical work, but even the scientific career for which she had so thoroughly prepared herself proved far from emotionally satisfying. Since the letters also reveal a wild, if vicarious, romanticism, it is also possible that she found it difficult to cut herself off from the full range of women's experiences, particularly from children, whom she found infinitely appealing throughout her life. Perhaps this helps to account for her eagerness to put the storms and uncertainties of youth behind her and for her fascination with the words of Robert Browning: "I shall know, being old."

The most surprising revelation of the early letters is the persistent strain of self-deprecation. Personal imperatives undoubtedly contributed to this self-doubt, for Alice Hamilton had exceptionally high standards, the consequence of her family's expectations and her own eagerness to excel. She also had models in the family to whom she compared herself unfavorably; her older sister Edith was intellectually precocious and her cousin Agnes almost oppressively good. But her self-doubt was not the product of a naturally morbid temperament. Rather, it was and is so common among women, even those who are apparently successful, that it must be considered in large measure culturally induced, rather than simply an individual difficulty. In the late nineteenth century, religious and cultural norms encouraged women to be self-sacrificing and humble. They often gained a sense of their own competence only after years of experience during which they repeatedly proved to themselves that they could indeed do a new task well. For those who remained single, lingering doubts about their course may not have been fully resolved until the approach of middle age.

The letters of the impressionable young Alice Hamilton are fascinating in themselves. They also highlight the achievements of the mature woman. She never entirely overcame the tendency to self-doubt. (In her early sixties, she called a laudatory article "a beautiful fiction which overwhelms me with shame. I feel myself a bluffer of the worst kind." And a few years later she observed: "You see, I simply cannot believe that I am a person of more than ordinary ability, though I know that chance has given me a more than ordinarily interesting life.") But she learned to master the emotional impulses that had earlier overwhelmed her and to utilize them in pursuit of her own goals. More than most individuals, Alice Hamilton did what she wanted to do. Without avoiding responsibility for family and friends, she followed her own course, even when it was not entirely clear. What is perhaps most remarkable is that she managed to move far to the left of her family's political values while remaining close to its members. In her mature years, she arranged her life so that it afforded both variety and the order that she considered conducive to well-being. If she did not seek honors and recognition, neither did she reject them. Of the small typeface in which her name appeared on the title page of her last book, she wrote: "It is all right to be modest, but one does not want to be anonymous."[14]

Alice Hamilton emerges from these letters a fully realized individual. Her professional achievements were matched by a rich private life. If this life excluded husband and children, it encompassed rewarding personal friendships and unusually close family ties. She took intense pleasure in life's offerings, both grand and simple, waxing eloquent about her garden, a bicycle ride, a concert, a walk through a bitter storm. Most important were the friends and relatives with whom she probed the human condition or shared quiet moments.

From the time she was fourteen, when she expressed a desire to move outside the tightly knit family circle, those who knew her well or casually remarked on Alice Hamilton's thirst for new experiences and her desire to know everything. She thought her curiosity sometimes bordered on impertinence. Certainly it carried her down avenues customarily not even glimpsed by women of her class and allowed her to approach new situations remarkably free of preconception. Her passion for the concrete and her power for visualizing help to account for the unusual interest of these letters. So does the exquisite self-consciousness that permitted her to convey the intensest emotions while maintaining sufficient detachment to observe the foibles of others—and her own as well.

Separated for long periods from those she loved, Alice Hamilton sought, by recreating the internal and external realities of her life, to eliminate the distance between them. Her sisters remarked on the charm and descriptive power of her letters. The stranger, reading them many years later, can share her experiences to a remarkable degree. Alternately humorous, gossipy, and despairing, the letters of Alice Hamilton convey the intensity of her desire both to know life and to make her own existence count. They also provide a glimpse into the lives of her contemporaries: of the unwed mothers and poor families whom she met in her student and Hull House days, and of the women of her own class struggling to make their way in the world. Although Alice Hamilton was in no sense a "representative" woman, her powers of observation and her fresh responses to people help to make these letters also a journal of her times.

I

THE HAMILTONS
OF FORT WAYNE

A LICE HAMILTON'S DEEPEST ATTACHMENT, from childhood to old age, was to her family. It was a privileged, cultured, and self-sufficient family with a sense of its own uniqueness that deeply impressed its younger members. Besides her parents and four siblings, in Alice's childhood the "home people" included the Fort Wayne households of three sets of cousins and that of her widowed grandmother and unmarried aunt, all of whom lived nearby. In later years, the Hamilton sisters and their closest cousins recreated the pattern in rural Connecticut by purchasing summer homes close to each other.[1]

Most important to Alice were her sisters. Born in 1869 in New York City (her mother's former home), Alice was the second of four girls born in the short span of six years. The sisters were close companions in childhood, and as young adults they missed each other even during short separations. They took turns pursuing their professional goals in the face of diminishing financial resources. They also divided up family responsibilities, although Alice, whose presence brought solace to others, and who was also the healthiest as well as a trained physician, did more than her share of nursing. In later years the Hamilton sisters often traveled, lived, and even worked together. When they spoke of their childhood, it was usually as "we" rather than "I."

By any standard they left a remarkable record. Edith (1867–1963), eighteen months older than Alice, became an internationally renowned writer whose vivid prose and timely connections between past and present made the classics accessible to a wide public. Intellectually precocious, as a child she was a "passionate reader" and "natural

11

storyteller." After graduating from Bryn Mawr College and studying in Germany, she put aside her plan to earn a Ph.D. in classics; in 1896 she became headmistress of the Bryn Mawr School in Baltimore, which she developed into a rigorous college preparatory school for girls. Only after she retired in 1922, at the age of fifty-four, did she begin her phenomenal career as a writer; by the time of her death in 1963, *The Greek Way* (1930) had sold about 775,000 copies and *Mythology* (1942) nearly 1,700,000. High-strung and moody as a young woman, Edith blossomed personally as well as intellectually in her later years. She also attained "dignity" and "manner," traits to which she had aspired as a young woman, striking others as almost regal in bearing.[2]

Margaret (1871–1969), two and a half years younger than Alice, also graduated from Bryn Mawr College, after which she studied biology in Paris and Munich and anatomy at the Johns Hopkins University Medical School. But she gave up her intended career in science, probably in large part because of the protracted sequelae of a broken hip, and followed Edith to the Bryn Mawr School, where she first taught science and then headed the primary school; shortly before her retirement in 1935, she became acting headmistress. Alice considered her the "quiet, stable, thoughtful one among us," traits she attributed to her sister's frequent childhood illnesses. Although overshadowed by her older sisters, like them Margaret had strong views on people and politics and remained vitally interested in both throughout her long life; some of her cousins' children and grandchildren found her the most companionable of the sisters.[3]

Norah (1873–1945), whose artistic talent manifested itself early, was according to Alice their father's favorite. After studying at the Art Students' League in New York, she spent two years in Europe, where she worked with James McNeill Whistler, among others. "Vivid and intense" as a child—"with her it was rapture or tragedy"—she suffered a severe breakdown in her mid-twenties and subsequent periods of incapacitation. Nevertheless she went on to become an artist, doing bold etchings, lithographs, and drawings of working-class and urban life in the manner of Käthe Kollwitz. She illustrated several of Jane Addams' books, as well as Alice's autobiography, and pioneered in art education for underprivileged children at Hull House and in New York City.[4]

The only brother, Arthur (1886–1967), some thirteen years younger than Norah, was nicknamed Quintus or Quint by his sisters after a neighbor rashly suggested to their father that he be called Primus. Like his sisters, he had a career in education. After earning a Ph.D. in romance languages at Johns Hopkins University, he became a popular teacher of French and then Spanish at the University of Illinois at Urbana; in later years he also served as adviser to foreign students there. He coauthored a French and a Spanish grammar and wrote *Sources of the Religious Element in Flaubert's Salammbô* (1917) and *A Study of Spanish Manners, 1750–1800, From the Plays of Ramón de la Cruz* (1926). The only one of the family to marry, he had no children.[5]

The circle of cousins numbered seventeen children, of whom eight were close enough in age to be playmates. "The Hamilton family Junior," as they called themselves, devised not only their own games but a code of conduct that set them off from "the grownups." They had boisterous good times playing endless rounds of Robin Hood, Knights of the Round Table, and the siege of Troy. The cousins were virtually everything to each other, for, except when they attended the First Presbyterian Church, the young Hamiltons did not associate with "outsiders."[6]

From earliest childhood, Alice formed exceptionally close bonds with two of her cousins, Agnes Hamilton (1868–1961) and Allen Hamilton Williams (1868–1960). "The three As," born within five months of one another, often led the others in play; they also got into the most scrapes and had their own secrets. Vowing eternal love and faith, as adolescents and even as young adults they looked to each other for inspiration, guidance, and support. Agnes encouraged Alice's choice of medicine as a career when her cousin's immediate family opposed it. And Allen, also preparing to be a physician, urged Alice to accept a position to which her family had raised objections.

Life for the Hamilton cousins centered on the family estate, which occupied approximately three city blocks. Three elegant houses—appointed with marble fireplaces, black walnut staircases, book-lined libraries, and lovingly tended French gardens—attested to the family's social standing. The Homestead, one of the first mansions in Fort Wayne, belonged to the family matriarch; the others—the Red House and the White House—to her two sons, Andrew Holman Hamilton and Montgomery Hamilton, the fathers, respectively, of Agnes and Alice.

Although Hamiltons of the second and third generations lived largely on inherited money—Agnes once estimated that twenty-five persons depended on it—they did not think of themselves as wealthy. They

13

had servants, of course, but since most of their inheritance was in the form of real estate, the Hamiltons always felt hard pressed for ready cash and several of the households suffered severe financial distress during the economically depressed 1890s. It was the family's intellectual and religious tradition rather than its wealth that most impressed its younger members. Among Fort Wayne's leading families, the Hamiltons were probably the most cultured. The women of Alice's generation valued intellectual achievement and moral purpose. They looked down on their leisure-class counterparts, whose chief goals were social success and early marriage, and disdained the artifice in dress and behavior favored by society women. Hamiltons of both sexes mastered Greek and Latin, as well as French and German, and often pursued such studies on their own initiative. They observed the Sabbath strictly, forgoing "worldly employments," reading Sunday books, and attending one or more church services. Teaching Sunday School was a family tradition. So was reading aloud, for young and old alike. The distinctive quality of the family vividly impressed a young neighbor who came upon one of the Hamilton cousins, surrounded by a circular bed of pansies, reading Dante in the original.[7]

Allen Hamilton (1798–1864), the patriarch who made this life possible for his descendants, arrived in frontier Indiana in 1823, having left his native northern Ireland six years earlier following a decline in the family fortunes. His rise was a classic American success story: a tale of driving ambition, entrepreneurial skill, and political sagacity. Starting out as a fur trader and retailer to the Indians, Hamilton shrewdly diversified his activities as Fort Wayne changed from a frontier to an agricultural community. Through a series of carefully staged partnerships, he prospered in dry goods, retailing, and flourmilling. But the bulk of his later fortune came from his success as a land speculator. Much of it was Indian land, for Hamilton profited from his friendship with John B. Richardville, chief of the Miami Indians, and used his position as executor of the chief's estate and guardian of his minor children for personal gain. After 1850, Hamilton shifted his business interests once again, this time from commerce to finance, subsequently devoting himself to railroad promotion and banking.[8]

As befitted a man of high ambition, Hamilton took a leading part in the public life of the new community: he held a succession of local offices; petitioned the American Home Missionary Society in 1828 to send a regular minister to Fort Wayne and three years later helped to organize the First Presbyterian Church; with one of his partners, he also imported from the east the city's first teachers. A man of strong family feeling, he brought his father, three brothers, and a sister to Fort Wayne and assumed financial responsibility for educating the younger ones.

In 1828, Allen Hamilton married Emerine Jane Holman (1810–1889), the oldest daughter of a prominent political family from Dearborn County, Indiana. Her father, Jesse Lynch Holman, was a judge of the Supreme Court of Indiana and subsequently a United States district judge, and her younger brother, William Steele Holman, later served as Democratic Congressman for thirty-two years. Allen and Emerine Hamilton had eleven children, of whom only five lived. They sent their sons, Andrew Holman (1834–1895) and Montgomery (1843–1909), to Wabash and Princeton colleges respectively (both institutions were bastions of Presbyterianism), to Gottingen and Jena universities in Germany, and to Harvard Law School. Their daughters, Mary (1845–1922), Ellen (1852–1922), and Margaret Vance (1854–1931), went to Miss Porter's School in Farmington, Connecticut.[9]

Like her husband, Emerine Hamilton was active in the community. A passionate reader who often lost herself in books, in 1887 she established with her daughters a free reading room for women so that the less privileged might enjoy the pleasures of the printed page. Deeply religious as well—her maternal grandparents had been Methodist missionaries on the Kentucky frontier and her father a Baptist preacher—she was a pillar of the First Presbyterian Church. (Her husband had successfully led a move in 1847 to grant women the vote in church affairs.) She gave away as much as one-fifth of her yearly income, much of it in response to direct appeals by the needy. Friends and her youngest daughter disapproved of these acts of charity, which did not distinguish between the worthy and unworthy poor. But she could not bear to turn anyone away from her door; during her last illness, a member of Fort Wayne's small black community who came to convey his respects, observed: "She has done everything for us." She also supported temperance and, more daringly, woman suffrage. Susan B. Anthony and Frances Willard stayed with her when they visited Fort Wayne, and Emerine Hamilton willed $500 each to Anthony and Lucy Stone for the suffrage cause.[10]

She survived her husband by a quarter-century, presiding serenely, if impersonally, over the Hamilton compound. Agile in mind and body, she enthralled her grandchildren with stories and poems. She was even tempered and able to ignore personal unpleasantness (much

15

of it no doubt occasioned by her difficult sister-in-law, Eliza Hamilton, who lived with her). Agnes made much of her grandmother's love of people and her interest in everything around her, which seemed to keep her younger in spirit than her own children. But Alice found her elusive and somewhat impersonal: "I always felt I was 'Montgomery's second daughter' to her." When Emerine Hamilton died in 1889, Agnes observed: "We feel like a great body with the head gone." In the next few years, family tensions surfaced.[11]

Despite, or perhaps because of, their many advantages, the children of Allen and Emerine Hamilton seem to have known little serenity or happiness. Only the middle daughter, Ellen (or Nell), who married Samuel Wagenhals, minister of the Holy Trinity English Lutheran Church, seemed to Alice to be both reasonably happy and congenial. Alice's oldest aunt, Mary, married Henry Williams, the son of one of her father's former business partners, from whom she later separated. Uncle Henry, a religious fanatic with a penchant for wild business schemes, was a disturbance to family harmony; even Agnes' dutiful mother tried to avoid him, and Alice considered him utterly detestable. Mary Williams clung to her older son, Allen, and her nieces thought her secretive and demanding. The youngest aunt, Margaret, who never married, devoted herself to her mother. Initially Aunt Marnie (or Marge) was a glamorous figure to her nieces, whom she entertained in her cottage in Farmington, Connecticut, and took on shopping sprees to New York. A patron of the arts, she remodeled her carriage house in 1892 for the Fort Wayne Art School, where several of her nieces taught. But she seems to have lost her purpose with her mother's death, and by the mid-1890s she described herself as a failure who resented the successes of others.[12]

Andrew Holman Hamilton succeeded his father as head of the family in 1864. As administrator of his father's estate, he spent the next eleven years tending the family's complex finances. His father's personal property alone amounted to $80,000. The real estate—the bulk of Allen Hamilton's holdings—was not appraised at the time the estate was settled, but the value of his fortune has been estimated at over $750,000. A Mason and a Knight Templar, Holman Hamilton was an active Democrat and served two terms in Congress (1875–1879). Subsequently, except for business affairs, he seems to have spent most of his time in his library. While impressing his family with the need for financial restraint, he spent large sums on rare books; at 6000 volumes, his was the largest library in the city. A recluse in his later years, he rarely ventured out of the house and took meals in his room when company was present. His family assiduously catered to his wishes, while he vetoed his daughters' plans for study or travel: "no" was their "usual answer" to any request to leave home. His wife, Phoebe Taber Hamilton (1841–1932), whose father was another of Allen Hamilton's partners, had been brought up virtually as a daughter by Emerine Hamilton following the death of her mother in 1847. She resented her husband's restrictions on their daughters and after his death in 1895 allowed them greater freedom. But she was nevertheless an anxious and rather possessive mother who preferred to have them at home.[13]

The second son, Montgomery, Alice's father, early departed sufficiently from family custom to alarm his elders. He distressed his parents by leaving Princeton in 1862 to join the Union army. Dispatched to Germany a year later to recover his health, he sought to allay his loneliness in fine wine, cigars, and good fellowship. Pursuing pleasure more than his studies, he felt it necessary to entertain—and lavishly—those less affluent than he. The demands on his purse were great, and his father and older brother found his repeated requests for funds disturbing.

In Europe he fell in love with Gertrude Pond (1840–1917), an unaffected woman in whom he sensed a kindred spirit. Her father, Loyal Sylvester Pond (1811–1881), a Wall Street broker and sugar importer with Southern sympathies, had sent his family to Europe to wait out the Civil War. Her mother, Harriet Sarah Taylor Pond (1819–1907), was a serene and kindly woman with artistic talent who shared her husband's conservative social outlook. Gertrude, the oldest daughter in a family of eleven, had charge of her younger siblings and managed them with ease. Montgomery Hamilton admired her spirit, her good sense, and her preference, despite exposure to fashionable city life, for the family's quiet home in Tarrytown, New York. She also impressed him with her knowledge of French and Spanish and her familiarity with modern literature, which surpassed even his own. To his family he made light of the affair, dismissing it as puppy love, but his feelings for Gertrude Pond suggest a more serious side to his character than his more customary tone of masculine bravado. After a two-year separation, he returned to Europe in 1866 and married her. His mother, though irritated by her son's extravagance and surprised by the pre-

17

cipitancy of the marriage, considered it altogether "for the best." But Montgomery's sister Mary thought him "demented" for marrying without money and feared he would "take all we have."[14]

In 1865, following his initial return from Europe and a year at Harvard Law School, Montgomery Hamilton had entered into the wholesale grocery business in partnership with Alexander Huestis, an older man experienced in the trade. The firm of Huestis and Hamilton, which purveyed such items as tobacco, fruit, coffee, spices, and oysters, lasted for twenty years. When the partnership was dissolved in 1885, the firm's liabilities substantially exceeded its assets and Hamilton expended $40,000 of his own funds to discharge its obligations. Not until 1894, after he had twice taken Huestis and his son to court, did Hamilton receive full payment of Huestis' half of this reckoning. In the meantime, the financial loss weighed heavily on his family.[15]

Like his brother, Montgomery Hamilton was active in the Democratic party in the 1870s and 1880s and served on the city council. He also became a director of the Hamilton National Bank, in which his father had been a major figure, and a trustee of the Fort Wayne College of Medicine. Although he apparently turned down requests to run for Congress and for mayor, he was more gregarious than his older brother: indeed, he sometimes extended luncheon invitations to passengers debarking from the New York–Chicago train during a change of engines. Talkative and argumentative, he held forth on such subjects as theology and the single tax.[16]

Montgomery Hamilton took a deep interest in his daughters' education, and in their early years he was their principal guide. (He objected to the public school curriculum—too much arithmetic and American history—while his wife disapproved of the long hours indoors.) Starting his children early on languages, literature, and history, for the most part he let them learn by reading on their own. Their only formal instruction was in languages: their father taught them Latin; their mother and later a tutor spoke French; and the servants and a Lutheran schoolteacher taught German. Montgomery had Edith and Alice memorize passages of The Spectator as models of style. Edith in her teens reported on the results to her cousin Jessie: "I flatter myself my style is getting quite Addisonian. I hope you keep all my letters; some day, you know, they will be all treasured up as the works of 'Miss Hamilton, the American Addison, Scott & Shakespeare'!" Their father, who had a passion for theology, also required his two oldest daughters to memorize the Westminster Catechism. He encouraged all his children to use his ample reference library to answer their own questions or those he set them. Alice's first research assignment was to find proof of the doctrine of the Trinity in the Bible, a question her father thought should be answered in the negative. When she wanted to study physics, he told her that everything she needed to know was in the *Encyclopaedia Britannica.*[17]

After the failure of his business, Montgomery Hamilton contemplated making a new start. He considered buying a sheep farm in West Virginia, where the family sometimes vacationed, or managing a branch office of a national firm. Defeated in his bid to become president of the First National Bank in 1887, he too retreated to his library. With his watch chain, fine cigars, and glass of port, he seemed to a young visitor an imposing if remote figure. He had a bad heart and drank too much, and neighbors sometimes had to assist him home. His daughters, mortified by his drinking, rarely spoke of it. By the time they reached young adulthood, they viewed their father as an ineffective but disturbing presence who demanded more attention than they wanted to give. So difficult was the situation by the early 1890s that their maternal grandmother apparently suggested divorce or separation. But although she took extended leaves from home, often to attend to Pond family matters, Gertrude Hamilton seems to have retained a fondness for her husband that eluded her high-minded and romantic daughters. Of her mother, Alice wrote in later years: "She did have some good times in her life—but not many." Yet shortly before her death, Gertrude informed Edith that she was no longer afraid because her husband "had been with her and told her it was nothing to be afraid of and he would be with her through it." To Edith it seemed very strange: "for years and years she had never thought of him as a help or one to stand by her. But it was a calming strength to her in her last hours of consciousness."[18]

During the early years of her marriage, Gertrude apparently got on well with her in-laws. Emerine Hamilton, with whom Gertrude and Montgomery lived for several years, had liked her future daughter-in-law greatly and considered the entire Pond family "one of the most affectionate" she had ever known. And soon after the young couple arrived in Fort Wayne in 1866, Phoebe Hamilton observed that the marriage had wrought an "almost miraculous" change in Montgomery. But in a family that kept to itself, Gertrude may always have been something of an outsider—perhaps partly by choice. Certainly in later years she was a source of contention. Used to catering to their

19

own kin, the Hamiltons evidently resented the attention Gertrude gave her own large family. They complained of the extended visits to Fort Wayne of various Ponds (among them Gertrude's mother, Harriet Taylor Pond, who lived until 1907 and was an important person in her granddaughters' lives) and of Gertrude's trips east to see her family. This family had its own matriarchal tradition: Gertrude inherited the valuable silver tea set that had belonged to her great grandmother and passed by custom to the oldest daughter. (Portraits of the five first-born women of the maternal line, including Edith, hung prominently in later years in Alice and Margaret's home in Hadlyme, Connecticut.) Gertrude also had an independence of spirit and a love of privacy alien to her more conventional relatives, and Alice thought her mother had shocked the Victorian Hamiltons by openly discussing such taboo subjects as pregnancy and childbirth. By the 1890s, Gertrude's relations with most of her sisters-in-law were severely strained; at least two of them seem not to have been on speaking terms with her.[19]

Gertrude Hamilton had a deep and lasting influence on her daughters. Alice was especially close to her mother and still missed her nearly two decades after her death. A warm and sympathetic person, Gertrude knew how to amuse young children and also encouraged them to have good times among themselves; one of her few requirements was that they take daily walks. She shared responsibility for their intellectual development with her husband. But where he liked the clarity and definiteness of Macaulay, Pope, and Addison, she read with her daughters *The Mill on the Floss* and *Adam Bede*. An Episcopalian, she preferred the Sermon on the Mount and the Psalms to the more sober religion of her husband. Despite her parents' conservatism and her husband's scorn for sentimentality, she had a capacity for outrage that expressed itself in anger at lynching, police brutality, and child labor, topics about which she heard a good deal in her later years from Alice.[20]

Gertrude Hamilton made it possible for her daughters to escape the fate of so many women: sacrifice of individual ambition to the family claim. Remarkably unpossessive, she encouraged her daughters to pursue their own goals and took pleasure in their achievements. Alice recalled her mother's admonition: "There are two kinds of people, the ones who say 'Somebody ought to do something about it, but why should it be I?' and those who say 'Somebody must do something about it, then why not I?' " Despite her difficult domestic situation,

Gertrude resisted the temptation to keep her daughters at home; she thought it a waste of talent that her nieces Katherine and Jessie Hamilton spent their lives caring for their mother. It was from her mother, Alice later wrote, that she learned that "personal liberty was the most precious thing in life."[21]

In contrast to Gertrude's daughters, who left Fort Wayne at the first opportunity, of the three daughters of Holman and Phoebe Hamilton only Agnes left home. Katherine (1862–1932), considered one of the most brilliant of the Hamiltons, had not been allowed to attend Bryn Mawr College. She read and studied on her own, tutored her brothers, and took part in Fort Wayne life as treasurer of the women's reading club, member of the library committee, and first president of the Women's Equal Suffrage League (1912). But she never found a real vocation, and she seemed unhappy and unfulfilled to many who knew her. (She apparently gave up a devoted suitor—in one version he was Jewish, in another Catholic—at her mother's insistence.) Jessie (1864–1960), an intimate of Edith Hamilton as Agnes was of Alice, aspired to be an artist and after her father's death attended (with Agnes) an art school in Philadelphia. But she too stayed in Fort Wayne, working on her own as an etcher and oil painter. Unlike Katherine, Jessie seemed content to lavish attention on her widowed mother. An unusually sympathetic person, she was probably the best loved of all the cousins. Agnes, the most ambitious of the three and also intensely religious, considered several careers, among them architecture and art, before settling on social work, an interest she shared with Alice. In her twenties she attained some prominence in Fort Wayne: she prompted the First Presbyterian Church to establish ties with a mission school in a poor neighborhood, was a leader in the women's club movement, headed the students' art league, and served as first president of the local YWCA. Finally, in 1902, she took up residence at the Lighthouse, a Philadelphia settlement, where she remained for thirty years.[22]

The Hamilton family had a deep and complex influence on all its younger members. If the family was more troubled than Alice Hamilton revealed—she made no mention of the dark side of family life in her autobiography, *Exploring the Dangerous Trades* (1943)—the Hamiltons imparted to their daughters that sense of specialness that characterized so many successful women of this generation. Standards and aspirations were high, competition intense. At one time or another, most young Hamiltons felt they had committed the unpardonable sin of being stupid. For the correlates of high aspirations and special privileges were feelings of unworthiness and fear of failure. Alice and Agnes, who as children had "mourned over the martyr days which could never come again" and as young adults aspired to high moral goodness, felt these pressures keenly.[23]

For all their underlying seriousness, the young Hamiltons were far from somber. In adolescence, as word games and musical evenings took the place of swordplay, Agnes feared that their "noisy proceedings" would shock outsiders used to tamer entertainments. If Alice and her sisters taught Sunday School and studied Greek, algebra, and chemistry on holiday, they also read *My Shipmate Louise, Romance of a Wreck* along with more serious novels, and preferred life in the woods to the refinements of home or society. Most of all, the younger Hamiltons took pleasure in each other. The cousins' love and trust of each other contrasted sharply with the sometimes discordant relations among their elders and the tensions these imposed on young people trying to make their way in the world.[24]

This sense of belonging to one another also made it difficult, initially at least, to move outside the family circle. At fifteen, Agnes could not understand why Alice and Allen wished to meet the "*very* stupid and silly" boys and girls in town. Edith, in her twenties, expressed relief at a young visitor's departure: "I don't believe there is any one in the world, outside of my own family, whom I want every day in the week." And throughout her life, Alice Hamilton felt "absurdly and wretchedly homesick" when separated from her family.[25]

In October 1886, at the age of seventeen, Alice left Fort Wayne for Farmington, Connecticut, to attend Miss Porter's School for Young Ladies. Founded nearly forty years earlier by Sarah Porter (1813–1900), the school was already a venerable Hamilton tradition. Three aunts, two cousins, and Edith had preceded Alice there; Agnes soon joined her, and Margaret and Norah followed. Despite a strong intellectual bent, Miss Porter—a devout Congregationalist—encouraged the acquisition of Christian character, liberal culture, and disciplined womanhood rather than learning for its own sake or preparation for a career. (Indeed, she thought that the young Mary Hamilton "overestimated intellectual excellence in comparison with moral graces.") But if the school did not promote intellectual rigor, neither did it tolerate snobbishness or frivolity. Sarah Porter had retired from active management of the school a few years before Alice and Agnes arrived, but she had a deep moral hold over her students that Alice attributed less to any formal teaching than to the pervasive influence of her personality, which embodied integrity and self-control.[26]

In her autobiography, Alice Hamilton left a bemused description of her years at Farmington. Rules dominated life at the school, but since Miss Porter did not believe in academic requirements, examinations, or grades, Alice chose subjects she liked, avoiding mathematics and science, about which she knew little (her father had not considered them important). Experiencing some of "the world's worst" teaching, students learned German literature and mental and moral philosophy by memorizing long passages without attempting to understand them. But the study of Latin and Greek and—a special privilege—Dante gave Alice genuine pleasure. She considered Miss Porter's sufficiently important that when family finances were low she postponed her medical studies in order that Norah might attend. Most of all, Farmington meant a wider outlook and new people with different upbringings. Like the rest of the Hamiltons, Alice made several life-long friends there.[27]

1 To Jessie Hamilton

Farmington
February 12*th* 1888

My dearest Jessie

Hallie and I have just been foiled in an attempt to go and see Mrs. Dow so she is sitting up here and we are all writing.[a] I wonder if you think it was very horrid of me not to write before and thank you for your letter, for it was such a very lovely one. It would have cured me entirely if I had not had the sense, with Agnes' assistance, to get over my blues a couple of days before I wrote that idiotic letter to you. I am going to keep yours as a remedy to have on hand when I get my next attack and I think I shall give it to Margaret to bring back here. It was very lovely of you to take so much trouble and I felt so ashamed and such a goose after I had written to you. I should never have dared to tell you all that nonsense, if we had been face to face. Please don't think I feel that way often, for I don't and I have resolved never to be any more. This has been such a very nice Sunday, though I don't know exactly why. Mrs. Dow told us at Bible Class that Ramabar was coming here on Tuesday. Have you ever heard of her? She is a splendid Hindu woman who is very remarkably well-educated and who is trying to raise money to start a school for Hindu women in India. So she goes lecturing all over the country and Mrs. Dow says she wears her regular Hindu costume and will only eat just what the high-caste do. And she is going to stay at Miss Porter's and come up here to some of the meals and just think how nice it will be for me at Miss Porter's table, watching her. Agnes and Theo went and had such a nice call on Mrs. Dow that Hallie and I were envious and we asked her after evening lunch, if we could go there but she had an engagement so we couldn't, but as she told us so she added "I had been wondering at not seeing you girls" and it made us feel so happy to think she really had been thinking of us that it quite made up for it.[b] It is delightful to go and talk to her about Miss Porter and hear her enthuse about her. I think she reverences her more than anyone in the world does. I wish you could have heard Miss Porter's talk this morning. It was about being truthful and honourable in little things. She spoke more mildly than she mostly does, but she said just exactly what would hit us all. She told us how the tone of the school was affected from year to year and how the traditions were handed down and she made us feel as if even the most insignificant one of us had some influence. I never could see before why it was morally wrong to break the rules, but she said that the deceit in it made it so. Agnes and I came up here afterwards and made no end of good resolutions, one result of which was that we undressed before we put out the light, for the first time this year. I suppose Agnes has told you that she has asked Evelyn to room with her unless Theo comes back. I am so glad that it is fully arranged that she is to room with one or the other of them, for Grace McLeod is wild to room with her and though she will not say so herself, of course, makes Hallie and Maud do it all the time and I am afraid that Agnes would think she ought to sacrifice herself and room with her and that would just spoil the whole year for her, for Grace is dreadfully uncongenial to Agnes and riles her all the time.

Please don't answer this. You have plenty to do with writing to Agnes and I did not mean to make this into a regular letter, only to thank you.

Very lovingly
Alice

a. Mary Elizabeth Dunning Dow, a former pupil of Miss Porter's, served as assistant principal of the school from 1884 to 1900, when she succeeded Miss Porter as principal. In 1903 she founded a preparatory school for girls that later became Briarcliff College.

b. Theo was Theodate Pope (Riddle), 1868–1946, later an architect, who founded and built the Avon Old Farms School for boys in Avon, Connecticut.

THE HAMILTONS' first sustained encounters with society took place at Mackinac, an island in the straits entering Lake Huron that was first a national and then a state park. The family summered there for many years in a house they built, and Alice later observed that she and her sisters loved the island "passionately, almost painfully." Fishing trips with their father gave way in adolescence to swimming, sailing, and tennis parties with their contemporaries. Sketching, novel reading, and long tramps in the woods alternated with serious study. Edith and Margaret read for their Bryn Mawr College examinations, while Alice tried to make up her deficiencies in chemistry and biology. For Norah, sketching was always a serious pursuit rather than a diversion. [28]

In the informal atmosphere of Mackinac, the young women received invitations to go boating, straw riding, and dancing. Chaperones accompanied them—the Hamiltons sheltered their young—but sometimes the "matron" was only an extra person their own age. Having been brought up so much by themselves, they found these occasions trying. In letters written between the vulnerable ages of eighteen and twenty-two, Alice's ambivalence about these social forays is apparent. Shyness, fear of rejection, and self-deprecation alternate with disdain and mock bravado. A letter written when she was eighteen captures the tone: "I do wish people knew how we dislike invitations. We no sooner tide one over, than another one comes." But later in the same letter she asserted: "I don't think four sisters ought ever to expect to go into society, nobody wants them. If I could only turn my three sisters into boys for the evening, I suppose I should have a very good time." [29]

Two years later, after Edith and Norah had lectured her on her "stiffness and primness" with boys, Alice announced that she was "deeply crushed" on a handsome neighbor. Both sisters had "taken fright" and "backed out" at once, she wrote Agnes with gentle, sardonic wit, and possible longing: "So I don't see how I am to get in any proper amount of sweet glances, passing greetings and all the rest

25

of it. I believe I have capabilities of lowness in me which only need activating to bloom out beautifully. Shouldn't you like to hear reports of my being wildly gone on some handsome youth and trying all methods to meet him . . . That is the sort of thing that goes on here all the time and why should I be better than my neighbours?"[30]

Like many intellectual women, Alice Hamilton had little in common with most men her own age. Both she and Edith professed to being bored by all the young men who paid attention to members of the family. At the same time, Alice idealized several older men and looked to them for intellectual and moral guidance. She preferred the company of those of any age or sex with whom she could talk freely— about religion, books, anything that mattered. In view of the social conventions of the day, such friends could rarely be found among the opposite sex. (Indeed, Allen Williams rejoiced when his fiancée treated him with the same forthrightness as Alice and Agnes did.) Few outsiders could match Alice's sisters and cousins for intellectual stimulation, high spirits, and secure affection. Nor did anyone else permit her to be so entirely herself.[31]

2 To Agnes Hamilton

Mackinaw[a]
July 6th 1890

My dearest Agnes

I feel like writing pages and pages to you to-night, not because I have anything particular to say, but because I feel very talkative. A half finished letter to Allen lies beside me and, if I have any brains left at the end of this, I shall finish it to-night, otherwise it must wait till to-morrow morning. I suppose you are just coming home from the Christian Endeavor now and making Jessie assure you over and over that you did not make a goose of yourself and that everybody could hear every word you said.[b] I wish I could have been one of them. I met Joe Bursley today and he said it was much cooler in Fort Wayne on Saturday, for which I am profoundly thankful. There has been a cold easterly storm here all to-day. We went to church this morning and then came home to find the Magees and Mr. Eagle here for dinner. My parents are certainly the most incorrigibly hospitable people I ever knew. It made a funny mixture at the table, Presbyterians, a Catholic, Episcopalians—the Magee girls are Episcopalians—and an

Agnostic—at least that is what Mr. Magee calls himself. And Mr. Eagle would talk theology. Papa was beautifully forbearing, Norah thought too much so. Mr. Magee makes me actively ill. You can't pick up a pin or turn your head or say "good-morning" without its reminding him of something he saw in Stockholm. I happened to say to Papa that the fire needed some wood and went to get it. As I passed out I heard Mr. Magee begin "In Stockholm they burn birch wood." So I went to the pile and picked out the loveliest birch logs I could find, all with the bark on. Mr. Magee looked quite crest-fallen when I put them on and said "Why those are birch logs." "Oh yes" I said "we have quantities out in the yard, we always burn it." The girls are not bad; Margaret is quite nice. Mary might be, but she goes in for the pretty too much. Not that she has the least claims to prettiness, except by contrast with her sister, but I suppose she has compared herself with her so often that she has begun to think herself quite a beauty. I feel painfully wall-flowery this year. If I go to a dancing party I am morally convinced that I shan't have a single partner. Norah and I went down to the Sheelys [?] with the Williams boys on Fourth of July evening. They have several young people visiting there so the room was quite full. I sat down at one end of it and talked to the least attractive girl and Mrs. Brooks, Norah sat at the other with five boys in attendance. It was painful but not so bad as when they all decided to go down to the docks to see the fire-works. Stan Brooks wanted to go with Norah, Wal Brooks was determined that he wouldn't go with me even if his own mother were the only alternative. And Jess, whom I can generally depend on, went back on me and took possession of Norah. So I was forced on poor Stan and he did his duty nobly, I must say. To be sure every time we stopped he would make a violent effort to take Jesse's place, but he was always foiled and had to come back to me.[c] I wish it were not so hard for me to talk to those boys, but it is. I am not interested in base-ball or tennis or their jokes with Rome [?] and Kate. College stories are a little better, but even those aren't wildly exciting and I find myself making a tremendous effort all the time to keep to subjects that interest them and to be interested in them myself. And I am afraid the effort is very apparent. Mr. Eagle is a blessed relief. He came up yesterday morning and in the afternoon Mr. Brown and we three girls and he took a long walk through the wildest part of the woods.[d] Mr. Eagle talked Woman's Rights and Prohibition to Norry and me and Mr. Brown discoursed on the single tax to Edith until we reached British Landing where we met Mrs.

27

Brown and as she detests Political Economy and politics, we dropped the subjects and recovered our tempers. I lose mine with Mr. Brown all the time, he is so dictatorial and illogical. Mrs. Brown took me home to dinner that evening—the others wouldn't go because they were so dirty; I was too but I didn't mind—and of course Mr. Eagle took me home afterwards. He wanted to talk theology. "Now why cannot we discuss these things fairly and calmly as we do everything else?" "But what is the use?" I said "we never could convince each other." "Not on the great points perhaps, but there are minor points which we might perhaps be convinced on." I said I could not by the wildest stretch of my imagination conceive of my being won over on any single point. But I know he has not given up. He is very nice, though, and it is so lovely never having to try to keep up the conversation and always being able to talk of things one really cares for. He is not as nice as Mr. Rowland, though. Edith has definitely invited Sairey [?] at last and I dread her visit more than anyone's. It is not to be till the last of August, for Edith says she will not care a bit for the little society we have up here. I rather wish she and Mr. Rowland could be here together.

Well I am going to return to Allen, so I must leave you, though I believe I have expended every idea I ever had on you and have none left for him. I think it is a shabby trick for him not to take my Nebraska class.ᵉ If it is only because he doesn't want to, then it is very selfish of him; if it is because he thinks he does not know how, then it is still worse. He ought to be ashamed of himself if he cannot teach them the gospel of Luke. Give him the "Flat-iron for a farthing."ᶠ I have a class here, just three boys, but the Harrisons will be here next Sunday and they will make it interesting.

Lovingly
Alice

a. Mackinac is pronounced "Mackinaw," and AH usually spelled it that way.

b. Christian Endeavor was an evangelical and interdenominational young people's movement founded in 1881 to encourage young men and women to lead a more vital religious life; it attracted millions of members.

c. Jesse Lynch Williams, a first cousin of Allen Williams, became a well known writer who won a Pulitzer prize for his play *Why Marry?* (1917).

d. Edward Osgood Brown, a Chicago lawyer and later a judge, was an ardent single-taxer whom AH credits with introducing her to "the literature of revolt." He also espoused the cause of the anarchists who were hanged for the 1886 Haymarket bombing in Chicago, a position contrary to that of the elder Hamiltons. His wife was Helen Gertrude Eagle Brown.

e. AH taught a class of boys, Agnes a class of girls at a sabbath mission school located in Nebraska, the poor district of Fort Wayne. The cousins referred to the school simply as "Nebraska."

f. *A Flat Iron for a Farthing* (1885), a children's book by Juliana Horatia Ewing, published in England, was a series of tales about the moral education of a boy whose mother had died. Its tone was not heavily religious.

IN JULY A MR. ROWLAND visited Alice in Mackinac. She had met him the preceding March in Athens, Georgia, where she had accompanied her mother, who was recuperating from an illness. Mr. Rowland, who came of an old Southern family in reduced circumstances after the Civil War, was staying at the same boarding house. Alice's letters from Athens expressed amusement at her unaccustomed popularity—which she attributed to the absence of competition—and surprise at the freedom with which Southern women accepted invitations from men they scarcely knew. She had preferred Mr. Rowland to another male acquaintance and had described him as "not a bit fascinating, but just as nice as nice can be" and "really good too." In a letter written during Mr. Rowland's stay at Mackinac, Edith humorously described the visitor's attempts to be alone with Alice and her sister's efforts to avoid a tête-à-tête. Her assessment substantiates Alice's own professed eagerness to have her visitor depart.[32]

3 To Agnes Hamilton

Mackinaw
July 13*th* 1890

My dear Agnes,

Mr. Rowland and I are sitting at the dining-room table writing home letters. Will it be a very great breach of hospitality if I say that I am beginning to invent forged telegrams from Athens, informing him of the disastrous state of the cotton market that absolutely requires his presence there? It is an awfully mean thing to say and I trust to you not to repeat it, for it isn't that he is not as nice as he can be and the easiest possible guest to entertain, but don't you know the feeling that you must not sit down for a half hour with a novel, because you don't know what your guest is doing, you must devise some way to spend the evening, you must have pleasant conversation at the table (though that is not as much an item as it would be if Papa were not so nice about it, he quite surprises us). You see we are used to getting up here with the whole long day before us to lounge, tramp, sketch exactly as we please, without ever planning ahead. I think Mr. Rowland would be somewhat amused, though, if he should read this, for we really have done almost nothing to entertain him. He has not been a bit well ever since he came, malaria I suppose, that is coming out with the change of air, so a sailing party we had planned was given up and several things we started to do, he backed out of at the last minute and went to sleep instead, so that he has really had the quietest possible time. I suppose it goes with Southern hospitality not to mention just how long your visit is to last, but it is a little inconvenient. How mean all this does sound! but then it is only to you. Don't go and read all around, please. We took Guy de Maupassant, the other day, Mr. Rowland and I, and went and sat down in a nice little spot on the way to Arch Rock and I read aloud one of the stories and then we talked. I suppose it all seemed perfectly natural to him, but I kept thinking how amusing we would think it, if we came upon a couple like that and what sarcastic little remarks we would make on the nice time they were having together and how we would have laughed over the scraps of conversation we overheard, which wasn't a bit more sensible than such conversations generally are. I am constantly amused at the positions I get into.

By the way, I ought to have done it at the beginning, I want to thank you so much, so very much for my hat. I haven't seen it yet, but I got your note this evening and I know it will be ever so pretty and it certainly is not a bit more expensive than I expected. Thank you very much. I hope it wasn't a melting day when you got it. I had a letter the other day from Allen and he says he sees almost nothing of you, that you are always either going down town or sketching. Norah and I have made nothing but outrageous failures so far, except one flower study apiece and one shore scene that she did. I am getting in despair. If I do not make a success soon, I shall simply give up. Edith and I read your and Jessie's letters aloud to each other, sitting on a stile in the government field. I wish Jessie would always mail hers, so that it should reach us the same day as yours, it is so nice to read them together.

Didn't you like the lesson to-day? I did, quite well. One nice thing about this class is that I can get off all my old chestnutty stories to them, without thinking up anything new. Today I began with a sketch of the state of the times, with Ben-Hur as usual for an illustration, then in the course of the lesson I could tell Sandy and the Cousin from India and they were quite fresh. Mamie [?] and Florence went into Edith's class this time and I am to have Bartlett next Sunday.

Very lovingly
Alice

As a young woman, Alice Hamilton consistently portrayed herself as a wallflower. After attending a dance, she wrote Agnes that although all the girls agreed that "it was simple blindness on the part of the men not to be attentive to me," that blindness was "quite incurable, unfortunately for me." Then twenty-two, she claimed: "I didn't wish myself in it all, as I used to . . . I feel now as if the time for that sort of thing were over." Such statements undoubtedly reflect her feelings about herself. But there is reason to doubt their accuracy. Certainly she was reserved in such social situations and, given the Hamiltons' penchant for sticking together, it was a brave young man who ventured to breach the ranks. But Alice took more of an interest in the social scene than did most of the Hamilton women and accepted invitations to occasional hops and other "dissipations," always suitably chaperoned, of course. Agnes, who avoided such affairs, marveled at her cousin's ability to socialize and yet stand so well in her work. In later years, Alice was considered the most attractive of the sisters, and many putative romances have been claimed for her. She was a beautiful young woman as well, and contemporary sources indicate that she was not without admirers. As in the case of Mr. Rowland, the evidence suggests that she discouraged them.[33]

For Alice Hamilton a social life remained a luxury rather than a necessity. Like other women of her generation, she believed she had to choose between marriage and a career. Highly motivated and eager to prove her worth, she made this choice while still in her teens. (A student at Miss Porter's believed that Alice would succeed in making something of her life, but doubted "if her bitter aversion to the married state continues more than three or four years.") It is unlikely that the choice occasioned an undue sense of sacrifice. Except for Allen Williams, the men in the Hamilton family were difficult and demanding and did not encourage a high regard for their sex. Certainly the marriages of her mother and aunts had little to commend them to an intellectual woman brought up to value independence and achievement. By the early 1890s, the older female cousins assumed they would not marry. Only the youngest of the eleven ever did. A full generation younger than Alice, Hildegarde Wagenhals believed that by marrying she was rebelling against a family tradition. (Of the impending marriage, Alice wrote in 1926: "The taboo is over at last.") Even Allen Williams initially determined against marriage, a resolve broken only when he fell desperately in love. Though it is unlikely that Alice Hamilton was greatly tempted by the prospect of marriage, it is possible that her self-image as a wallflower and her rush to middle age helped to reduce any conflict occasioned by her choice and to protect her from the possibility of dangerous entanglements.[34]

II

MEDICAL TRAINING
1890–1894

ALICE HAMILTON DECIDED to become a physician sometime in her teens. From the start, she planned to combine medical work with a career of humanitarian service, a goal that prompted many women physicians of her generation. She and Agnes took their religion more seriously than some of the Hamiltons, and Alice had initially hoped to become a medical missionary in Persia (because she doubted that she "could ever be good enough to be a real missionary"). But by the time she left Miss Porter's in 1888, she gave as her future address "Corner 375 St. & Slum Alley," a change of place that points to the increasingly secular nature of her mission.[1]

In her autobiography Alice Hamilton claimed she had chosen medicine "because as a doctor I could go anywhere I pleased—to far-off lands or to city slums—and be quite sure that I could be of use anywhere. I should meet all sorts and conditions of men, I should not be tied down to a school or a college as a teacher is, or have to work under a superior, as a nurse must do." In addition to identifying the three leading professions open to women in the late nineteenth century, her explanation reveals several lifelong traits of character that were apparent from an early age: a desire to be useful, a love of adventure, and an intense need for independence. But in a society where a woman's eagerness to be of service might easily be transformed into self-sacrifice, the goals of autonomy and service were not always compatible, and as a young woman Alice Hamilton felt this tension keenly. Medicine proved a more problematic choice than her retrospective explanation might suggest.[2]

Her decision to become a physician was an unusual, but by no means unprecedented, act. The late nineteenth century was in fact something of a golden age for women in medicine. Elizabeth Blackwell had received her medical degree—the first by a woman in modern times—in 1849. By 1870, there were 544 registered women physicians in the United States, and twenty years later, when Alice Hamilton was preparing herself for a career, more than 4500. At first they trained mainly at one of the women's medical colleges, most of them short-lived, founded in the latter part of the century. Some also went abroad, principally to the University of Zurich, where women made up almost half of the American student population. During the 1890s increasing numbers of women studied at coeducational medical colleges in the United States, where they sometimes constituted a large proportion of the student body. At the University of Michigan, for example, fourteen of the forty-seven graduating physicians in Alice Hamilton's class in 1893—nearly 30 percent—were female.[3]

Still, Alice Hamilton's later claim that she had not considered herself a pioneer underestimated the commitment a woman needed to sustain a medical career. In her own case she had also to overcome the initial opposition of her family. At a time when the Hamiltons were selling off lots in order to meet interest payments, Alice and Edith decided they must become self-supporting if they were to lead "a wide and full life." But Edith's pursuit of the classics better accorded with the family's patrician aspirations than did Alice's choice of medicine, then a less prestigious profession than now. It is unlikely that the Hamiltons, who refused to discuss even an impending confinement, considered the practice of medicine a fit career for a well-bred woman. Edith, whose opinion in Agnes' view counted most, actively opposed her sister's choice and let Alice know that she considered science a "disgusting" pursuit. Nor did Alice receive support from Miss Porter, who, she informed her mother, "treats the whole affair as if it were an amusing childish whim that I shall outgrow" and "think[s] me a bluggy-minded butcher" for wanting to observe an appendectomy. Only Agnes (and probably Allen) remained true. Agnes wrote in her diary that she "would rather Alice be a physician than any thing else. I care more for that than I do for being an architect myself."[4]

After leaving Farmington, Alice Hamilton spent three and a half years in Fort Wayne preparing for a medical career. She had not only to make up her deficiencies in science, but to convince her father of her seriousness of purpose. She studied chemistry and physics with a local high school teacher, worked at biology in Fort Wayne and Mackinac, and then in 1890 entered the Fort Wayne College of Medicine, a "little third-rate" school where she recalled mainly studying anatomy. One of six women in a school where, according to Agnes, most of the men belonged to the "roughest class," Alice evidently impressed her professors. One of them gave her extra work, allowed her to assist at operations, and took her to meetings of the local medical society. Another physician let her prescribe for a charity patient. So did a druggist, for a fee. She also gained practical experience by nursing an occasional convalescent. Some of her cousins thought she was getting into "queer society."[5]

During this period, Alice joined her cousins in activities more traditional for a woman of her class. She went to the theater, studied Greek, and participated in informal clubs devoted to dramatics, German, cooking, and sketching, some of them initiated by the Hamiltons themselves. From the start she and Agnes took a special interest in the underprivileged, and they soon began to follow the flourishing new literature on social problems. Shortly after their return from Farmington, they asked a woman who worked at the reading room established by their grandmother to show them the poor section of Fort Wayne; they subsequently taught at a sabbath mission school there, an enterprise in which other Hamiltons also joined.

Although others considered Alice exceptionally devoted to her work, she sometimes berated herself for insufficient zeal—particularly compared to Edith, who entered Bryn Mawr College in 1890. In August 1891, when she had not yet taken out her microscope, Alice wrote Agnes: "I have waked up to the sad fact that I have no love for intellectual employment . . . Why there are small boys who care enough for such things to work at them every spare moment and here I sit around and fritter away my time and am too lazy to do what should not be work to me at all." At this time she planned to go east to complete her medical training, probably at the Woman's Medical College of the New York Infirmary. But finances were tight, and autumn found her still in Fort Wayne.[6]

In March 1892, Alice Hamilton entered the medical department of the University of Michigan as a special student. This change of plans, probably the result of financial considerations, gave her access to one of the best medical educations then available in the United States. In the early 1890s Michigan was in the forefront of the revolution that would transform medical education from a short and haphazard affair,

35

with little exposure to clinical or laboratory work, into one that required a lengthier and more rigorously scientific training. Under the leadership of the new dean, Victor C. Vaughan, the medical department had recently introduced a four-year graded curriculum (one of the first schools to do so) and had also added a number of outstanding young professors to its faculty. From these men, imbued with the values of German science and committed to research, Alice Hamilton received excellent instruction in the basic sciences. Vaughan himself had developed one of the earliest courses in biochemistry and also taught a comprehensive course on hygiene which featured laboratory work as well as lectures. In addition to Vaughan, Hamilton's teachers included John J. Abel, a highly trained scientist who transformed the traditional study of "materia medica" into the independent and experimental discipline of pharmacology; Frederick G. Novy, who established at Michigan what may have been the first laboratory course in bacteriology in the country; and William H. Howell, whose physiology lectures commanded the total attention of Michigan's sometimes rowdy students. Hamilton also worked closely with George Dock, professor of the theory and practice of medicine and clinical medicine and physician to the university hospital, who greatly expanded clinical training opportunities for Michigan students. A former assistant of William Osler, Dock introduced laboratory instruction into the clinical program, a major advance over traditional medical education, which forced students to sit through the same lecture course two years running and offered little clinical or laboratory instruction. As late as 1890, most schools still taught clinical medicine principally by lecture and few offered students opportunities to examine patients.[7]

During her first term, Alice Hamilton studied obstetrics and gynecology, the theory and practice of medicine, surgery, materia medica (pharmacology), descriptive and surgical anatomy, physiology, embryology, chemistry, toxicology, and urinalysis; much of the instruction took place in the laboratory. On her own for the first time, she responded initially in what would become predictable fashion: with self-doubt. If she never succeeded entirely in shaking off this tendency—it reappeared in any new situation—it did not often debilitate her. Her native ability and eagerness to prove herself carried her far, and experience invariably made her more confident, as even she came to recognize. Albert B. Prescott, director of the chemical laboratory and the author of several texts on qualitative and organic chemistry, took her true measure. He praised her "strong intellectual power," "sterling purpose," and "unconscious spirit of helpfulness," aptly concluding: "With a somewhat severe distrust of her own abilities she is found to possess incisive determination."[8]

The environment of Ann Arbor was different from any Alice Hamilton had known. She was surprised, and initially disturbed, by the freedom allowed young women—her parents had even insisted on watching as she crossed the yard at night to visit her cousins—and by the irreligious atmosphere of the university town. But she liked the security of rooming with Dr. Prescott, an elder in the Presbyterian church, and his wife, Abigail Freeburn Prescott, both of whom she greatly admired. She was soon delighted to be introduced around town as "Mrs. Prescott's Miss Hamilton."[9]

4 To Agnes Hamilton

Ann Arbor
March 6th 1892

Dearest Agnes,

I have just been out to mail Mother's letter. Yours will probably have to wait until I go to the College to-morrow morning, so if you do not get yours as soon as she does, don't think it is because it was not written on Sunday. You and Madge have probably just come home from Communion service. I wonder if you are up in Jessie's room talking things over and wondering what I am doing and if I am very lonely. Well I am not, not very. Things are so interesting that I have not time to be. Only I shall have to talk all Summer without stopping, to make up for these long silences. I went to Communion service this morning too. Mrs. Prescott took me to church and I stayed to communion afterwards. It was a nice service, not like ours of course, but much nicer than at Farmington. It made me feel as if I belonged to church and I was glad it came on my first Sunday. Mrs. Prescott is very lovely to me, but I try to keep out of the way as much as possible. She is a thorough lady and the doctor is dear, but I fancy they have not very much money and they never took a "roomer" but once before. The house is exceedingly tasteful and shows so much cultivation, especially in the pictures and the lovely harmonies in the walls and curtains and there are any number of queer old things that look as if they came from Europe. My room is very pleasant. It is a south-west one, the southern window looking out on a big oak grove

37

across the road and the western one, a broad window just above my writing table, looking way across the valley to "Germantown" on the other side. This part of the town is filled with professors and is not very thickly built up. It remind[s] me very much, as indeed all the town does, of Cambridge. It is much larger than I expected. In some ways it isn't at all like a college town. Of course you meet swarms of students everywhere, but you are not stared at at all and you can go wherever you please and meet other girls everywhere in the campus, in the buildings and everywhere. So it is ever so much nicer than in Cambridge.

To-morrow I am going to my first lecture, on Materia Medica, at half past nine. Then comes one on Surgery, then one on Obstetrics and that finishes the morning. After that I suppose I shall drag what is left of me to dinner and study during the afternoon, if I have any courage left to study with. I have not the slightest idea what my standing will be, whether I shall find myself utterly deficient or pretty well advanced. It is so queer to be one of so many and of such very little importance. I am absolutely nobody, for the first time in my life, with no family name or reputation to fall back on, just one of the multitude with no more deference shown me than any of the others. I saw three "female medicals" on Friday when I was with Mother and more forlorn, micky looking specimens you couldn't find in the overall factory.[a] But I met a nice one this morning, a Miss Bishop and she promised to pilot me around to-morrow.[b]

The meals are going to be very nice. I don't mean the eating part, which is quite good though, but the people. I sit at one end with one man on one side and five on the other. Mrs. Hertel sits next to the one man, then come three girls and one more man.[c] One of the girls is foolish, one is quite pretty and the other ordinary-looking, but the two last seem very nice. Several of the men do too. When I come to know them better I fancy I shall enjoy my meals very much. They are all literary students and I can imagine their horror when Mrs. Hertel told them they were to have a "medical" among them.

My back is tired writing or I could go on for ever so much longer. I wish I had somebody with me just to tide me over to-morrow. After that it will not be quite so bad.

Very lovingly
Alice

a. "Micky" was a derogatory term derived from the slang word for Irishman.

b. Frances Lewis Bishop, whose nickname was Fanny, was a classmate of AH's and became one of her closest friends at the medical school.

c. AH ate in a boarding house run by Emily Hertel.

5 To Agnes Hamilton

Ann Arbor
March 20*th* 189[2][a]

Dearest Agnes

I have just come home from a missionary meeting at Newberry Hall, a medical missionary meeting. It wasn't bad but very, very small, considering the number of students there are. Christian Association work is looked upon here just as it is in every college town, a man loses caste as soon as he goes into it. That is the last meeting I shall go to to-day for I stayed to Bible Class and that takes up enough of Sunday. Miss Stoner wished me to go with her to a meeting way out at the hospital and I know she thought me a back slider for refusing.[b]

You and Margaret and Taber are out at Nebraska now, just in the middle of the lesson.[c] I shouldn't mind being there with you, walking home afterwards all together and going in to see Jessie and eat some orange cake. Why it seems so much more than three Sundays since I did it all. Madge says Taber gets on so beautifully with my boys. It is awfully good of him to do it.

Your letter was late in coming this week. I got it yesterday afternoon when I came home from my various callings and receptions and waited until I had dressed for tea before I opened it, so as to read it peacefully and leisurely. Please go on with Ada Hamilton when you write next.[d] I am much interested in her.

I have decided that if I were not a physician—I mean going to be— I should like to be a professor's wife in a college town. They seem here to have such very nice times together and they are such lovely, intellectual women and of course their husbands are the most fascinating part of all. There are three whom I am very much gone on, all married, and the one wife whom I have met is charming. Everyone says Dr. Abel's wife is too, but I haven't seen her yet.[e] He is fascinating. Everybody, I mean all the girls, commiserate [with] me because I cannot come back and graduate here next year. It is a fact that I cannot, for they require the last two full years to be spent in this school, but even if it were possible, I don't know that I should wish to come here. In the first place the girls in Philadelphia will be nicer and then I know a woman's school will be nicer too. To be sure the men here are respectably behaved, but there is a sort of constant aggressiveness on the part of the women and half-veiled ridicule on the part of the men, that one feels all the time. As for thoroughness, I am not sure about Philadelphia for I haven't seen their examination papers, but New York is not as thorough in the fundamental branches. Of course that would not make any difference next year, for the last year is almost entirely practical in both places. Dr. Prescott tells me that they have acknowledgedly the finest physiologist in the country here and I can easily believe it.[f]

Next week my laboratory work begins and, what is worse, my class in Physical Diagnosis. It was bad enough to have to percuss people at home with Dr. McCaskey, but to do it here with Dr. Dock and the whole class looking on is much worse.[g] Because I always find I am more ignorant and scared than anyone else here.

I went over to see Miss Rich and Miss Reilly yesterday afternoon. They are the two Freshmen at our table. Miss Reilly is one of the prettiest little things I ever saw, strikingly so here, for Ann Arbor abounds in homely girls. It struck me as so strange the way those two girls live. They don't look a bit over seventeen years old, and yet there they are living as independently as if they were women of thirty. They have a landlady, to be sure, but she hasn't any more to do with them than the chambermaid has, and they receive men in the evening and go off to evening parties with them, men they never met till they came here, and no mother around or anybody. I don't like it a bit. I would no more have a daughter of mine come here, unless I came with her, not under any consideration. It is not nearly as nice as a boarding school in so many ways. The girls are not silly about the men, but all their good times, all their excitement, is inseparably connected with them. About the lovely times that girls have just among themselves, they know nothing. They talk of girls having much or little attention, just as society girls would, and it is just society on a small scale. I think there are very many objections to it and I don't see a single advantage it has over an exclusively girls' college.

I am growing fonder and fonder of Dr. and Mrs. Prescott. It is the prettiest thing to see them together, for they are as devoted as if they were in their honeymoon. She is quite talkative and he, very silent, and he sits by and beams fondly on her while she talks. And this morning, when she didn't go to church, he came in and kissed her good-bye before he went. I don't think I ever saw so old a husband and wife as they, so very devoted to each other. I love to see a thing like that, for one is apt, without knowing it, to grow a little skeptical about such feelings lasting.

What with meditating between the lines, I have been three quarters of an hour over this and must end it.

Very lovingly
Alice.

a. This letter is incorrectly dated 1891.

b. Cora Lane Stoner was a classmate of AH's at the medical school.

c. Taber Hamilton (1876–1942), the youngest child of Holman and Phoebe Hamilton, became an engineer and had a long career with the Pennsylvania Railroad.

d. Ada Hamilton, not a relative, was one of Agnes' Sunday School pupils.

e. Mary Hinman Abel, an advocate of cooperative housekeeping, had a career as a nutritionist.

f. The outstanding physiologist was William H. Howell.

g. G. W. McCaskey was the Fort Wayne physician with whom AH had studied Materia Medica. He was also attending physician to Agnes, Jessie, and other Hamiltons.

THE REFERENCE to the "half-veiled ridicule" by male students was unusual. Both at the time and later, Alice Hamilton commented favorably on the no-nonsense relationship between the sexes at Michigan. In her next letter, she informed Agnes that she had spoken "too emphatically" against a girl's life there, although she still believed that a women's college had many advantages over a coeducational one. But although she made several close female friends and observed that "as far as I am concerned, this might be a girls' college altogether," for the most part she looked down on her fellow female "medics" as lacking in the intrinsic marks of gentility she considered essential for true womanliness. Certainly she felt a sharp discontinuity between her previous social standing and her new status as a medical student. As a Sunday school teacher of the poor, she was accustomed to being a patron. Now for the first time she found herself "one of the patronized," and was mortified to think that the gracious faculty wives, with whom she identified, took great pains "to make you feel that they think you are very worthy, good sort of girls." One of the wives, Anne Janet Howell, also perceived the incongruity between Alice Hamilton's person and her situation, and later told her daughter that she "never got over her astonishment and pleasure at seeing such a beautiful young girl in that group of rough medical students."[10]

Despite Alice Hamilton's doubts about her standing and her recurrent fear of failing one "Star Chamber" or another, she did exceptionally well in her studies. In May the medical school faculty voted to allow her to graduate "when she has completed the work," thus shortcutting the normal two-year residency requirement. (This special arrangement permitted her to graduate after three semesters and forestalled a plan to transfer to the Woman's Medical College of Pennsylvania, where she expected to complete her course in shorter time and at less expense.) She also joined the journal club initiated by John J. Abel to allow the best students to keep abreast of the European scientific literature. Along the way, she almost qualified for a bachelor's degree, falling short in college standards only in mathematics.[11]

During her second year in Ann Arbor, Alice Hamilton served as an assistant to George Dock at the university hospital, one of three seniors so honored. Dock had a reputation as a stern taskmaster and on one occasion she observed that he had "squelched" her completely for not knowing the natural history of mycelium: "I think a word of praise from him would upset me for weeks, but I have never heard him give one to anybody and don't believe I ever shall." But despite Dock's stern medical manner and his sarcastic remarks about religion, Alice Hamilton revered him. Because the university hospital was a teaching rather than a general hospital, many of the cases were of obscure origin, and as an intern she was to feel deficient in her knowledge of practical therapeutics and even of ordinary obstetrics (she claimed she had never seen a typhoid case in Ann Arbor). But since Dock had his students conduct microscopic and chemical analyses as well as physical examinations before reaching a diagnosis, like other Michigan graduates, she learned how to work up thorough clinical and laboratory reports. During her senior year she also served on the obstetrics ward and in the neurology clinic, learned to use the ophthalmoscope and to give electrical treatments to nervous patients, and did laboratory work in bacteriology.[12]

6 To Agnes Hamilton

Ann Arbor
October 2nd 1892

Dearest Agnes

There is no hospital meeting this afternoon, so I am not going out, but have settled myself for the afternoon, to read Hypatia and write to you.[a] I am going to assume that you saw my letter to Madge, as I am quite sure you must have, and so, not to repeat myself I shall skip my first two days and only talk about the time after. My Chemistry is off my mind at last, and it was taken off in the nicest way you ever knew. Friday evening I came home from supper and found the Doctor out on the porch alone, looking at the moon and evidently very forlorn because Mrs. Prescott had disappointed him and had not got back that afternoon. He asked me what I meant to do and I told him that I meant to cram Organic all evening, upon which he looked a little disappointed and said that he had thought of asking me to go driving with him. Of course I told him that I would cut fifty examinations rather than miss that, so we went. He drove us way out in the country, and you can imagine how lovely it was, soft and warm and moonlit, the air heavy with the fragrance from the swamps and fern thickets, the deepest stillness everywhere and the dearest man in the world in the buggy beside me. When we got back he proposed that I should come into the sitting-room and talk over a few things in chemistry with him and he might clear up some points for me. Naturally I was only too glad to, and I asked him all the tough points and he explained them, asking me a few questions, but only incidentally, and then when I rose to go, he informed me that I had passed a very good examination and he would give me credit for it. Wasn't it the loveliest way to hold an examination? I think we both felt that it was a little too informal though, so we agreed that I should go to his office the next day and take a short formal one. Well I went, but he only kept me a few minutes and I got on all right. I think it was so dear of him. Fanny [Bishop] took me out to the Hospital yesterday morning, after I left Dr. Prescott's and there I had the most depressing time. I cannot help it, the very atmosphere of that place scares all the courage out of me and I know I shall never get used to it. The way the Seniors go calmly around and do physical diagnosis and examine fluids and give electricity, why it simply amazes me. I know I shall never dare do any of it. Dr. Warthin gave Fanny and me a nice red-faced Deutscher to examine, and we spent about an hour percussing him and deliberating whether one side of his chest expanded more than the other.[b] When Fanny had decided that there wasn't any difference between them, and I had decided that I shouldn't have seen it if there had been, we left him and went to examine his expectoration to see if there were any tubercle bacilli. Fanny taught me how to do that and I managed it all right and was rejoicing over the little red lines I had found until I remembered what it meant to the poor man; that sobered me. That is the trouble with this sort of work. I don't mind chemicals or frogs even, or cats, but when it comes to living, feeling people, then I grow frightened. I wish they would give us dummies and manikins. Dr. Dock was not there, but I met a Dr. Whinery, who is Dr. Warthin's assistant, as Dr. Warthin is Dr. Dock's.[c] I don't know exactly whose we are, all I know is that the one who gets me will get precious little assistance. Dr. Whinery is a foolish looking youth, like most of the assistants and instructors. I wish they would not choose infants of such very tender years and such blushing, diffident manners. It makes one doubt their knowledge and ability.

After dinner yesterday I went and called on Helen Dryer and Artie Chapin, but I did not find Artie at home. Helen is nicely settled now, with a sweet little thing for a room-mate, and as the Delta Gammas and Gamma Phis are both doing their best to get her, I don't believe she will be forlorn at all. Mrs. Dryer is here still.[d] She seems to take the greatest care of Helen, not only getting all her arrangements made, but planning out all the day for her, and even investigating the different fraternities, to see which would be the most desirable. I suppose Helen will have to think and act for herself for the first time, when her mother goes.

Mrs. Prescott came back at five in the afternoon. It is beautiful to see the Doctor hover around her, and watch her and limp off to get her a foot-stool, or a cushion or a shawl just for the pleasure of doing something for her. It is as if they had been separated for months instead of days. In the evening Joe Quarles came over. He is a nice, fresh, sweet-tempered looking boy, just a little awkward but nice and natural. It was quite a surprise, when Mrs. Prescott asked him to play for us, but after a little hesitation he went to the piano and played something of Beethoven's, played beautifully too. One does not expect such things from an ordinary boy and he looks very ordinary. His mother did not come, for which I am thankful. Mrs. Prescott has come for me to go out walking, so I must stop.

Lovingly
Alice

a. *Hypatia: or, New Foes with an Old Face* (1853) was a novel by the English clergyman Charles Kingsley about a female Neoplatonic philosopher in Alexandria who was stoned to death by a band of monks. It illustrated Kingsley's Christian Socialist philosophy, in which AH and especially Agnes were then much interested.

b. Aldred Scott Warthin, a demonstrator of clinical medicine in 1892–93, later became professor of pathology and director of the pathology laboratory at Michigan. He also had a teacher's diploma in music.

c. Joseph Burgess Whinery, who had received his medical degree from Michigan in 1892, served as an assistant in internal medicine in 1892–93.

d. Alice Mary Peacock Dryer was the wife of Charles R. Dryer, a physician and geographer who had taught chemistry and toxicology at the Fort Wayne College of Medicine and with whom AH had privately studied chemistry. Helen Dryer was their daughter.

A WEEK LATER, Alice informed Agnes: "I have got over my fear of the Hospital, for as usual it is only the first step that frightens me; after I have done a thing once I can always feel sure of it." Though an apt statement of her characteristic response to new situations, her optimism in this case was premature. Nearly a month later she observed: "The work grows more and more fascinating, but I grow more and more afraid of it." Some of her best letters capture the panic that medical students (and interns) experience during their early encounters with patients. Although she later claimed to be more fascinated by laboratory work than by patient care, these letters reveal her intense involvement with patients and her determination to save them. Of one deformed "Poor House child" whose neglected condition she contrasted with the "fuss" made over the least ailment of her brother Quint, she wrote: "I believe if I were established and had a practice and money, I would be silly enough to adopt him." In fact, Alice Hamilton never entirely overcame her fear of hospital work, which she found "just a little bit hardening." Even though she recognized that nature had more to do with most recoveries than medical skill, she evidently could not—or did not wish

45

to—acquire the necessary detachment to take these responsibilities in stride.[13]

The early letters about her hospital work further reveal the disjunction between Alice Hamilton's sheltered background and the realities of medical practice. Some of her patients were unmarried mothers, a condition that she, like other proper Victorian young women, had been taught to regard as the ultimate disgrace. She found herself unable to offer spiritual comfort or moral guidance to these patients, although this was an aspect of medical practice to which she had previously aspired. In part this was a matter of reserve and embarrassment, and, she thought, a reflection of the family's inability to talk about difficult matters. But her reticence also stemmed from her awareness of the irrelevance of her own experiences to those she would help. As she came to know these patients personally the moralistic tone diminished, but she never overcame her resentment at the inherent inequity of the situation: "And to think that those men have no more care or responsibility or suffering because of their sin than if they never had done it. It seems so horribly unjust."[14]

7 To Agnes Hamilton

[Ann Arbor]
January 22nd 1893

Dearest Agnes

I mean to begin this letter to-day, but it may not get finished till next Sunday, unless I decide to make it very short, for there are no odd moments any more which can be filled in with writing letters. The mid-years come in three weeks and everyone is putting on extra steam and cramming early and late. We have petitioned for extra quizzes too, which adds to the rush. I am just as sure as sure that I shall be conditioned in at least three things, but there is no use groaning over it, or talking either. There is a case out at the hospital now, whom I am rather worked up about. It is one of the obstetrical cases, her baby will come in a few weeks. She is an exceedingly pretty girl and seems to be much better bred than the other girls in the ward, with whom she has nothing to do at all. I had not noticed her much until she was referred to our department for an inflammation of the joints. Then I was sent to take her history which she gave in a defiant

46

sort of way, with her eyes on the novel she was holding, and with just as short answers as she could. Dr. Dock discovered that she is infected with a disease that bad women very often have and that renders her approaching confinement very dangerous, almost surely fatal.[a] It seems so dreadful, for she is so young, so utterly alone and has probably been so wicked. I told Kate about her and she exclaimed over the beautiful opportunity I had of getting at her, but, Agnes, I cannot get at her at all.[b] I simply don't know how. I would give anything to be able to reach her, and not only her but the other women in the ward too. There are three more, one whose baby is born and whose husband has deserted her; and two wretched looking girls, who might have been chambermaids in some fourth-rate boarding house. They turn away when I come in and seem to shun any notice, yet I am sure anyone who knew how could be a comfort and a strength to them. You know I told you when I was at home how provoking it was that the nurses would not let us have anything to do with the patients, except in the line of our work. Well since Fanny and I have been so much with Dr. Dock for the last month, the nurses have come to look upon us with much respect and now I can go in the wards where I please and talk to whomever I wish and I just wish I could not. For then I could put the blame on someone else and now it is all my own fault. Sometimes I think that if I were used to speaking plainly and having people speak plainly to me, it would come easily, but don't you think that the way we all have of never putting into words things that we feel at all deeply about, is a disadvantage in a way? It simply is a mental and physical impossibility to me to do what comes as easily as can be to some people.

To change the subject, has that very rude cousin of ours ever acknowledged the—I forget just what it was that we sent him—but whatever it was? Because he has not to me and if he has to you, please write and tell him to divide his gratitude evenly.

This has been a more Sabbath like Sunday than the last two. Fanny and I decided to take turns staying out at the Hospital and to-day was my turn, but fortunately Dr. Dock came early and I got away a little before twelve. When she announced that she meant to go to church he said in a nasty little way that he was glad his staff was not neglecting its spiritual interests and he hoped that she would worship for the rest of us. I wish he were a little more reverent. I went the rounds with

47

him and then stained specimens for him to examine, until just about twelve, when I had time to skip down to the church for Dr. Prescott's Bible class. Mr. Gelston, the minister, is ill and three of our professors have offered to take the Bible Class off his hands for the next three months. Dr. Prescott has it this month and his subject today was the limitations of human knowledge in the field of chemical science. It was very beautiful and so many came to hear him that the room would not hold them and they had to move into the church. After dinner I went with Kate to see a Mission school that has been started down in lower town, where she has a class of girls. On our way back we stopped at Mrs. Lombard's, for she had asked us to come in and have a cup of tea. Mr. Bourland and Mr. Budgett, the nice Englishman who assists Dr. Lombard, came in by and by, and we had a very nice time.[c] Mrs. Lombard's rooms are charming, the tea was real tea, which Mrs. Hertel's is not, and there were macaroons and chocolates besides. I am so glad to have Mrs. Lombard learn to know Kate, for they will like each other so much. One of the Delta Gammas told me to-day, that she was perfectly sure Kate had refused Dr. Warthin last Summer. I am rejoiced to know it, but I wonder at his forgiving her, as he seems to have done.[d] This letter began with an announcement that it was to be very short, but it seems to be lengthening out someway. I must write to Mother now, though.

Very lovingly
Alice.

Who do you suppose could have sent me "Japanese Girls and Women" *anonymously*. It was sent from Putnam's and there is no card or name anywhere. I am mystified.

    a. Presumably gonorrhea with gonococcal arthritis.

    b. Katharine Louise Angell, one of AH's closest friends in Ann Arbor, completed her medical training at the Woman's Medical College of Northwestern.

    c. Warren Plimpton Lombard was professor of physiology and histology. His wife was Caroline Cook Lombard.

    d. Katharine Angell and Aldred Warthin finally married in 1900; AH remained friendly with them.

8 To Agnes Hamilton

[Ann Arbor]
Sunday February *19th* 189[3][a]

Dearest Agnes

I am waiting anxiously for your weekly letter to come, because I know it will tell me all about the dreadful time at Aunt Nell's. Of course they have written me from home about it, but they don't know the details as you do. I don't know when I have been so shocked as I was when I got Madge's note about it, Thursday evening. I could not believe it, and that evening at supper when they were all talking to me, and all that night while I was cramming, it kept coming up in my mind that it really was true, the baby really was dead and had been buried that very day.[b] It is so sad that it is beyond words. Madge says you have been with the children all the time. It must be pitiful. Only I am glad that it was an inevitable kind of sickness, one that no care on her part could possibly have helped. You will tell me all about it, won't you? I keep thinking of how beautiful Aunt Nell was at Grandma's death and I imagine she must be that way now.

Do you want me to tell you about my baby cases, my girls, I mean? I don't know why I always write to you especially about them, but you know about them now, so I shall take it for granted you are interested in them. Little Miss Jackson's baby came a week ago. I had got her some flannel and muslin and she had made clothes for it and enjoyed it so much. The day that it came, when she was first taken sick, Fanny and I took her in some little dresses and jackets which one of the faculty ladies had given us and she was as delighted as a child over them. Fanny was in that section, I wasn't, so I didn't hear till the next morning that the baby was born dead. Then I went in to see her and she told me all about it and cried so because it hadn't lived long enough for her to hold it even a minute. To all of us it seemed such a good thing that it should die, but of course she loved it. Yet she couldn't have kept it, for she is going to live with her brother and he would never take her if he knew of this. I asked her how she came to get into trouble and she began to cry again. Then she told me that she had not meant to be bad but they were to be married very soon, the house was all ready and she didn't think there would be any harm and then, just a little while after, he was killed in an accident. Poor little girl, she is only twenty now. Fanny says she is sorry the baby died, for she thinks it would be a restraint on her afterwards always, but I don't believe she will be bad again.

As for Bell Willoughby, I finally made up my mind that if I was incapable of helping her I would find somebody who wasn't. So I told Cora Stoner all about her. Cora is about the finest girl in the class, if not in the whole department, a real Mrs. Vroom sort of a Christian and she took up the case in earnest and managed to break down the barrier.ᶜ Then we heard the story and such a sad one. She is a Canadian girl, who has been sewing in Detroit for the last year. Her people are all in Canada and know nothing of this, and she says she would die rather than have her father know. She says he is the dearest father a girl ever had and it would kill him to know that she had disgraced him. She kept saying "If only I hadn't done it, if I only hadn't. I never had any pity for girls who did such things before, and now to think that I have sunk as low as they." Miss Stoner asked her if she had ever known Christ and she turned away in the saddest way and said "Yes I used to once." She never says anything about the man, except that he has disappeared, she does not know where. We all dreaded her confinement very much, for that infection that I told you about makes her liable to blood-poisoning, besides giving her a swollen, very painful and entirely helpless knee. Last night just after supper Dr. Whinery, Dr. Dock's interne, came over for me, telling me to hurry for she was in labor. I went at once and found her already in the room and most of the students of the section there. They arrange those things very nicely here. A curtain falls from the ceiling down to the woman's waist and we girls are the only ones who go behind it. I stayed with her almost all the time, and it was very lovely to have her cling to me and feel that I could help her a little. Everything went off well and she has a big, hideous boy baby. But I can assure you that while I sat there and held her hands and heard her moans I thought pretty murderous thoughts of that man. This morning she was looking pretty as a picture but oh dear! what is she to do! With her helpless leg, and that baby. Of course she isn't anywhere near out of danger, will not be for a week, but the doctors are all surprised at her strength now, so she may pull through. I don't know whether or not to advise her to tell her father. Fanny say he ought to know, Cora says she cannot keep on living a lie, Dr. Dock says "why break the old man's heart?" and she herself says "what right have I to put my misery on him too." Well perhaps death will be the easiest and best solution.

50

Nothing else seems to have happened to talk to you about, for this has been examination week and I have not had time to eat and breathe, much less sleep. It is good to have it over, but dreadful to think that a whole semester is gone and only one left.

Lovingly
Alice

a. This letter is incorrectly dated 1891.

b. Winfred Wagenhals, son of Ellen Hamilton and Samuel Wagenhals, died on February 13, at the age of sixteen months.

c. Mrs. Vroom, whom AH knew in Mackinac, could best even Montgomery Hamilton in religious argument.

9 To Agnes Hamilton

[Ann Arbor]
Sunday March 5th 1893

Dearest Agnes

I am beginning this letter right after tea, so as to be sure to finish it before I grow sleepy. Margaret and Norah and Mother and Edith ought all to have letters written to them this evening, but they will have to wait until yours is written. I am very sorry that I had to skip last Sunday's letter, but there doesn't seem to be a place where I can squeeze in an extra fifteen minutes unless I take it away from sleeping time and that I don't dare to do until all the sleep lost during the mid-years is made up. Your last letter smote my conscience, as yours generally do. Edith's letters always make me feel how I am idling away my time and wasting my opportunities and how utterly ambitionless I am. Yours make me feel how many chances I have to do something worth doing, how much you would do in my place and what a wicked thing it is for me not to do it. And I realize that I have been writing to you as if my heart were bound up in these things and as if I thought and planned about them all the time, while really I live in such a rush that it is only once in a long time that I ever stop to think at all and when I do, it is much more apt to be about some purely selfish thing than about Bell Willoughby or Miss Jackson or anybody whom I could do some good to. But this last ten days I have not been allowed to see them at all, for my surgical case broke out with erysipelas and Dr. Martin forbade me to go into the obstetrical ward as long as I was attending to him.[a] So I did not see Miss Willoughby until yesterday, when Dr. Martin dismissed her and she became entirely Dr. Dock's patient and he, like a sensible man, let me go in to see her again. Her knee is much worse, poor girl!, and it would be impossible to move her as she is now. Yet after Dr. Martin had dismissed her, there seemed to be nothing to do but to have her go to the poor-house. You see these patients are the only ones who do not have to pay, they are taken free of charge and kept for a certain length of time before and after their confinement, but when they are dismissed as well and strong, they must go or begin to pay as the others do. We heard about it one morning, Friday I think, and worried over it down in the laboratory until Dr. Dock came out. Then we told him and he came to the rescue. He said he had some money, given him in trust for such cases, and that he would keep her here for some weeks yet. He is the finest man. He may try to make himself out cold and sneering and heartless and looking on his patients as purely pathological specimens, but really he is, inside of him, a simply splendid man. So she is to be here until she can go home to Canada which I hope will be soon. Some country people near here have offered to take the baby and she will let them. She asked us about it, for she has begun to love the ugly little thing, and I think she half wanted us to tell her to keep it. But we would not advise her one way or another, for we couldn't tell her to give it up and we could not tell her to keep it, when she has no way of supporting it and we can't help her do it. She told me yesterday that the man had promised to marry her, which I was glad to hear, for it makes it a little better.

I am mystified about Allen Williams. The only way I can account for his calmly ignoring it, is that the candlestick never reached him. Because you know, even if he thought it was hideously ugly, he would have to write and say something about it to us. Mother wrote me that Aunt Mary had told Papa Allen was not coming home for the Spring vacation, much to my disappointment, for I should like very much to see him then.

I wish I could write you something interesting about my work, but it is all so much the same to an outsider, from week to week, that there doesn't seem anything new to talk about. Yesterday was a fascinating day to me, but I am afraid if I told you about it you wouldn't see the fascination at all, for it was only examining blood and percussing lungs and staining slides for bacteria all morning; and in the afternoon staining and mounting specimens of cancers. But of course I can't make it sound as interesting as it is. And to you, I suppose, the human side of my work is all that appeals, while more and more I find the other side appealing to me. Hospital work is just a little bit hardening.

Mrs. Lombard had us stop in for afternoon tea again to-day and we had such a nice time. I spent all yesterday evening with her too. She is so very charming, like Mrs. Dow, only with a sincerity and intolerance of people and things she doesn't like, which are not Mrs. Dow at all. She tries to make me talk about home things and tell her all about you girls and I try to, but it seems so unnatural that I simply give it up. I love to have people tell me things and can listen all evening to their talk about their home life and their home people and that is just what she wants me to do, but when it comes to doing it myself about my own self I cannot, it's a physical impossibility.

I have meditated and dreamed over this letter until it is eight o'clock and I must stop and write to somebody else, so good-night.

Alice

a. James Nelson Martin was professor of obstetrics and diseases of women.

10 To Agnes Hamilton

[Ann Arbor]
April 8th 1893

Dearest Agnes,

This is Saturday evening, but I am sure of it to myself, while something always seems to happen on Sunday to keep me from writing you anything but the shortest, sleepiest notes. Talking of sleepiness makes me remember that I am to go to bed very early tonight and to sleep till all hours tomorrow morning. Fanny is to dress my cases, in return for a like service some Sundays ago, so for the first time since Christmas I shall have a long Sunday morning nap. The cause of this unusual craving for slumber on my part, is that I have been dissipating for the last two nights, and of course having to get up just as early in the morning. I know you are thinking that it is very wrong for me to dissipate when I need all my strength and so perhaps it is, but I cannot help seizing the chance of being human and like other people now and then, of getting rid of the odour of iodoform and the society of pathological specimens. This time the temptation came in the form of a Wagner concert, which I had no real misgivings about,

53

and the Junior Hop, which I virtuously refused. The concert was the most wonderful I ever heard. Oh dear! if I could only describe it to you, if I could make you see Isolde dying, Elsa dreaming at her window, Siegfried listening to the birds, the Valkyre rushing through the battle with their wild cry of triumph. You may say you know nothing about music but you could not have helped seeing it all, it was so vivid, more vivid than words. I wanted to clutch someone and hold tight to keep me from springing from my seat sometimes, and then again I wanted to bury my face in my hands and have no one near me, no one speak to me, only to be alone with the music. It was Seidl's orchestra, the greatest in America except the Boston Symphony, and I liked it better than that.[a] That was Thursday night, and the Hop came Friday. I was repenting bitterly of having refused to go, but of course it was too late and I consoled myself by thinking that any one really wrapped up in her work, as I ought to be, would not care anything for such things. Which did as much good as such general reflections usually do. But on our way home from the concert,—I went with Dr. Ward—he told me that he had just made up his mind to go and asked me to go for the first part anyway, though there would not be time to fill out my card or do anything decently and in order. It was a very irregular way of doing things but I thought I would, and really I am very glad I did, for it was the prettiest sight, I would not have missed it for anything. Imagine a great gymnasium, the room a seventh of a mile around, the ceiling very high, with great wrought iron beams, and around the three sides a gallery, filled with spectators. The floor covered with white canvas, the walls draped with the college colours, a raised platform at one end, with a canopy and divans for the matrons, and a grand march of two hundred couples. And the girls were most of them charmingly pretty and so were the dresses. Even with almost four hundred dancing at the same time there was plenty of space, and one could promenade around the hall without any danger of being run into. They had two bands and the dancing was tantalizingly pretty. Fortunately for me there were some of the younger instructors and professors who were not dancing, so I was well looked after, but I only stayed till after supper and then went home, under the impression that it was about one, to find it was quarter past three. The other girls got home between five and six. It was four by the time I got to bed and as I had to be up again at seven I begin to feel sleep hungry. We had a lecture on Medical

Jurisprudence this afternoon and I almost dropped off while he was discoursing on the legal side of malpractice.

Perhaps you do not realize it, but exactly a week from tonight I shall be in the bosom of my family, gladder to get there than ever before. A note from Mother this afternoon speaks of Allen Williams' vacation ending Tuesday. I suppose that means that he is at home and will be gone long before I reach there. How exceedingly provoking. Vacation has been so unreasonably late with us.

This is my last letter till the thirtieth of April, next Sunday I can sit and talk and be talked to instead.

Very lovingly
Alice

Try to meet me Saturday afternoon, won't you?

a. Anton Seidl was conductor of the New York Philharmonic.

DURING ALICE'S early years away from home, her principal confidante was Agnes; sometimes she wrote her cousin of doubts or feelings that she initially kept from her immediate family, mainly to spare her mother needless worry. The cousins' shared interest in social problems was a further bond. Indeed, for some years Agnes constituted something of an ideal in this regard for Alice, who contrasted her cousin's success in social service and religious work among the poor in Fort Wayne with her own inability to reach patients. Agnes did make her mark there as a young woman. Her most conspicuous success was in gaining community support for the Nebraska Sabbath Mission School, the fate of which had long concerned the cousins. Six volunteers had responded to Agnes' speech at a meeting of the Christian Endeavor Society late in 1892 by volunteering to teach at "Nebraska." And the following January, leaders of the First Presbyterian Church agreed to purchase an old church building for the mission and made plans to hire a minister and to hold regular church services there.[15]

Of Agnes' achievements, Alice observed: "I think that, take you altogether you are the finest girl that this wicked world contains just now and, if I am not better for having known you, it is my own fault." In the next few years, as she became increasingly interested in

55

her scientific work and less committed to the personal moral and religious endeavors that still inspired Agnes, Alice frequently contrasted her cousin's virtues with her own selfishness and inadequacies. But just as Alice viewed Agnes as a model of moral goodness she would never equal, so Agnes idealized Alice, to the latter's extreme discomfort. To Agnes, still struggling to find her vocation, it seemed that Alice had found a unity of purpose so that "her work, her charities, her studying, her reading, her coming in contact with others all are towards one aim." Alice, whose sense of equity was always strong, felt increasingly distressed by the contrast between her own good fortune in being able to study medicine and the restrictions and "iron-bound plans" imposed on Agnes by her family.[16]

Although religion remained a frequent subject of discourse between the cousins, Alice's religiosity diminished somewhat during her Ann Arbor years. When she first arrived, she frequently attended several Sunday services (six on at least one occasion), but during her second year she not only went to fewer meetings but sometimes allowed laboratory work and other "worldly employments" to encroach upon the Sabbath, which she had been brought up to observe strictly. Having once contemplated becoming a medical missionary, she felt oppressed by the too-exemplary conduct of a missionary-to-be. Nevertheless, she taught a Sunday School class in Ann Arbor's poor district, the "Lower Town," and took this responsibility seriously enough to intercede with the superintendent on behalf of one of her charges. She also attended at least one church social at which she and her protegées joined in games of "drop the handkerchief" and "pussy and mousie" with neighborhood men. Following "heroic efforts" as pussy, she caught "the biggest and most awkward of the men," after which she retired because she was "in mortal terror of being made mousie to some man's kitty and that would have been much worse."[17]

The years at Ann Arbor were among Alice Hamilton's happiest. She liked the work, revered her professors, and enjoyed her first taste of independence away from the protected home environment. She loved the exposure to new ideas, some of them "crazy, of course," but all of them interesting. Despite her fears of falling behind in her work, she found time for occasional "dissipations," including afternoons in the woods hunting for wildflowers and hickory nuts and at least one "school-girl spread" for which she and a friend stewed oysters on hospital bunsen burners. Five weeks before her graduation in June 1893, she wished she could remain an undergraduate forever.[18]

She found it difficult to think of herself as a physician and was disconcerted to be addressed as "Doctor." Soon after graduation, she instructed Agnes that "if you care anything for me," she was never to put anything but "Miss" on envelopes. In her autobiography, Alice Hamilton indicated that she had settled on a career in science rather than in practice while still a medical student, but had decided to take a year of hospital training to avoid becoming too one-sided. Toward the end of her life, however, she attributed the decision to the agony of watching a patient die following childbirth. In either event, the choice was consistent both with the values of her Ann Arbor professors and with her own fear of clinical work.[19]

Internships for women, more limited in the 1890s than opportunities for formal medical education, were mainly available in women's hospitals. When one of the most prestigious of these, the New England Hospital for Women and Children, had no place for her, Alice Hamilton accepted an internship at the Northwestern Hospital for Women and Children in Minneapolis.

11 To Agnes Hamilton

Don't read this aloud till you have read it to yourself.

> Northwestern Hospital
> Chicago Avenue & 27*th* St.
> [Minneapolis]
> [July 16, 1893]

Dear Agnes,

I have been here just thirty hours and I feel as if I had more to talk about than ever in such a short time before, except the first day at Farmington. In this time I have been through the various stages of bewilderment, utter despair, then a slight ray of hope and now temporary cheerfulness, which will probably soon be succeeded by despair again. But I am going to tell you all about it from the beginning. I left Mackinaw Friday noon with Edith and Papa. Papa had to escort me as far as the junction where I was to take the express, for that part had to be gone over in a freight train and he was afraid that the caboose might be filled with rough lumbermen. It was lovely having Edith go too, for it divided up the good-byes and kept me from making a water-spout of myself on the dock, as I am sure I should have done otherwise. It was a very big wrench leaving Mackinaw, for I have been having such a perfect time there for the last two weeks and felt as if the Summer were only just beginning. Indeed when Papa and Edith put me in the train at Trout Lake and departed, I quite succumbed and was thankful that there was nobody to see me but the porter. It was only a night's journey to Minneapolis, I reached here yesterday morning at half past eight. It seemed the cleanest, freshest city I ever saw, but I was chiefly impressed by the feeling that there was not a single soul in the whole big place who knew me or whom I knew. So you can imagine my pleasure when I heard myself called by name and turning saw Belle Foster, an Ann Arbor girl, and an intimate friend of Fanny's, who happened to be waiting for a train. She told me how to reach the Hospital and sent me on feeling much better for the sight of a familiar face. The hospital is a big, dark-red building, out in the suburbs, with plenty of fresh air and space about it. I was shown into a rather forlorn little reception room and was told that Dr. Everitt, the resident physician, was making her rounds of the wards and I should have to wait.[a] I waited ten minutes, growing more and more frightened as time went on. Then the door opened and a neatly dressed woman, with a strong, sensible face, came in, greeted me very kindly and offered to show me my room. I followed her up the stairs and found a very nice, large room, well-furnished, with a big window and a writing-desk and a luxurious closet. As my trunk had not come, I asked her to show me over the Hospital and she did so. We went from room to room, while she pointed out the patients, explained their ailments and treatment and, incidentally, my future duties. She would say "The surgical and gynaecological treatments are given by the interne every morning, the electrical in the evening. The nurses are expected to go to the interne in any emergency, night or day. The interne makes the rounds with me in the morning, by herself at night. She takes the history and makes abstracts of the treatment of each patient. The obstetrical work is entirely hers." At this point, while I was wondering if I could stand any more, she pointed out a woman to me. "She expects to be confined any time and will be your charge." I gasped. "Dr. Everitt," I said "do you mean that *I* am to manage the case, when it comes off"? "Certainly" she said. "Of course if it should be a complicated case, beyond your powers, you could send for me." Complicated case! why I shudder at the thought of the simplest one in the world. I had far, far rather

58

amputate a leg. We ended up at the pharmacy, where she put the last touch to my despair by saying that, as there was no pharmacist, the interne was expected to put up the prescriptions. So besides all the other methods of killing, I am to have the chance of poisoning my patients too. While we were in the pharmacy Dr. Hood dropped in for a moment.[b] She is, you remember, the real head of the hospital, the visiting physician. She is tall, thin, worried-looking, with a kindly face and a nervous manner, not at all the sort of a woman I expected. And she graduated twenty years ago, only fancy! I must have been looking rather limp, I suppose, for she spoke to me very kindly and reassuringly. I imagine I shall grow quite fond of her and never shall be as afraid of her as I am of my self-reliant, matter-of-fact little superior, Dr. Everitt. She, Dr. Everitt, is an object of wonder and admiration to me. Not of unmixed admiration, for she is desperately unscientific, her surgery is slovenly, and her lax methods in physical diagnosis would turn Dr. Dock's hair gray, but she is so decided, has such confidence in herself, such calm, authoritative ways with the nurses and such a cheery indifference with the grumbling patients, that I constantly envy her. She looks upon me as a very doubtful experiment. Philadelphia training is very different from Ann Arbor, and really, in spite of my helpless feeling just now, I am glad that my school was Ann Arbor. All the accurately careful, elaborate work that I have been taught to consider so important, is ignored here, and I am expected to make off-hand diagnoses, rapid prescriptions and meet emergencies without losing my head. None of which I can do at all. And there is no laboratory, the microscope is not as good as my own, they diagnose tuberculosis without looking for the bacilli, and I don't believe they ever heard of the plasmodium of malaria. But there are eleven cases of typhoid fever here, besides several consumptives, paralytics and Bright's disease, so I ought certainly to be able to do good work, even without my trusted chief to oversee me. Dr. Everitt very kindly gave me all of Sunday to myself, so I have had to-day off, and have spent my time dreading to-morrow, when my duties begin. I look with terror at the confinement case whenever I pass her. Suppose she should do it to-night. There is no earthly reason why she should not. I have read over all my lectures on the subject, but dear me! everything will fly from my head in a moment.

My meals are sort of funny and sort of forlorn. The dining-room is in the basement, a pleasant room, with two long tables, at the head of one of which Dr. Everitt sits, the other I preside over. Dr. Everitt greets the nurses as they troop in, in her usual cheery, business-like way, sharpens her knife vigorously and carves as if she had done it all her life. I sit meekly in my place, speak when I am spoken to, and make an utter mess of the carving. If you knew how it rattled me to be called Dr. Hamilton! It seems always as if they did it in ridicule and forces the sense of my own ignorance on me. And they all do it. The nurses treat me as if I were a teacher and they boarding-school girls. I went up on the third floor, their floor, last night to take a bath and they were making very much more noise than they ought. As they caught sight of me, I heard a loud whisper "Goodness! There's the new doctor" and immediately the nightgowned figures scuttled off to their rooms. I felt like Miss Cowles.[c]

It is seven o'clock Sunday evening. Dr. Everitt has just made the evening rounds, and dismissed me for to-day, so I am free from the possibility of mistakes unless an "emergency" should arise in the night. If one does I believe I shall jump out of the window. I cannot imagine any other course of proceeding which would save me.

It is simply lovely out. I wish I had somebody to stroll up and down with. Last night Dr. Everitt came for me and took me for an hour's stroll, telling me all about the afflicting "ladies' committees" and confiding to me that Dr. Hood was old-fashioned and very un-scientific. I groaned in spirit, as you may imagine, for if Dr. Everitt thinks her unscientific what must she be?

Do you know, I am hovering on the verge of a very babyish and senseless fit of homesickness? I think I shall take a good dose of Epictetus to brace me up.

Good-by my dear. I hope there is a letter coming for me this week.

Lovingly
Alice

a. Ella B. Everitt, a graduate of the Woman's Medical College of Pennsylvania in 1891, was superintendent and physician in charge of the Northwestern Hospital for Women and Children.

b. Mary Gould Hood headed the first medical staff of the Northwestern Hospital in 1882. In 1893 she was senior physician.

c. A teacher at Miss Porter's School, who was responsible for turning out the lights at night on AH's hall.

12 To Agnes Hamilton

[Minneapolis]
Sunday July 23rd 1893

Dear Agnes

It is a very hot afternoon and I have come down to the drug-room as a last refuge. It is a clean little room, with a north window and there is a large table to write on. My room has the western sun and has been unbearable since two o'clock. This hospital has such wide halls, with big windows at either end, that it is always cool somewhere. It is visiting hour and the wards and halls are full of sympathizing relatives and friends. Just outside my door I can hear the poor old Bright's Disease pouring out her complaints and discomforts to a worried-looking old man, her husband I suppose. They all seem to do that, so that I really pity the visitors.

That blank means a few minutes stop, for I ran upstairs to see how a little new patient was doing. It is a baby with cholera morbus, who came in last night, all gray and cold and gasping.[a] Now she is warm and more natural-looking and was taking her peptomized milk nicely, when I saw her. It is perfectly beautiful having her come in for now I can watch Dr. Everitt's treatment and apply it to my dispensary patients. Did I write you about them? No, I couldn't have, for I wrote last Sunday and the first one didn't come till Monday. Well my dear they are the most terrifying things and I have had seven of them. Perhaps you think seven not very many, but just wait till you have them, and five of them children! The first one was the worst, a little six months old baby, and its face and hands were blue and cold and it gasped for breath and couldn't retain even water. I diagnosed the form of Summer complaint all right, but I couldn't prescribe all by myself on the spur of the moment, why it frightened me to think of it. So I went to Dr. Everitt and she laughed at me, but wrote a prescription and gave me a lot of common-sense advice for the mother. *She* took sick babies as coolly as she took her breakfast long before she graduated. I have worried and worried over that baby for the mother promised to come and tell me how it was getting on and she never has. I am so afraid it is dead. That same day two women came and I managed better then, for I was quite in my element examining them and I coolly told them to come back the next day for the medicine. Dr. Everitt laughed at that too, but I felt more easy about it than if I had had to run down to the drug-room and put them up at once. Well then came another baby, with bronchitis this time and as, Dr. Everitt was out, I had to manage her all alone. And yesterday the mother brought her back with summer complaint. Oh dear, I wish they would not bring me babies. I worry over them inexpressibly. This one was a little Swede—they are all Swedes or Norwegians or Danes—and the woman who came to interpret was a Swede with a different patois and they could only half understand each other. Now fancy getting anything into or out of such people. I just know she will put the Ipecac into the milk and give drop doses of the lime-water. Another baby with the same trouble—little Sigrid Engell—and three surgical cases, who only needed dressing, make the eight for this week. I wonder if the next will bring as many. If only they are grown up, I don't care so much, though they all frighten me.

Don't you ever write me a letter like your last one. Talk about making me feel better, it made [me] feel unutterably blue. It simply showed me what a beautiful, high idea you have of my work and what a low one I have myself. For I don't think of it as a mission of healing at all. In the first place I don't feel as if I were healing. When a surgical case heals up, or a typhoid goes home well I feel as the backers of the victorious man in a prize-fight feel. They have been watching the fight and helping and encouraging their champion and he has conquered, but he has done the fighting, not they. And when they don't get well, I don't find them at all less interesting, often more so. Indeed Agnes, you mustn't misjudge me so, and fancy that because I chose to study medicine and was fortunate enough to have a father who would let me, I have proved that there is something in me. I simply have placed myself in a position that will show if there is or isn't, and day by day I am finding out that there isn't. And some day you will find it out too, only I hope I will die first.

Dr. Everitt has been in here, sitting on the little slip-ladder and talking about the baby. She thinks it is developing brain symptoms. I went to look at it again and it is growing gray, the way it was this morning, and it turns up its eyes in the most ghastly way. Poor little atom! it has nobody to care. Its mother left it and the woman who brought it, told us to "drop her a postal if anything happened."

Do you know I feel better than I did last Sunday? I am not as ridiculously homesick and I am not afraid of the work much, except the babies, and Dr. Everitt is exceedingly interesting and growing very companionable. She loves to sit around for ten minutes at a time and gossip or rather talk medicine and tell me stories of her student life. But around the wards and in any of our work, we are just as much superior and subordinate as ever. She is very nice to work for. She is broad-minded and fair and can be trusted to tell me at once, if she has any criticism to make, so that I need not be fancying that [I] am doing wrong. I am awfully busy most of the time, and my spare moments I spend on "Diseases of children," and planning out what I should do if I were called on in some emergency. Dr. Hood took away my breath, the other day, by informing me that Dr. Everitt means to resign on the eighth of August and the new resident, a Philadelphia woman, does not come on until the first of September. So there are three weeks in which I am to run this hospital alone and unaided. Did you ever hear of anything so frightful? Of course Dr. Hood or Dr. Fifield can be called on at any time, but that does not mean every five minutes all day, and what shall I do?[b] Even as a matter of time I feel appalled, for every minute almost is taken now, and if I have the nurses' training and the diet kitchen and the admission of patients and accounts and ladies' committees to look after besides, I shall become insane. However, I don't mean to think of it until the time comes. Then I shall pray for no babies.

I think you might write me a longer letter this time. And don't talk about my work in that way again, for it makes me feel like a wretched hypocrite.

Lovingly
Alice

a. Cholera morbus or summer complaint was a severe form of diarrhea, which struck mainly in the late summer or early fall. In the late nineteenth century it was frequently fatal to infants.

b. Emily W. Fifield was attending physician at the Northwestern Hospital.

ALICE HAMILTON soon learned that there was an opening at the New England Hospital for her after all and set about finding someone to replace her in Minneapolis. (The substitute, a Michigan graduate of 1892, brought out her worst snobbishness: "She is Irish and Catholic and rather third class, but then, as I told Dr. Hood, you don't need a lady for practical hospital work.") Of her reasons for leaving, she wrote Agnes: "The work is fine, but it is a pretty lonely, desolate sort of a life, and a winter of it would be very hard. Boston will be so much better."[20]

She started her internship on the maternity ward of the New England Hospital for Women and Children in mid-September. Although she would have preferred working on the medical ward, she began the year with high expectations.

In her first surviving letter from the hospital, Alice's enthusiasm for meeting people of contrasting backgrounds and politics is apparent. The object of her interest was her fellow intern, Russian-born Rachelle Slobodinsky (1869–1946), a graduate of the Woman's Medical College of Pennsylvania. She remained a good friend and, with her husband, Victor S. Yarros, a journalist, lived at Hull House from 1907 to 1927. Rachelle (or Rachel) Yarros, a specialist in obstetrics and gynecology, was an early leader of the birth control and social hygiene movements. She taught at the University of Illinois Medical School and in 1926 became professor of social hygiene there. Yarros later recalled that one of Alice Hamilton's most striking characteristics when they first met was her "desire to know" and her repeated request to "*Tell me everything.*"[21]

13 To Agnes Hamilton

N.E. Hospital
Roxbury
September 27 [18]93

My dear Agnes

Here it is Thursday and I have not written my Sunday letter yet. Really I am so lazy that I do not even muster up energy enough to write. It is late in the afternoon and I have just come in from a very delightful solitary walk. It was cold and cloudy all day, but now the sun has come out just before setting and is making up for his absence by being wonderfully and almost unnaturally brilliant. I wish you could look out of my window just now, for I know you would exclaim "Inness!" at once.[a] The sky is all dark gray except down in the west, where there are long gold streaks with the sun in the middle of one of them. The rays strike down on some trees just below me and a bit of lawn, making them a most impossible orange gold, while just behind them rise tall trees and two sloping hillsides covered with a thick gold mist, as if they were miles away instead of being within a block or two. I have been wandering over Franklin Park, and trying to imagine myself back in Farmington. It is very like Farmington in many ways: steep little hill sides with oaks and sumach bushes, growing beside and on the bare gray rocks which look as if the grass had peeled off and left them naked. And in the distance the Blue Hills look very like our Farmington hills. I have quite a good deal of liberty now, for the next two cases that come in are Dr. De Hart's and I do not need to have anything to do with her cases until just at the end.[b] Our work comes so by fits and starts. Now we had five cases between Friday night and Sunday morning and we have not had one since. So we are either distractingly busy or have nothing to do. Saturday night was very exciting, for we had five patients come in, two for Dr. De Hart and three for me. My three all came off and one of hers, so we really had a very rushing time. Her case and one of mine were instrumental cases, mine an especially interesting one, for the baby was almost dead and the woman had haemorrhage twice and almost died. I am growing quite to like obstetrics and not to be as terrified over a case as I was at first. Dr. De Hart knows quantities about such things and I get her to instruct me in New York methods, while Dr. Slobodinsky gives me the Philadelphia standpoint. Dr. Sloboda,—as we have to call her for short—is the most interesting thing. I keep wishing and wishing that you could meet her. She is only our age, but she has lived through more than we will have when we are sixty. Yet through it all she is a light-hearted, natural girl, with something almost childishly simple about her at times. You have to question and question her to get out anything about herself and I am afraid sometimes she will think me impertinent, but it is so interesting that I cannot help it. Think! she began to be a Nihilist when she was thirteen years old. That is, she began to study in secret, read forbidden books and join secret societies, and that, she says, is the beginning of Nihilism. She used to study, unknown to her parents, with a scholar who had been banished from the University where he was studying, just the day before he was to take his Ph.D. and sent to this town where she lived. And one morning when she went to take her lesson with him, as usual, she was informed that the police had come for him during the night and taken him, no one knew where. That, she said, made an impression on her that she never forgot. Many of her own friends have disappeared in just such a way. Some they have traced to Siberia, one was hanged, others she has never heard from. Of course her

65

parents were bitterly opposed to her doings, and no wonder, for the police used to come to her father every now and then and tell him that, if he did not stop her going to those young people's clubs and talking so imprudently, they would simply have to arrest her. Finally matters came to such a pass that it was a question between banishment of some kind and flying the country, so she decided to come to America. She seems to have had money enough for the asking, but she said she had no right to use her father's money while she constantly disobeyed him, so she only took enough for her passage. She came with a number of enthusiastic young Russians. It is amusing to hear her tell about their intense enthusiasms and extreme ideas. She laughs at them herself now. They would not eat anything but the coarsest food, would only have one dress at a time, and would use no luxuries that everybody could not have. And they scorned all little courtesies and ceremonies, addressing each other in the bluntest, plainest fashion possible. As soon as they reached America they began to hunt for work. She became a factory girl in a large house in Rahway, New Jersey, and worked there for two whole years. Of course she was well-educated enough to be able to take a better place, but she wished to get in among the working classes. And she does know them thoroughly. She was bitterly disappointed in America though. She headed a strike in her factory and she says it was one of the bitterest and cruellest experiences of her life, for it failed utterly. Finally she decided she could not even help people when she was slaving from morning to night to get the bare necessaries of life, so she accepted the offer of some Russians over here to educate her in medicine. And that is how she came here. In her vacations she has nursed in poorhouses and has had the most interesting experiences. She believes that the only way in which people can reach the working classes is by living right among them and not letting them know for a while that you are not one of them. She knows that she can never go back to Russia, but she is so hopeless over the condition of things there that she does not care. I wish you could hear her talk for I can just imagine how intensely interested you would be. Here I have been writing pages on her instead of doing some useful reading. I must stop and tape [?] a brace.

With much love
Alice

a. George Inness was a prominent American landscape artist of the Hudson River School.

b. Florence De Hart was a graduate of the Woman's Medical College of the New York Infirmary.

THE MOVE to the New England Hospital (located in Roxbury, a twenty-minute "electric" ride from downtown Boston) brought Alice near to Allen Williams. He had been away from Fort Wayne for some years, first at the Chauncy-Hall School and then at Harvard College. In 1893 he was studying medicine at Harvard and lived at the home of William H. Ladd, principal of the Chauncy-Hall School, and his daughter Emily J. Ladd, who had taught there. Emily Ladd was considerably older than Williams, and their relationship was problematic. It was altogether a difficult period in Allen Williams' life. Alice found him "as beautifully and distantly courteous as any acquaintance might be" but felt she no longer really knew him.[22]

14 To Agnes Hamilton

[Roxbury]
October 15th 1893

Dearest Agnes

This will be just an apology for a letter, because I am so sleepy that I can barely keep my eyes open, and yet I haven't the face to go to bed at eight o'clock. It is partly because I was up all last night with a case and partly because I went off for a long blowy, windy walk with Allen this morning, and the two together have made my eyelids so sticky that I did not dare go church-hunting again, for fear if I found a church I should go to sleep during the sermon. This Roxbury is a positively heathenish place. Last Sunday evening Dr. Ingram and I went to five different churches and found every one shut up and dark.[a] Finally we found a little chapel and went in and heard some Gospel hymns and anecdotes about the minister's dead children. But isn't it strange not to have any evening service! Allen and I had the most beautiful walk this morning. We went to Arnold Arboretum, the tree park that belongs to the Bussey Institute. It is a very beautiful place. We climbed a hill-side that looked like the one I always remember in the American Artists, you know, with a white thunder

67

head in the sky above it. When we reached the top we found a most marvellous view on three sides of us, Boston and a faint glimpse of the harbor on one, the long range of Blue Hill with the tiny suburbs at its base, on the other, and low, rolling forest-covered hills on the third, while the Arboretum lay behind us. It was a clear, windy day, with light clouds passing rapidly across the sky and making shadows chase over the landscape. Allen has been particularly nice lately. He came out the other evening, to bring me some books, only I had a case on hand and could not more than speak to him. Two interesting cases have come off since I wrote you last. Somehow all my cases seem to have something unusual about them, while Dr. De Hart has case after case of bouncing children and healthy mothers. My little negress had a seven months' baby and made herself so interesting over it that ten doctors assisted at the performance. Not that they did anything, except two or three of them, but they came to see it. Her baby is the tiniest thing you ever saw. It weighs only two pounds and three quarters, while most babies weigh from seven to ten pounds. It looks like a poor little unfledged bird. I keep it rolled in cotton-wool and laid in a clothes'-basket with hot bottles around it. It is fed from an eye-dropper. Everybody insists that I cannot save it, but I am determined to, and as it has weathered through three days, I don't see why it should give up now.[b]

My case last night was a very lively one. Dr. De Hart and I were roused soon after we had gone to bed by a hack driving up to the door and the sounds of shrieks and groans proceeding from it. We slipped into our clothes and went down stairs, to find a middle-aged woman in an almost raving condition and a middle-aged man, her husband, in something very like hysterics. We hurried her into the confinement room supposing that we had only a few minutes left and then discovered to our disgust, that there were several hours more anyway, and that everything was as serene and normal as could be. At first we thought the woman was temporarily insane, as they some-times are, but Dr. De Hart went out and interviewed the weeping husband and found that she often was this way, it was only "her nervousness." So there was nothing to call it but hysteria, and I never saw a worse case. She raved, she moaned, she attitudinized, she gave up the ghost with many tender dying messages to everybody, she besought us in almost profane language to kill her on the spot. She would lie still, apparently unconscious, then suddenly start up with a shriek, clutch me wildly by my arm or throat or hair, with her eyes starting from her head, and gasp out "Death is at hand!" or some such thing. At first I tried soothing and sympathy, but she only grew worse. Then I sent the nurses out, tied a long towel to the foot of the bed and told her to clutch it when she wanted to, and Dr. De Hart and I sat down at a distance and talked about other things, telling her, when she shrieked, that every woman who had a baby had the same thing to bear. Well it did seem callous, but it was all we could do. She quieted down wonderfully, and began to weep and lament the hard-hearted-ness of single woman. "How hard you look" she said to me. "Don't you think you are very cold-hearted?" I said, no I didn't think I was at all. "Ah" she said "we never see our own faults." She was a very irritating case, as you can imagine, for she hadn't the sense of a baby and wouldn't help herself a bit. The baby came at a little after five and when I told her that it was a dear little girl, she calmly remarked that she did not care whether it was dead or alive, so long as she was safely through it. There is maternal instinct for you! I don't know anything that repels me more from a woman than the sort of animal selfishness they sometimes show. Here it is eight pages, and I said I would only write a short letter. And it is quarter to nine, so I can really begin to undress and get into that blessed bed.

Very lovingly
Alice

a. Maria P. de Boiij Ingram had been a medical classmate of AH's.
b. The baby died the following day.

15 To Agnes Hamilton

[Roxbury]
October 26th 1893

Dearest Agnes

My Sunday letter has waited over till Thursday this time and I think I shall have to keep it till next Sunday and send the two installments off together. I should really like to know how I can be expected to keep the fourth commandment if all of the babies will persist in coming on Sunday. It has grown proverbial here now, our Sunday babies. Not that they always come on the day itself, but they come Friday and Saturday and Sunday evenings, so that we are so sleepy on Sundays that we cannot do anything but take naps. I had a thirteen-pound one

Saturday night, the most enormous baby I ever saw. It is really a monstrosity.

Allen did not come out on Sunday, but Saturday afternoon he took me to see the Dartmouth and Harvard foot-ball game. Such fun! I enjoyed it immensely. Of course it was not a really exciting game, for the score was thirty-six to nothing, but some of the playing was excellent, and they were not only playing Dartmouth, but Yale indirectly, for they were trying to beat her score of twenty-eight to nothing the week before. There was a large crowd which I should have enjoyed watching, if I had not been too interested in the game. And it was nice to see the Yard again and recognize all the places. Allen is really so very nice. I wish I were quite sure that he really wants to come traipsing out here all the time. Now with you I should feel calmly assured that it was as much a pleasure to you as to me, but one never knows what is going on inside of him.

Sunday, October 29th

Allen and I went to Trinity this morning. Of course we were late. I find that that is the penalty of going any place with him. But we entered just in time to hear the Te Deum sung most magnificently. Of course you remember the church. Really I believe it satisfies me more than any church I have ever been in, as a building I mean. And to-day the sunlight came and went in the most beautiful way, making a fresh pleasure every time it lit up those glorious windows. The service was exactly as it was in Phillips Brooks' time, with him left out.[a] His successor reads beautifully but I don't care at all for his preaching. His [sic] is like Bishop Davies in his faculty of saying truistic things in a most impressive way, that imposes on you for the moment, until you think it over and find that there was nothing there after all. We walked almost all the way home, for it was such a lovely day. I half meant to go to church with Dr. Evans this evening, but I found that Dr. De Hart wished to, so I gave up to her with rather a feeling of relief.[b] I went with her last Sunday evening and proposed to go again only because I was afraid she saw that I did not like her church. She is a very good and rather dense girl, and she enjoys preaching of the Mr. Northrop description. The Baptist Church here in Roxbury is a refined and rather more intellectual edition of ours at home, with Mr. Northrop cropping out all the time.[c]

We had no Sunday babies this time, indeed we have not had a single case for a whole week. Do you know I believe I shall let myself out to you to-night and just tell you what I think of this old place. I have kept it in six weeks now and it simply is boiling over, yet I won't write Mother about it because it would distress her so to think that I am disappointed, and if I should write to the girls she would probably see the letter. So I shall take you as my safety valve and pour out all the abuse that has been piling up for six weeks against this narrow, petty, squabbly, idiotic place. Just fancy! there are exactly as many patients here as I had all to myself in the Northwestern and there are six internes to look after them. Do you wonder that we sit most of the time with folded hands, or that we have indignation meetings daily, and denounce the idiotic women who run this place. Why under the sun it has the reputation of being the best training place for internes I cannot conceive. It must be because it is so very old and so many prominent physicians have had their interneship here. Oh it is such a petty little place! There are officials on officials, none of whom have half enough to keep them busy and all of whom, therefore, try to get the work into their own hands and thus give rise to squabbles innumerable. We have enough officials and internes to run a large hospital and we have an amount of etiquette and red-tape that would overstock Bellevue. We were each presented, on coming here, with a list of printed rules which we were to obey and which we diligently set ourselves to study, until we discovered that the list only covered about a quarter of the existing rules, the rest of which are not printed and only discovered when someone has been unlucky enough to violate them. They issue appendixes and revised editions of the rules occasionally, but no list has ever been long enough to comprise them all. Most of them date thirty years back and were made for internes who were only prospective medical students, but they are still kept up and handed down to us graduates. The logic of the managing doctors is incomparable. You ask them why you must do such and such a thing "Oh because it is the rule." "But why was such a rule made?" "Because one must have rules." So they make rules in order to have rules and keep them because they have made them. As one of the old internes, now a practicing physician told me, "one never discovers how much wickedness and rebellion there is in one till one has had a year in the New England." And I never thought before that I was tenacious of my own dignity or jealous of my rights, yet here I find myself growing quite red-hot when I am treated—as we all are—like a raw school-girl, reproved before my own patients for a bit of Ann Arbor heresy, or pitched into by the Irish scrub-girl for not putting away my things

71

in the drug-room. The Northwestern has quite spoiled me, I am afraid. Really though, I would stand it all if they would give me work. But to feel that I am simply losing a year which I cannot spare, sitting around and reading text books, when I need practical work, kept by idiotic rules from using even all the inferior opportunities which I have—well it is simply maddening, that is all. And the visiting physicians are so bland and patronizing, and so convinced that there is no hospital like the New England and no advantages like ours. Someday I know I shall speak my mind out to them, for it is growing too full not to spill over sooner or later. I won't put it down to the fact that they are all women, for so they are at the Northwestern, but they are narrow women, women who study gynaecology and obstetrics and know absolutely nothing else, who are in a state of self-distrustful antagonism to all men doctors, and who escape discovering their own inferiority merely by avoiding their superiors. I think if they knew the severe criticism to which their work is put by the meek young internes under them, they would be utterly amazed. As it is they never suppose we could have an opinion of our own. It irritates me to think that there is not a man medical graduate in the country who would accept so inferior a position as this, yet here are we, who know just as much as men students, obliged to accept places where we must divide with six the work that is only enough for two. Why I was rash and impulsive enough to leave that excellent place and come to one that is living on the ashes of its former reputation, I cannot imagine.

There, I have let out some of myself and I feel better for it. This place makes me feel as if I were tight-laced and must burst my whalebones for a good, long, big breath. So if I elope or come sailing home or do something suddenly awful, you at least will be somewhat prepared.

Alice

a. Phillips Brooks, rector of Trinity Episcopal Church in Boston, had been one of America's most popular and respected preachers in the post–Civil War era.

b. Sarah Evans, an intern, had received her medical degree at the Woman's Medical College of New York Infirmary.

c. AH probably meant Stephen Northup, minister of the First Baptist Church in Fort Wayne.

ALICE HAMILTON's diatribe against the New England Hospital was the first of many. Unaccustomed to such rebellious feelings toward those in authority, she sought to justify herself to Agnes, who had disciplined herself in the art of obedience. She also needed to convince herself. In subsequent letters she complained that the rules and the ways in which they were enforced gave her no authority whatever, which "reduces me to a slightly lower level than an ordinary nurse. Yet I notice that if anything goes wrong with a patient I am blamed, and blamed in the patient's presence too." She also charged that patients with serious diseases were not admitted because they might die "and that would make the Hospital report look badly." By mid-November, she decided to inform her mother of the situation, claiming that she had "not been even ordinarily prudent" and might be "sent home decapitated, drawn and quartered."[23]

What of the validity of her complaints? The New England Hospital, founded in 1862, had been considered a model women's hospital, but evidence from the 1890s tends to support Hamilton's charges. In 1891, Bertha Van Hoosen, then the resident physician, provided a detailed critique of the organization and working of the hospital. Van Hoosen, later a prominent surgeon and a founder of the American Medical Women's Association, considered the position of the interns "unendurable": because they lacked authority and were entirely subservient to the resident, neither patients nor nurses respected them. Most serious, she believed, was the failure of the hospital to provide interns with essential practical experience. And in the fall of 1894, two attending physicians were asked to resign following a complaint brought by the resident physician. At that time a member of the Board of Physicians declared that the hospital was in a "state of 'dry rot' " and that strong measures were needed to halt its deterioration.[24]

16 To Agnes Hamilton

[Roxbury]
Tuesday December 5*th* 1893

My dearest Agnes

I have a case in and as soon as the nurse has prepared her, I must go down and look after her, so this will be an interrupted letter. Still, as it promises to be a slow case, I can probably finish this in the intervals of attending to her. My dear, you have attained to your quarter of a century since I wrote to you last, haven't you? It makes me feel as if I were tottering on the verge of decrepitude, when I think

73

that we are really twenty-five. It is the greatest step of all since we were twelve and twenty.

Again I have let Sunday slip by without writing to you, this time out of sheer laziness. I could not get off in time for church in the morning, so I telephoned to Allen not to come, as Dr. Slobo wanted the afternoon. He is so very scrupulous about always coming on Sunday, that I was glad to release him once, for I know that sometimes it must be a bore. We had had snow in the night, which turned to rain later on, and I clothed myself in my mackintosh, rubber-boots and tam and went for a long, lonely, slushy walk. It was horrid in the streets, but very beautiful when I reached the Park. I don't believe that any day is really ugly in the country. I met nobody but one solitary Park policeman, standing like a picture of misery, with the rain dripping from his rubber cape and looking at me as if he thought I must be insane. I enjoyed it very much, as much as I can a solitary walk, for you know I do not enjoy my own society very enthusiastically. But there is nobody here, except Allen who accords with such a walk. Dr. De Hart is nice, but, to use Rose Red's expression, one is always "toppling over the edge of her mind," and Dr. Slobodinsky would talk individualism and Herbert Spencer all the way.

Wednesday—

I had to stop there and attend to my patient, so I put my letter up on the mantel-piece and Dr. Slobo came in and dropped it into the baby's bath-tub, giving it this wept-over appearance. But I shall send it all the same, even if it is illegible. Your letter came in the course of the morning, but I could not read it then or indeed until almost nine o'clock that evening. It was one of my very few busy days. We had three cases, two of mine, and one of Dr. Slobo's. My first one was dreadfully tedious and tiring, and it finished just in time for me to help Dr. Slobo with hers, and then to go on to my second, who kept me until after midnight. At the end I was so tired that I dozed between the pains. But they were very satisfactory and I would willingly have every day as busy. I have read your letter through three times already, and shall keep it to look at every now and then. I believe that with the first part of it I do agree, on thinking it over. Yet when I wrote, I was thinking of some of Allen's incomprehensible ways, which have puzzled all of us, and wondering if it would not have been better if he had known, as I should know, that a complete, frank account of every thing would be demanded by his family and that he

74

would have to show that commonsense motives had ruled him, rather than being allowed to shroud the whole in impenetrable mystery. But as to telling him of his faults, there I do think you are right. I never should do it with one of my sisters or with you girls, except in case I should think you had done something which I couldn't understand and which, for my own peace of mind, I should have to have cleared up. Then I would ask you, knowing that your explanation would set things right. But you know how at home we make fun of each other's small absurdities and criticize what we say and do. It is that sort of thing that seems so impossible between Allen and me now, though I can remember when it was as natural as could be, years ago.

As for the rest of your letter, I will acknowledge that I have done a great deal of unnecessary grumbling, but, my dear, do you think it is best to take quietly and without protest, things that you know are wrong, absurd and unjust? I truly believe that if the internes who have been coming here for years, had made a strong stand against the treatment accorded them, that the chiefs would have seen that they must improve matters. As it is, whether from politic desire to stand well with the authorities, or from lack of courage to face an angry woman—and I must confess that it is not a pleasant thing to do—the internes have submitted to everything and contented themselves with abusing the place after they were safely out of it. Now Dr. Smith is detested by every girl who has worked under her, so much so that she is famed for it in both the schools (Philadelphia and New York), yet until Dr. Slobodinsky came, not one of them had ever dared tell her that her behaviour was unbearable. The result to Dr. Slobo is that she has a sworn enemy in Dr. Smith, but if one or two more would follow her example it would have its effect.[a] And really some of the treatment here ought to be exposed before the State Medical Society as outrageous malpractice. I know that I have promised them my services and am under a solemn contract, but I consider that they have not fulfilled their part of the contract. They promised me good experience and teaching and they are giving me neither. I know it is only for a year, but think how much a man can learn in one year's interneship. Now don't think that I am protesting against what you said, I am only enlarging upon my own letter. And do please write me just what you think of all that I tell you. Sometimes I really am puzzled to know how far it is right to resist and how far I ought to submit. For you see it is not all the treatment of us internes that we resent, that is a small matter, but our chief's ignorance is really great

75

enough to be dangerous. We do not claim to great diagnostic powers ourselves, but when Dr. Harrington tells a woman with beginning melancholia that there is nothing the matter with her but unreason-ableness, when she refuses to notice heart disease that we all can diagnose, when she keeps poor working women here week after week without ever finding out what is really the matter, why then it becomes a very nice question as to whether it is right to hold one's tongue.[b] I have been learning "Rabbi Ben Ezra" by heart lately and lines of it come to me now and then "Young, all lay in dispute; I shall know, being old." One of these days I suppose I shall look back and see where I was right and where I was wrong. All that I am sure of now is that I am not altogether either.

Lovingly
Alice

a. Mary Almira Smith, who had received her medical degree from the University of Zurich in 1880, was attending surgeon at the New England Hospital.

b. Harriet L. Harrington, attending physician in 1893, was one of the physicians asked to resign the following year.

ALICE HAMILTON was delighted when she began work at the hospital's dispensary in Boston late in December. She had looked forward to this work even before her arrival, and the relief it afforded from the strife at the hospital made it doubly satisfying. Dispensary service also allowed her to indulge her love of adventure while "slumming" in immigrant and black neighborhoods; by the end of February she counted up eight different ethnic quarters she had visited.

17 To Agnes Hamilton

29 Fayette Street
Boston
[ca. December 28, 1893]

My dearest Agnes
    I have just finished reading your letter and now, until my clinicians come, I can put in a few minutes in trying to answer it. You were an angel not to speak of my long silence, for you must have thought it

strange that I let two successive Sundays go by without writing, no three, only this letter will do for last Sunday. Instead of giving you my excuses, let me tell you what I have been doing for the last two weeks and then you will see that writing has been almost an impossibility. Three Sundays ago, the first time I missed, I spent the whole day between a woman very ill with pneumonia and two babies with convulsions. One baby almost died and I was up all night with it, dividing myself between its crib and its mother's bed. The baby got better, but the mother got worse and I was with her day and night until early Saturday morning, when she died. My dear child I cannot tell you what a terrible time that was. She was the sweetest woman, young, lovely-tempered, with the most utterly devoted husband and relatives that any woman could have. They, the relatives, were people that we would all have thought coarse, vulgar, uninteresting. But death-beds do bring out people in their own characters, so that all their externals are forgotten. This red-haired, vulgarly-dressed husband, who looked like a fifth-rate commercial traveller, showed himself self-controlled, brave, utterly unselfish and most deeply loving. And the loud-looking mother and sisters-in-law, overdressed millinery clerks, the silent old Scotch father, the dried-up, shabby old mother, they all seemed heroes and heroines to me, when I watched them around that death-bed. She was quite different from them, a fair, delicate woman, looking like a mixture of Madame Elizabeth and Jessie's "Mistress Anne Page." Of course in a hospital it is not usual to allow the patient's relatives to stay with her, but she was in a room alone, and my chief— who with all her ignorance has a womanly heart—let the husband and all of them practically live there. On Friday she began to sink and at noon we thought she was dying. And my dear, the heart-rending scene! all the worse because they were so quiet and obeyed my slightest word so implicitly. She rallied then, sank again and rallied again. The second time I picked up heart and, as I was quite worn out, I let Dr. Slobo relieve me and went to sleep from eight till twelve. That is, I meant to sleep till twelve but at nine they called me and then I knew that all hope was over. The poor things, who had gone home so full of encouragement, had to be sent for again, and all gathered around her bed-side again. I thought I could not stand another minute of it, but my dear I had eight solid hours of the most heart-rending work I ever did. I could keep up as long as there was anything to do, but when it became plain that anything more meant only useless tor-

77

menting I stopped, and then I just broke down and had to fly to my room. My dear she was conscious to the end, to the last moment, and turning always, even when she was choking to death, with the same smile to her husband, that big, ugly, red-haired drummer.

Well the next day I left Maternity to come down here, so glad, so glad to see the last of that place. Here life began very busily at once. All morning I had clinic, that is, patients coming here to be treated, all afternoon and evening, till midnight, I was paying calls, down alleys, in cellars, up attics, over saloons, everywhere in the slums of Boston, which, after all, are not very bad slums, and very picturesque and interesting. It happened to be a sloppy day, pouring rain, ice and slush and mud on the sidewalks, rivers in the gutters. I rather enjoyed having it so, it was such delightfully typical slumming. My calls were not all paid until nine o'clock and then, just as I was going to bed, a frantic husband came to the door, begging me to go see his wife, who was having a haemorrhage. I knew they were respectable, though they do live over a "Deutsche Billiard Stube," in the fifth story, so I called the nurse and we went. It was midnight before I thought it safe to leave her, and I stumbled home so dead tired that I almost slept on the way. The next day, Sunday, I made calls all morning and in the afternoon rested up in Allen's room. Miss Emily [Ladd] asked me to dinner, and then to tea, both of which I thankfully accepted, though I had to go home in between. The next week went just the same and I never even found time to write home to Mother.

This letter has been interrupted by a clinic, then by a distracted mother with a dying baby, who wasn't dying at all,—they never are when you are routed out at ten P.M. and particularly tired—and now it must be stopped because I am too sleepy to keep my eyes open. Mother will have to tell you about the why and the when of my going to Plainfield, unless you can wait till next Sunday to hear.

Good-by my child. Your last letter was an unusually dear one. If you knew what a comfort they are!

Lovingly
Alice.

18 To Agnes Hamilton

[Boston]
January 22nd 1894

My dearest Agnes,

These are prescription papers, but they are all I have and if I don't snatch this time to write to you, there is no telling when I shall be able to. It is a pity that just the time when I most want to write to you, I cannot. All the time I am seeing things and doing things that I tell you in imagination, describe my amusing and pathetic patients, and all my worries and reliefs over them, but somehow the descriptions do not get on to paper, probably because at such times I am generally tramping the streets or sitting beside some woman and trying to keep awake until she does something. I believe hereafter when I go out to a confinement case at night, I will put pencil and paper in my bag for I sometimes have hours then with nothing to do. Tuesday night I had an all night case in the funniest family. They are negroes and real down-South, Carolina negroes. There was the husband and the husband's sister and three neighbour women. They treated me with a profound and awful respect, but they talked and laughed and "jawed" among themselves with perfect freedom. The sister-in-law was a typical Topsy and she looked like a picture, sitting on the edge of the bed, swinging her bare black legs and eating a huge piece of smoking [?] cake, which the husband passed around for refreshments. He was a solemn darkey, who wore his hat all the way through, and was very dignified and careful to use long, many-syllabled words. It was his first experience of the kind and the women all took pains to instruct him on the various ways in which to find out whether it would be a boy or a girl, and the bearing that his dreams and her dreams of the night before, had on the case. He was much impressed, because it seemed that he had dreamt of something to eat, the night before, and that was a sure sign that a disappointment awaited him. Poor man! it turned out true, this time, for the child was still-born, and they both felt dreadfully over it. They live in a house where I have had a serious case of pneumonia on the ground floor, and a case of nervous break-down on the middle floor, so I am quite at home there, and it is amusing how each family has confided to me that the other family thought I was only a school-girl and couldn't have been through college, until they gradually discovered that I was older than I looked. One woman has delighted my soul by thinking me older than Dr. Evans, who is every day of thirty. My coloured patients are always nice about obeying me and treating me very respectably, but one is never sure about what sort of a place one is getting into, in the negro quarter. Now in the Jewish quarter, filthy though they are, there are no bad houses anywhere.

January 25th

Another clinic morning. And your letter has come in the mean time, but I could not answer it before. So the hard times are coming heavily down on Fort Wayne as well as on the eastern cities. It seems strange to hear you talk of so many men being out of work. We always used to feel that a family with an able-bodied man in it were poor only from laziness or drunkenness. Times are terribly hard with us. It is the greatest exception when I find a family whose father is doing steady work. All the Jewish quarter makes its living pressing and cleaning clothes, and they are all out of work. And the cab-drivers and waiters and cooks among the negroes, the fruit-men among the Italians, have almost nothing to do. Often I find the man at home taking care of the children, because his wife can find work better than he can. Boston is a very charitable city, fortunately, and the charities are systematically managed too. When I find a destitute family, I can refer them to the "Associated Charities," and in two or three days I get a letter telling me what they have found about the family, what they mean to do for them and just what help they need. And if I find people really hungry, I can get food from the Christian Association for the mother and children and send the fatl er over to a Rescue Mission, with a ticket for a good hot meal. Then my consultant for the month has a fund that I can draw on for coal and for invalid broths, or eggs and milk. And another doctor has a place to which I can send for clothes and bedding. So that my work becomes more and more mixed up with the work of the charities, in a way that makes it much more delightful.

January 29th

If I keep this sort of thing up, I shall have a serial extending over all my Dispensary service. Somehow I cannot find time to write even as much as I could at the Northwestern. And Sundays, when I always expect more time, I am either at church or with Allen or making calls. Then Mrs. Lombard's being here has taken my spare moments lately, for she is just near enough to make me feel that I can go there often and far enough to take me some time to go. At first she was out in Jamaica Plain, with the Welds, and I had one very delightful evening with her there, Saturday before last. And this Saturday evening I dined with her at the Oxford and went afterwards to the Music Hall, to hear the Boston Symphony. I don't think I ever enjoyed more keenly little things like that. Instead of making Dispensary work seem harder, they make me come back to it so rested, and keep me from getting tired and impatient. I think I could stand the hardest sort of a life, if there were only change in it. Yesterday was a very delightful Sunday. I had had a baby in the course of the night, so I did not feel much like getting up early, but Allen was coming for me to go to church, so I hurried through the few necessary calls before half-past ten. We went to King's Chapel, the dearest, queerest old church I ever have been in. The pews are square little rooms, with seats running all around them, so that a part of the congregation sit with their backs to the clergyman. It is almost entirely undecorated, except for quaint old memorial tablets in Latin, and there is only one place where the windows are stained glass. Above the pulpit they have three exquisite windows, but the rest are small-paned white glass, looking out on to the business blocks around and to the old grave-yard below. Oliver Wendell Holmes sat up in the gallery to our left.[a] I wondered if he felt all eyes upon him when he came in. Perhaps he is so used to it that he never thinks of it. The service was adapted from the Episcopal and I hate adaptations, so I could not enjoy it much. They have simply left everything as it was except when the Trinitarian idea comes in, then it is changed, even in the Te Deum. Mr. Moxam, the ex-Baptist, who has just left his church, because of his preaching Eternal Hope, preached the Sermon.[b] It was good, but like all Boston sermons, not wonderfully good. I went home with the Ladds and Allen and, as we had an hour before dinner, I settled myself with pen and paper to write to you, but Miss Emily and Allen came in and discussed the "Heavenly Twins,"[c] abstract moral principles and their own and each other's failings, so that I didn't even make a beginning of a letter. Miss Emily spends her time in criticizing and arguing with Allen, but she is awfully fond of him, much much more so than he is of her. Indeed if she were younger and he older, it would be rather tragic, I fancy. After dinner I meant to go off and pay another call, but I lingered, talking to Mr. Ladd, until Frank Sever came in, and then just as he was going, Phil Savage came.[d] Have you heard Allen talk much about him? He seems to be one of his very dearest friends and

81

I had heard Allen talk so much about him since I came, that I was very glad to meet him, and still more glad to find him very delightful. He is decidedly handsome, to begin with, and has just that touch of half shyness, half nerve [?] that I like in a man. And he impresses me as being very strong. He is the son of my old enemy, Minot Savage, the man who wrote that wretched article in the North American on the orthodox belief in eternal punishment, an article that made me uncomfortable and upset for months and months, until it forced me to settle the question in my own mind.ᵉ I suppose I ought to be grateful to him, but I cannot get over the intense dislike I have to the man's very name. Allen wants me to go hear him some day, but I don't feel like it. Phil is taking his first year in the Divinity School at Harvard, but he cannot be a Unitarian, especially as Radical a one as his father is, neither can he subscribe to the Westminster Confession or the Thirty-nine Articles, so Allen says he is in quite an unhappy, restless state. I fancy he will end in liberal Episcopalianism. It was after six when he left and then I had to dash off to the slums to see my woman of the night before. It was a funny contrast, the dark little alley, the dirty entry and stairs, and the crowded little room, smelling of kerosene and onions. I came back to supper—I simply live on the Ladds Sundays—and after supper Allen and I went into Cambridge to see a Mrs. Demerit [?], the chaperone of the Browning Club, who has been very nice to me, the few times we have met. There I met Algernon Tassin, who informed Mrs. Demerit [?] that I had been brought up with him in Washington and was much disgusted when he found I had not.ᶠ I do not fancy him at all, less than any friend of Allen's I have ever seen. He has joined a fifth-rate theatrical company, in order to get the training to help him in writing plays, and whether it is lately acquired or not, he has a second-rate dramatic manner. There is an underlying vein of coarseness and a very thick overlying layer of conceit, that I disliked exceedingly.

This letter is making up in its length for its tardiness, but it will come to an end now.

Let it stand for the three letters that ought to have been written.

Lovingly
Alice

a. Oliver Wendell Holmes, Sr., essayist, poet, and physician.
b. Philip Stafford Moxom, a popular Boston preacher, had resigned from the pulpit

of the First Baptist Church in December because of doctrinal differences; he later became a Congregationalist.

c. *The Heavenly Twins* (1893), a novel written by Frances Elizabeth (Clarke) McFall under the pseudonym Madame Sarah Grand, is critical of the plight of women, who are often kept in ignorance, denied an outlet for their talents, and expected to conform to the wishes of their fathers and husbands at whatever cost. The heroine, who learns on her wedding day that her soldier-husband has led a dissolute life, refuses to live with him as his wife.

d. Frank Sever and Philip Henry Savage were Harvard friends of Allen Williams. Sever became a banker; Savage, after a year at divinity school, became a serious poet who published two books of verse before his untimely death in 1899.

e. Minot Judson Savage, a prominent Unitarian clergyman and head of the Church of the Unity in Boston, was an adherent of Darwin and of the Higher Criticism. The article that so upset AH was probably "The Inevitable Surrender of Orthodoxy," published in the *North American Review* in 1889; it attacked such major tenets of Protestant orthodoxy as the infallibility of the Bible, the depravity of man, the atonement, and eternal punishment.

f. Algernon Tassin, founder of the Browning Club (a literary club for Harvard and Radcliffe students), later had a career in the theater before becoming a professor and writer.

19 To Agnes Hamilton

[Boston]
February 9*th* 1894

My dearest Agnes

Ever since your letter came I have been struggling to get time to answer it, but until this evening I have not had ten minutes which I could call my own. In the meantime I don't know how many imaginary letters have been going to you; some taking everything back and promising that I would never think or speak of such a thing again; others arguing the point out, and others provoked at you for feeling more over it than I meant you to. But now that I actually am writing, I do not know just what to say. If I could only see you and talk it over, it would be so different. In the first place, I shall not "keep your letter and read it over" for it would make me exceedingly uncomfortable if I did. In the second place I will not drop the whole subject forever. You may be quite sure that whenever you do anything that I don't understand or that seems strange, I shall write and ask you about it. You can tell me frankly that you are not allowed to explain and then I shall be satisfied and know that there is an explanation somewhere, but I mean to ask anyway. That is the difference between my way and Margaret's. She was as reserved as the rest of you and the result was that, instead of letting a few frank words clear everything up, she never asked you about anything, but kept it in and brooded over it until she began to misinterpret everything. Then there is another thing that I don't agree to in your letter. I don't by any means think that a person who cannot keep a secret is as bad as a secretive person. It is deplorable when anybody lets other people's secrets leak out, but I doubt if it causes as much trouble as the other fault to others, and I am sure it isn't as disastrous to one's character. Nor am I sure that people have a right to tell us things and then demand secrecy when that secrecy may put us in most trying positions. Look at Aunt Mary and Allen now: the uncomfortable place she puts him in. He knows that the Ladds are wondering over all this mystery and speculating as to where she is and why she is hiding, and he is forbidden to say a word.

My dear, I don't suppose, living as constantly in the home atmosphere as you do, that you realize how different it is from that of the rest of mankind. Whenever I come back home it seems to close around me like a lot of choking, cobweb chains, all this caution and mystery and reserve about nothings. It is always "Now don't repeat this for anything," "I wouldn't speak of this before so-and-so," "There is no use asking about it, for they will tell you they don't know, anyway," until I long to brush the whole web away and make people speak out openly, no matter what, rather than choke themselves with all this secrecy. And it is that that I could not bear to see in you, because it is not you nor your nature a bit, and would be one fault, where I have not seen any. Now am I hurting your feelings? My dear child I hate to risk doing that, and yet I must tell you what I think and how I feel about it. The longer I live the more I value openness and transparency, and even if it does hurt a little sometimes, it never can hurt deeply because it is an open hurt and can be openly treated. Don't think for a minute that I could ever doubt your love and trust. If I did I should simply give up. And don't talk about your needing me, for I need you twice as much. Of course in big things I could trust you to the very last, no matter what disgraceful thing I ever did, but Agnes, it would fret and trouble me to think that you were condescending to any littleness, and little secrecies are littlenesses. You say Madge began all this. No, she did not. It was Aunt Nell's baby dying when none of us knew he was ill unusually, that started it. And then she added the little things to it, instead of asking you frankly about them. Now do you think me all wrong and have I hurt you? Write and tell me.

With all love
Alice

THE EVENT or events that elicited Alice Hamilton's passionate letter of February 9 are not known. But by the mid-1890s there were numerous signs of tension between her immediate family and some of the other Hamiltons. Their Aunt Margaret openly disapproved of Alice and her sisters, whom she considered self-centered, extravagant, and, especially, neglectful of their father. She expressed these views to her other nieces and sometimes precipitated unpleasant family scenes. Phoebe Hamilton too, although eager to maintain good relations between the households because of the friendship between the cousins, frequently commented on her nieces' failure to attend more closely to the needs of their father, who was often left alone in Fort Wayne. In February 1894, Mary Hamilton Williams was in the Boston vicinity but refused to allow Allen to disclose her whereabouts to Alice. And somewhat later neither Mary Williams nor Nell Wagenhals seems to have been on speaking terms with Gertrude Hamilton.[25]

What is clear is that as Alice Hamilton had an opportunity to observe different ways of life, she was becoming increasingly distressed by the "choking," secretive, and moralistic atmosphere at home. In November she had written Agnes of her fondness for Emily Ladd, who had "a way of speaking to the point in an un-Hamilton way that simply rejoices me. For one of our family's most unfortunate traits, I think, is our dread to call a spade a spade, and an over-sensitive way of walking around and around a subject." Though she associated these characteristics with the older Hamiltons, she feared they were cropping up in the younger generation as well. Letters from Margaret and Edith to Jessie Hamilton in the spring of 1894 suggest that they believed that their cousins were under instructions not to discuss certain subjects with them.[26]

Alice's letter of February 9 (or the one preceding it that has not survived) may have been the one for which she apologized more than a year later: "Do try and think that it wasn't really I that wrote that letter, but a Mr. Hyde transformation of me. In a way I think I must have been temporarily insane." She had in fact apologized at the time: "I will promise not to speak of it again, but to quietly accept the fact that we are different in some ways and that probably my difference troubles you fully as much as yours does me."[27]

Conventional notions of womanhood died hard in the transitional era in which Alice Hamilton lived. A graduate physician with almost two years of hospital work behind her, she was in her mid-twenties keenly aware of the contrast between her professional training and the requirements of polite society. She caught herself about to discuss a confinement with one of Allen's friends, only later concluding that "very likely he would not have been shocked at all . . . but still I don't want to get in the habit of talking medically to non-medical people." Her medical and dispensary experiences were making her less judgmental, more able to accept people on their own terms, and her night calls took her to seamy neighborhoods and even to houses of prostitution. Yet she still accepted the social conventions that restricted women of her class, and observed after attending the theater with a fellow intern: "She is almost forty, you know, so I can go anywhere with her." Even as she struggled to free herself of the "family caution," she maintained that women should be shielded from knowledge of the sordid. She reported that Emily Ladd had been studying "questions that most unmarried women in her class of life know nothing of," but to Alice it seemed that "a girl should know all or almost nothing." Thus medicine, though in some respects a false start, permitted Alice Hamilton to acquire a knowledge of life that she felt most women should not have. By choosing to become a physician, she had given herself permission to see and know more of life than she could otherwise have done.[28]

When Alice Hamilton's dispensary service ended in mid-March, she returned to the hospital, this time to the medical ward. In November, she had written that she would "commit suicide, or elope, or be expelled" if she were sent there, and by the beginning of April she wished she could break her contract. At the end of the month, she asked the hospital's Board of Physicians to release her from the remainder of her service, pleading "strong family demands which make me most anxious to be at home." She added that they might also find it convenient to let her go because she believed that the number of interns was to be reduced in the near future. The Board accepted her resignation, "not because of the reasons stated . . . but because in the opinion of the board the best interests of the hospital will be served by her withdrawal." No explanation accompanies this tantalizing notation in the Board's minutes.[29]

Shortly after Alice returned to Fort Wayne in mid-May, "family demands" did become pressing. Her sister Margaret, a student at Bryn Mawr College, suffered a severe hip injury in a carriage accident. Alice left to nurse her sister and then brought her home. The following fall, Margaret returned to college, but Alice remained in Fort Wayne.

III

"I SHALL KNOW, BEING OLD":
CAREER AND FAMILY
1895–1897

Having decided against medical practice, Alice Hamilton sought to prepare herself for a career in science. This was by no means an easy task for a woman of her generation. The movement for formal admission of women to American graduate schools was just beginning. Several of the leading universities opened their doors to women in the early 1890s. But few of them offered advanced training in bacteriology, then a new field still closely allied with pathology. Even for men, a career as a research scientist was a new and precarious objective. In addition, Alice Hamilton had to balance her professional aspirations against family obligations of the sort that rarely interfere with the careers of men. In the 1890s, when the Hamiltons had more than their share of troubles, she was keenly sensitive to the competing claims of public and private life.[1]

She returned to Ann Arbor in February 1895 as a resident graduate and assistant in the lab of Frederick G. Novy, an important figure in early American bacteriology. She was also supervising the convalescence of her sister Margaret, who was initially confined to bed, her leg in a pulley. Despite the worrisome situation, Margaret improved steadily, and Alice took pleasure in her own success in juggling her obligations as "sick-nurse, cook, bacteriologist, semi-society girl." Some of the aura of her undergraduate days had already worn off. On seeing Dr. Dock again, she observed that she "felt dreadfully disappointed and flat . . . I cannot imagine that that boyish, slightly embarrassed and noway extraordinary man is the one whose praise I used to long for above everything and who could make me happy or wretched by a word." She felt no longing to be back at the hospital, but had "a feeling of joy that I could get back to my beloved germs." At this time, she hoped to obtain a Ph.D. from Johns Hopkins University, the premier school for training in the medical sciences. But although anyone with a medical degree could take graduate courses there, the university did not award doctorates to women. (There had been one exception in 1893.) In view of "all the difficulties," William Howell, who had joined the Hopkins faculty, advised her not to try for a degree, which he did not consider essential for her branch of work. But Hamilton still wanted to study bacteriology, either at the University of Wisconsin (where she may have been offered an assistantship) or, preferably, at Hopkins.[2]

In June, Edith Hamilton underwent a laparotomy for removal of an abdominal mass that was causing her excruciating pain. The operation was dangerous, but Edith's Ann Arbor surgeon considered it the only way to save her life. Alice, who did not trust herself to assist at the event, observed: "Everything that I have ever borne or ever dreaded sinks into insignificance beside it." But the surgery went well and Edith made a rapid recovery. In July, Alice wrote Agnes that she had been unable to pray or give thanks, for "it seemed so firmly fixed in my mind that all had been planned and decided long ago and planned for the best and wisest, that whether for death or for life God was good."[3]

In the fall, Alice and Edith Hamilton (who held the Mary E. Garrett European Fellowship, Bryn Mawr's highest honor) sailed for Germany to pursue their studies in science and the classics. There they encountered open prejudice against women for the first time. The universities did not officially enroll women, but foreign students could make their own arrangements for study with a professor or department. (In fact, the first two women—one American, the other English—had received Ph.D.s from German universities only the summer before the Hamiltons' arrival.) The sisters both gained permission to study at the universities of Leipzig and Munich. In Leipzig the professors allowed them to attend classes but told them they should be considered "invisible." At the University of Munich, less used to women students, one of Edith's professors escorted her past the large crowd that had assembled to watch her enter her first class, and seated her on the lecture platform where she need not sit near the male students, some of them Catholic seminarians. A female presence was evidently less disturbing in the laboratory than in the lecture hall; Alice was surprised and on the whole pleased by her reception there.[4]

But she was never allowed to forget she was a woman, and her training did not equal that of her male peers. She had hoped to work in Berlin with Paul Ehrlich and Robert Koch, but neither of these giants in bacteriology saw fit to take her on. Leipzig, where she had gone to study gross pathology, barred women from attending autopsies. In Munich she wanted to investigate the role of white cells in combatting infection, but the professor, Hans Buchner, would not permit her to conduct the necessary animal experiments. Instead, she worked on a problem of spore-formation and on a new bacillus that Buchner thought might be a companion to the cholera bacillus. Her findings in both cases were negative and the bacillus "turned out to be nothing but a big, fat nonentity." Only in Carl Weigert's lab in Frankfurt, where she spent the long spring vacation, was she treated as an equal. There she worked with Ludwig Edinger, one of the founders of the science of comparative neurology, on the olfactory system of bony fish. She developed a lasting friendship with Edinger and his wife, Anna, a social welfare leader and feminist.[5]

Taken as a whole, the year was disappointing. Not only did her experiments fail, but Alice Hamilton later claimed that she learned nothing about bacteriology that she had not known in Ann Arbor. Like other American students who flocked to Germany, the Hamiltons found themselves in an atmosphere at once more restrictive and more liberal than any they had known. They resented the patronizing attitude of German men, who never tired of asking: "But who will darn the stockings if women are going to be bacteriologists?" The lack of manners also repelled them. Men frequently pushed women off the sidewalks, and one "great blond Siegfried" lifted Alice right out of her front-row gallery seat at the opera so that he might sit there himself. Nor could the Hamiltons tolerate standards of gentility that prohibited young women of their class from working even when marriage was also out of the question because they had no dowries. Yet Alice rather liked the relative freedom that characterized relationships between men and women. The men in her lab spoke freely about their fondness for drink and the resulting physical distress. They even invited her to accompany them to a farewell evening of drinking "strawberry bowl." Although she declined, she reassured her mother that the invitation was not an insult, as it would have been at home. She was also impressed at how much good work got done in the laboratory and at scientific meetings despite an informality that would have been considered unseemly in the United States.[6]

The year in Germany sharpened Alice Hamilton's political and social consciousness. Still a frequent, although not an uncritical, churchgoer, she was amazed by the indifference of most Germans to religion. She found their attitude—"outwardly devout, inwardly atheist and hypocritical"—utterly repulsive. Most disturbing was the pervasive anti-Semitism. "I am becoming an ardent champion of the Jews," she wrote Agnes after she had made some Jewish friends. She could not understand why the Edingers, among others, calmly accepted the insults and restrictions on the professor's career, which she found intolerable.[7]

Alice Hamilton's letters in her twenties record her struggle to carve out an independent and satisfying life without shirking family responsibilities or alienating those she loved. This was a difficult task for a conscientious woman raised in a culture and religion that measured a woman's worth principally by her success in fulfilling family obligations. Viewing her education as a special privilege—unlike Allen Williams, for whom it had been "so much a matter of course"—she felt impelled to prove herself worthy of her opportunities. There were few models to guide her. Puzzled herself about how to live, she was acutely sensitive to the dilemmas of other women. She attributed their struggles to the confusion of living in a transitional era, when old standards were falling away but no universally accepted new truths had yet replaced them.[8]

Alice Hamilton frequently hinted that the parental home threatened a woman's health as well as her ambition. While maintaining that married women should give up their professional aspirations, she criticized single women who sacrificed their careers (or even their fiancés) for the sake of their parents. She undoubtedly believed that Agnes (who remained in Fort Wayne caring for her parents and whose health broke down following her father's death in May 1895) needed such admonitions. But she was also shoring up her own defenses against becoming too absorbed in family affairs, which reached crisis proportion in the 1890s. This was the period when Alice's maternal grandmother wanted her daughter to leave her alcoholic husband. Moreover, Margaret, Edith, and Norah Hamilton each suffered a serious illness, during which Alice took major responsibility for their care.

Alice feared that the decision to follow her own path might impair her relationship with Agnes. In family debates about the obligations of women, Alice and her sisters stood for "individualism" while their

91

cousins leaned toward "socialism," which to Agnes meant a woman's ultimate loyalty to her family. From Germany, Alice implored Agnes, who was bound to hear family matters discussed "from the other side," to have confidence that she and her sisters acted only "after long thinking and consulting, yes and praying." Agnes, who had observed in her diary the summer before that she and Alice "seem quite far apart in our opinions," assured her cousin that despite their differences each believed the other acted from her conscience "and so complete trust stays always between us." Alice had not yet received this letter when she returned to the theme, this time mainly in connection with Allen Williams.[9]

20 To Agnes Hamilton

Albertstr. 44 III
Leipzig
January 14*th* [1896]

Dearest Agnes

I meant fully to write to you while I was in Dresden, but there was no time to write to anybody and I had to try to write home, though I did not succeed much. You can imagine, can't you? how full the days were and how we felt we must see all that we could because who knows if we ever would be there again? We used to be so tired that we had to sleep late, very late, so that we hardly ever finished breakfast before it was time for the gallery to open. Then we would be there till three o'clock, then go to dinner and the rest of the day was spent in wandering over the city or seeing people. And the evenings we spent at the opera whenever we could. It was such a delightful week. The three days in Berlin were not as thoroughly delightful. We were quite tired by that time and it was either very cold or rainy, and we had to hunt up our professors, and the city is so big that you take forever to reach places, and we could not find as nice places for our meals. And then we had to hurry. Berlin really must be ever so much fuller and more interesting than Dresden, but Dresden is dear and old and very quiet, with beautiful views of the river and, of course, the gallery. We only could go twice to the gallery in Berlin, because it was closed on New Year's. I was delightedly surprised over it for I had not thought we would find very much there. But it is very rich in the early Italians and early Dutch. I did not think I could care for

Albrecht Dürer or enthuse very much over Botticelli, but I could here. The Dürers are infinitely better than the Dresden ones. His portrait of his wife is there and several studies of heads. And the Nativity of the two van Eycks with the choruses of singing boys and girls on either side is here. The Rembrandts are poor, and there are almost no Van Dykes, but the Franz Hals are fine, six or seven of them at his very best. As for the early Italians, there are, as I said, splendid Botticellis; one Madonna especially is perfect. I think you must know it. The child stands on her knee with a dear little grave face and his hand with two fingers uplifted. There are Ghirlandajos and Filippo Lippis and Rossellis and then a vast number of the very early ones which I can't think beautiful, only interesting. And a wonderful Andrea del Sarto. I don't know why I am writing all this long list to you, for you can't make the rooms rise up before you and see the pictures as you read it and they won't be anything to you but a list. Yet if you had been there you would have enjoyed it more than I, for you used even to like those hideous Arundel pictures, which I still detest and I think more now than before.[a] We have been back ten days now, but it seems no time at all while the vacation lengthens itself out to weeks and weeks. Everything is just the same in the laboratory, except that my nice little American neighbour has gone home to be instructor in the University of Texas. Dr. Lange still imitates all the instruments in the orchestra and they all continue to envelop me in tobacco smoke until the people here say that everyone will think I have been smoking too. I hope the next laboratory I go to will be as amusing. It might easily be more dignified but I don't believe that any better work will be done. It amazes me to see how much work they manage to get done and yet always they have time to stop and eat and drink or sit around and gossip. Germans think it fatal to go for more than three hours without eating. They eat at eight, at eleven, at one, at four and at seven and if they go to the theater they eat there or after they have got home. Everybody does. I cannot get in the eleven o'clock meal, for I do not care to send the diener off for sausages and beer as the men do, but I get in all the others. Yet they say dyspepsia isn't as common here as in America. Certainly American stomachs could not stand it long.

I wonder if you have written me about the Christmas doings. Your last letter came just as I had sent mine off, so I am going to think that another is on the way and will come when this one has been sent. Norah's and Madge's were not as satisfactory as if they had seen more of the boys. Allen Williams sent me a book, at least I imagine it was he though there was no name in it. It is very amusing: Alvan Sanborn's sketches of tramp-life.[b] I had read some of them in the Forum already.

I wonder if I shall always have this unsatisfied feeling about Allen, a feeling that we might be such dear friends and that we never will be. I do love him so dearly that he is a great loss to me. I was wondering to-day just what the reason is. It isn't because he will not write to me. I don't believe that if you stopped writing to me I should have that feeling. Nor is it because I don't feel that he is fond of me, for I do. It is because I know that he judges me and condemns me in things which he can only know by indirect and one-sided evidence. It is no use saying I cannot know what he thinks. It is easy enough to see every time he comes in our house. And I resent it bitterly. He ought to trust me, as I trust him in things which I cannot know as well as he does. I know that when anything comes up, even if you do not agree with my decision you do believe that I make it as justly as lovingly and conscientiously as I can, even if your conscience cannot approve of it. But with Allen I always feel that he has judged me cruel, unjust, hard and wrong. And of course he must always go on thinking so and I must always go on feeling it. It is that that makes it so impossible for me to imagine ever going to him for sympathy. I don't mean comforting speeches by that. I don't believe you ever made me one all last year, yet you have no idea what an inexpressible comfort it was to feel that you were there, you and Jessie, and that you knew without needing to talk it over and that you cared and made the trouble your own. I don't see how people who must bear things alone ever live to bear them, or how people can bear being mistrusted and disapproved of by those they care for. I can take Allen's disapproval and want of sympathy quietly enough now, but it is only because he has made me grow used to it and because I have the girls and you and Jessie. But I ought to have him too and it is very bitter to think that I never shall. The old childish friendship that he clings to as if it were the only possible thing seems to me worth very little now. We aren't children any more; we can't sit on the top of Grand-ma's carriage and talk about what we are going to be when we grow up. And I have not the least yearning to go back to that time with you because I know that the friendship we have now is worth a hundred childish ones. Instead of mourning over the impossibility of our ever being children together again, as he does, I mourn over our friendship not having grown up with us.

I don't know what made me write about all this just now. I am sure you would much rather have had a letter of doings, but I have not been doing anything much since we came back and for some reason or other I have been thinking of things. Is there anything in the world more utterly unprofitable than going over the last six years and wishing you had lived them differently? I suppose everybody does it, though, but I don't believe everybody has big decisions, important turning-places to regret. At any rate if *you* look back and regret, you cannot regret as much, for it can never be as bad to think of neglected opportunities as of neglected duties. Yet I am not sure I do really regret. Sometimes I think I do and then individualistic principles reassert themselves and I think I was right. So perhaps if I had it to live over again I should be where I am now; perhaps at home without any medical education or hospital training or Europe trip. It is just "Rabbi Ben Ezra": "Let age speak the truth." "I shall know, being old." Only one rather dreads the knowledge, I think. What must it be to feel that one's whole life has been a mistake.

I have had letters from Edith [Trowbridge] and Evelyn [Noyes Saltus] and Kittie [Ludington] since I wrote you last, and have heard about Grace McLeod and seen the announcement of the marriage of Goosie Schultz.ᶜ I am so glad she succeeded at last. Grace I heard about from one of the girls at Mrs. de Soto's, a Troy girl who knows Grace very well. She said that Grace's first engagement was very unhappy. The man was very attractive and Grace was tremendously in love, but he was dissipated and drank more and more all the time until Mr. McLeod told Grace that the engagement must be broken off. Grace refused and they had a very unhappy [time] over it but she held out until it grew so bad that at last she told him he must give up it or give up her. And he chose to give her up, very fortunately for her, don't you think so? A promise kept for a couple of months would have been quite enough to make her defy her father and marry him and then he would have been free to do as he pleased. The man she has married is quite different, not at all a charming, society fellow like the first, but a quiet, grave, steady man, about twelve years older than she. She said that Grace was very much broken up after the first one, not a bit like herself, but now she was quite her old self again. I hope she isn't in every way the same again. Such a thing must almost inevitably make a person older and stronger.

Evelyn's letter was a brimmingly happy one. She is established in a little house of her own. Rollin is working somewhere all day long, whether in the Beaux Arts or not, I don't know, and she is keeping house, studying French, going to lectures on literature and Art and— having Rollin. Which last is the largest part by far. She went over all the wedding, dwelling on each detail with such loving pleasure, and ending with "It was perfect in everything. I could not wish for a change." She is happy, very, very happy. She spoke of meeting Chester Aldrich "a friend, or is it a relative of yours"? Funny, when I don't even know the man by sight. She liked him very much and said she had heard from him a rumour of my coming abroad, which of course she didn't believe. Both she and Kittie told me that Ethel Saltus was just recovering from a severe attack of typhoid and that she is to have a baby in February. Laura, Mamie Brinckerhoff and Ethel Saltus with babies! It ought to make me feel very old but it makes me feel young and ignorant beside the women who have had experiences I shall never know anything about.

Kittie's letter was very sweet. She spoke of beginning another winter of helping her mother, in a very unselfish way. I wonder if there is some reason why she never has married. Did I tell you that Susan Duryee told me when I was there that she had always imagined Kitty cared for a New York man who used to be a great deal with her and now is not any more.[d] Somehow I cannot but think Kitty could not reach the age of twenty six without falling in love, can you? Of all girls she was the most susceptible. Edith [Trowbridge]'s letter was written on her way from Spain to Italy and was full of a tirade against Americans who come to Europe to "fritter away years in useless pleasure- and beauty-seeking", and against me for saying that it seemed so much more appropriate for her to be studying the Florentine and Venetian schools, than to be taking lessons in nursing and going to Working-Girls clubs. Well I shall apologize humbly when I answer it, but I must say I am afraid I don't know the Edith of to-day. She isn't the girl I knew at the Estate and in Litchfield. Did you write her that Aunt Marnie was in Florence? I would if I were you, for she is almost sure to run across her and she would think it so queer that neither of us had spoken of it.

I ought to keep this letter till I have something really worth saying to put into it, but it is so long since I have written at all that probably you would rather have me send this and write a better one later.

Perhaps something very exciting will happen next week and then I will save it for your letter.

Very lovingly
Alice

Notice this beautiful German envelope.

a. The Arundel Society of London issued prints of works by early masters; the most famous were lithographs in color.

b. *Moody's Lodging House and Other Tenement Sketches* (1895), by Alvan Francis Sanborn, was an example of a popular genre of the period.

c. Farmington friends. Katharine Ludington (1869–1953), who became active in the National American Woman Suffrage Association and was later a vice president of the League of Women Voters and president of its Connecticut branch, remained a lifelong friend.

d. Susan Rankin Duryee (later Fahmy), a Farmington friend of the Hamiltons, became a missionary in China for the Dutch Reformed Church.

IN JULY ALICE wrote her mother that she planned to remain in Germany until October so that she could finish an experiment she had begun in Frankfurt. Instead, she returned in September with Edith, who had a position as headmistress of the Bryn Mawr School in Baltimore.

21 To Jessie Hamilton

Plainfield [N.J.]
September 5th [1896]

Dearest Jessie

Your letter came only to-day and Agnes' came day before yesterday, so I ought really to be writing to her instead of to you, but a letter from you is such an unusual honour that I can't help hastening to acknowledge it right away, while a letter from Agnes, which ought to have been written at least two months ago, may wait a while for its answer. I am very glad my family thought that I was to be left alone and desolate and wretched in Frankfurt, for it has made them all write to me much oftener. Yours was the last letter of condolence, though, and after this I cannot expect to be sympathized with over my loneliness.

I am so glad you all liked Dr. Slobo[dinsky]. I thought you would find her interesting but it was so long before I myself discovered her lovableness that I did not think you would have time to. I think that she must have changed, as you said, and the really best things about her must be more on the surface than they used to be. But her rampant materialism used to often revolt even me and I thought you and Norah especially would shiver over her often. I think it was extremely nice in you all to like her so much. Mother wrote me a very graphic description of the studio party, even down to the part of the room in which they all sat, so I have visualized it all and imagined just what every one said and did. I never knew before what a power for visualizing I have. I can see so vividly the scene in which every one of the letters have been written, even down to your timothy grasses. And next summer, next July, seems a long way off. Don't you know how, when you are away from home, you keep putting by in your mind all sorts of things to tell the home people, things too slight to put in a letter, and now I keep thinking that way off next Summer I shall have forgotten them all and Germany will have sunk so far into the background that I shan't have anything to say about it. And all the details that I want to ask about, things that people have forgotten to explain in their letters, will have slipped from all your minds and you won't remember what I am asking about. I don't think I ever in my life longed to be in a place as I long to be in Mackinaw now.

Do you know I am quite in love with my country? I didn't expect to be. The Americans on the "Werkendam" were so queer and awful that I was afraid I would find other things and other people queer and awful too. The first delightful thing I found was the Customs House official. He was so jolly and informal and human that I quite loved him. The next thing was the conductor on the train, who helped us with our bags and asked us so respectfully for our tickets. In Germany the conductor always treats you as if you were an Anarchist in disguise and he wanted to let you know he had found you out. Then we had hardly left the city when I saw another most delightful thing, a waste field with a tangle of golden-rod and milk-weed and beyond it a field of corn. And there was a bit of autumn red in the woods now and then, and there were so many rambling, untidy, wild places, where everything grew as it wanted to and nobody interfered to make it tidy and profitable. And the air was clear and brilliant, no soft blue distances anywhere, everything quivering with light for miles away. Then the girls in the little stations looked so nice and trim and pretty. Not that they were new to me, for Holland was swarming with them, but they were unusually numerous. Indeed I got the impression that Plainfield was inhabited by girls between fourteen and twenty-five. And perhaps the very nicest thing I saw was a clean, smooth-faced youth, in white duck trousers. I hadn't seen him for a whole year, not once, and I loved him immediately. It is a pity that we cannot send one over to the Berlin Exhibition and stand him under a glass case as our most distinctively American product. Altogether I feel waves of patriotism sweep over me. America seems just as nice as she did when I used to think of her over in Germany: when I sat over my forlorn breakfast, for instance, or when I got into one of those awful beds or when I passed a group of bloated-looking, slovenly, swaggering German students. Somehow I fancy that in a great many ways I should like Holland better. And yet I had one big disappointment there. Afterwards it was not so bad but at first I was dreadfully disappointed. It was Scheveningen. Now, when you think of Scheveningen, don't you see a great stretch of sandy shore, with rolling sand-dunes in the distance, a stormy gray brown sea with Mesdag boats and ships,[a] a wonderfully picturesque fishing-village, stretches of brown nets spread out to dry, and delicious peasant costumes everywhere? Well then just imagine my feelings when I stepped out of the car and saw myself confronted by a row of enormous, hideous summer hotels, far worse than the "Grand;"[b] a beach covered with basket chairs and gay little tents, lemonade stands, merry-go-rounds, curiosity shops and beer-rooms, and a crowd of fashionably-dressed people everywhere. I could have sat down and wept. Well we hurried away from the hotels as fast as we could and by and by got to a place where the beach was comparatively clear and we could climb up and sit on the top of one of the sand dunes and look over to sea without being distracted by the summer people. And the grass on the dunes was very gray and the sea and the sky and the sand were all somehow softened down so that the same note came into them too and you felt that you were looking at a landscape that had been beaten by rain and wind and sun until it was all faded to gray, like the little towheaded, brownfaced children who play out all day with no hats on. But we still felt that there must be some remains of old Scheveningen: the Mesdags couldn't have invented it all. So we went down past the hotels and cottages and gradually we found the streets growing cleaner and the houses

99

smaller and tidier and we came upon women and children in sabots and sometimes the women had the little close caps with funny wire arrangements at the sides. It was quite satisfyingly clean, everything, street, gardens, paths, were paved and scrubbed and shining, but it was not picturesque at all. The houses were of brown brick with white woodwork, all exactly alike and all looking brand-new, though I believe they are really very old. Altogether it simply did not compare for beauty or for interest with a North-Germany or a Rhineland village. And of course summer cottages were elbowing their way in everywhere. Isn't it a pity? I felt as if things abroad always stayed the same, but Scheveningen has been spoiled within the last five years.

Aunt Annie is going to send this to the station so I must stop and write just a line to Agnes.[c]

Alice.

a. Hendrik Willem Mesdag (1831–1915) was a Dutch marine and landscape painter.
b. The Grand was a large hotel in Mackinac that the Hamiltons considered crass and touristy.
c. Annie Boorman Pond was the second wife of AH's uncle Charles-Frederick Pond.

SOON AFTER her return, Alice learned that Allen Williams had become engaged to Marian Bartholow Walker (1874–1930), a student at Radcliffe College. Because the wedding was to be postponed until Marian completed her medical training, the couple had decided to keep the engagement secret. But Allen kept a promise the three As had made in adolescence: to inform one another first if any one of them became engaged. Alice's reaction to the news reveals her continued affection for her cousin as well as a deep romantic strain. In the spate of letters that followed, she let Allen know that she did not think Marian could do justice to both marriage and career. He teased her with being an "unconscious hypocrite" because, caring only for her work, she would never have to make the choice she would impose on others. The couple finally married in 1901, shortly after Marian received her medical degree from Johns Hopkins. They both practiced medicine, first in Hartford and later, after Marian contracted tuberculosis, in New Mexico, Arizona, and California.[10]

22 To Agnes Hamilton

East Quogue [Long Island]
September 12*th* [1896]

Dearest Agnes,

Allen's letter came in the same mail with yours. He had written to Frankfurt, but hearing I was home he sat down and wrote another letter, which is certainly proof of a change of heart if anything could be. My heart just stopped short when I read his first words, for in spite of my guess I hadn't really believed it. Oh I do want to come over and spend to-night with you, to be by ourselves for hours and hours so that I can talk and wonder and speculate and lament a little and rejoice a great deal. I don't know why I say lament, for there is nothing to be sorry about. Were it you or Jessie or one of my sisters it would be different, but for years I haven't felt any possessorship in Allen and so I can't feel that I am losing him. And it just warms my heart and fills me with gladness to have the boy turning into the sort of a man that I wanted him to be, to have him doing an impulsive, unpractical, youthful thing. More and more lately I have dreaded so his growing cold-hearted and self-centred and old-minded. And this letter! Why it is sixteen pages full of the freshest, dearest, most romantic nonsense. Whatever Marian Walker may be she has brought out the very best there is in Allen, she has saved him from all I feared for him, and I do feel grateful to her. Why he talks like the heroes in Black's novels, like Willy Fitzgerald and George Brand and Frank King. I am so glad. Do you know I always imagined that if he ever married he would marry a girl like Lina Farnam or perhaps Grace Duffield, somebody who loved him tremendously and whom he liked in a cool sort of way and whom we would all be afraid of and feel that she looked upon us as a lot of queer old maids. Now, little as I remember Marian Walker—I had forgotten her very name—I remember that she was not a particle imposing in any way, or swell, or even very decidedly pretty. I have an impression of a middling tall, rather plump, fluffy-haired girl, with pretty eyes and a singularly frank, unconscious manner. But no intuition warned me that here was my future cousin and I turned from her to absorb myself in the Mesdags and Artzes and Mans, for it was a private view of the Dutch paintings from the Fair.[a] I am glad you are glad, and that you told him so. I wish you could have written me as soon as you heard, for now your wonder has cooled down and you can't go over and over it as I want you to and you have even reached the stage of wondering how long it will last and that I will not let myself do. I have been proved to be mistaken in Allen once, for I always thought him incapable of passionate love and now I shall make myself believe that he is capable of constancy too. Agnes it musn't pass over, it is too sweet and dear and fresh and cunning. His letter! I half laugh and half cry over it. Would anybody believe that that introspective, slightly cynical, critical, over-cultured fellow could be so naive, so unconsciously trite, so deliciously young! "Something has happened which has wholly and entirely changed me." "Love is really quite different from anything that has been written about it." "And then I made the discovery that a person may be quite happy in loving one who cannot return his love, just content in caring for her and knowing that she knows it." Isn't that cunning? And then the account of the courtship. "I kept saying hypocritical things to put her off the scent and she never suspected." Of course she didn't. Girls never do. "Finally I gave up and realized that I cared more for two minutes with her than for anything else in life." "It was an immense surprise to her and she was tremendously sorry for me." "There were *other people* who were not so considerate (he is speaking of how he kept everyone in the house from suspecting it) and did bother her a lot"—Of course Marian hated it. Girls always do dislike such affairs so much.—"On thinking it over we cannot help believing that Marian had begun to care by that time, though nothing makes her so angry as to say that she did not know her own mind. So conventional and improbable!" Oh Agnes how he would have laughed at such a letter two years ago. Is he so much younger than we? or is it just because love is rejuvenating and makes trite old things seem real and fresh? I think almost the funniest part is his account of the effect the announcement had on the two mothers. It is just like poor Traddles and his Sophie in "David Copperfield" and these poor children seem to take it just as seriously. Allen says very sadly that Mrs. Walker became extremely ill and every subsequent sight of him brought her illness back. As for Aunt Mary, she was quite as much overwhelmed. Well I do feel sincerely sorry for her, but I cannot see how anybody who truly loved Allen could help but rejoice. The one thing about it that troubles me, they seem to take very calmly, and that is the long engagement. If I were Aunt Mary and Mrs. Walker I would make it possible for those two to marry within a year's time. It is all nonsense that a man may accept an allowance from his mother while he is single, but must never marry until he can support a wife. Why can't the two mothers put together that which they give the children already and let them spend it together? The one bit of sense in that mass of disgusting nonsense "The Woman who did" is the principle insisted on that money considerations ought never to interfere in marriage, that every young man ought to be free to take to him the woman he loves while he is still in his fresh, enthusiastic youth, in the time when Nature meant him to mate.[b] Wait till he is thirty or thirty-two and the freshness, the wonder, the selflessness is gone from his love. It steadies down to a good sound affection, but it isn't like "love's young dream." And I do think it such nonsense Marian's studying medicine. That is the fault of the transition period in which we live. Girls think now that they must all have professions, just because they are free to, not realizing that the proper state of society is one in which a woman is free to choose between an independent life of celibacy or a life given up to child-bearing and rearing the coming generation. We will go down the path of degeneration if we lose our mothers and our home-life. We can easily get on with fewer professional people. I don't mean that she ought not to take up whatever studies she chooses, but she ought not to choose a work which is in its nature absorbing, which cannot be laid down and taken up again. It isn't as if the world were in crying need of doctors, it has too many already. I suppose she might say that she has a right to follow her own tastes as well as Allen. Well, let her do it then, let her study medicine, but if she practices it simply means either avoiding the burden of maternity or fulfilling its duties imperfectly.

Isn't it queer to be thus solemnly discussing Allen's future wife? Do you actually realize yet that love has come to one of us? And don't you feel a little afraid when you think of it? I said that I would not let myself be afraid, but I can't help it. Oh I am so glad there are two of us. I couldn't hold all this in possibly, and yet I couldn't pour myself out and say what I really think to Allen. He says Marian is to be in New York this winter. I will certainly go to see her and I'll write you pages and pages of description. I was so thankful that Edith wasn't here when your and his letters came for you know how bad I am at keeping things.

He has not said one word to me about his degree either. Can Emily [Ladd] have been mistaken? She spoke of it as a well-known fact and I can't think that she was simply going on the strength of that paper, yet neither can I believe that he would have said not a word about it.[c]

I am here at Quogue with Aunt Louise and Uncle Fred.[d]

The telegram telling of Flory's death has just come.[e] I am going at once to Elizabeth, that is, to-morrow early, it is now late evening. The telegram said "Remains brought to Elizabeth Monday."

Oh there is trouble in this world!

a. David Artz was a nineteenth-century Dutch painter of rustic and domestic scenes in the seventeenth-century manner. Frédéric-Henri Mans was a seventeenth-century Dutch landscape and figure painter.

b. *The Woman Who Did* by Grant Allen, a science writer and novelist, caused a sensation when it appeared in England in 1895. The heroine rejected marriage in principle as degrading to women and preferred to live with the man she loved without legal sanction.

c. Allen Williams did not receive his medical degree from Harvard until 1902.

d. Louise Hélène Loizeaux Jennings (1847–1938) was the second wife of Frederick Charles Jennings (1847–1948) whose first wife, Annie Beatrice Pond, was Gertrude Hamilton's sister. Jennings, a tea importer, was a devout member of the Fundamental Plymouth Brethren Church. Although they considered their Uncle Fred something of a religious fanatic, the Hamilton sisters remained close to the Jennings family, particularly to their Aunt Louise.

e. Florence Pond, a first cousin, had died of typhoid fever.

ALICE HAMILTON spent the year 1896–97 at the Johns Hopkins Medical School, where she was relieved to be accepted "without amusement or contempt or even wonder." She worked mainly with Simon Flexner, a young pathologist (later head of the Rockefeller Institute), who helped her publish a paper on tubercular ulcers of the stomach and another on a case of neuro-glioma of the brain. At Hopkins she also got to know some of the great figures in American medicine, including William H. Welch, then professor of pathology, and William Osler, the legendary professor of medicine whose clinic she sometimes attended.[11]

Alice and Edith Hamilton were both fascinated by a young English medical student who spent six weeks in the boarding house where they lived. To Alice, Bonté Sheldon Amos, who belonged to the "young revolutionary circle in London," represented "advanced individualism, a revolt against accepted ideals of self-sacrifice and an exaltation of individual liberty." Though generally approving, as was Edith, Alice believed that Miss Amos and her friends Bertrand Russell and his wife Alys Pearsall Smith went too far in their "exaltation of physical existence, of sexual passion, of the prime instincts of man."

Alys Russell, a graduate of Bryn Mawr College, was then on a lecture tour of the United States with her husband. She took as her official topics temperance and woman suffrage, but she shocked many by endorsing free love.[12]

23 To Agnes Hamilton

1114 McCulloh St.
[Baltimore]
December 6*th* [1896]

My dearest Agnes,

I truly meant to write to you on your birthday, but I was busy all day and in the evening we had a full dispensary, so the day passed and the point of my letter was lost, and I probably should not be writing to you now, if it were not for that very nice letter of yours that came Wednesday. I am so very, very, glad that you are getting better. Mother's first letter said you were and now if you acknowledge it or rather think it worth while to tell me so, then it must be true. And you were really quite forebearing not to add triumphantly that you were probably much better than if you had followed all my enthusiastic schemes and gone kiting off to Jerusalem, Madagascar and North and South [illegible]. I don't know exactly why Allen resented my letter about professional married women. I hadn't the least idea he would. I had told him that I would not, in Marian's place, think that I ought to take up two professions at once and that wifehood and motherhood ought to be considered a career worthy of the best efforts of the most intelligent women. And then he wrote back a very provoking letter, assuming that I took my stand on the old conservative ground and held views just like Miss Elizabeth Porter's. So I wrote back and explained my position just as I did to you and treated the case simply from a general economic standpoint, without even thinking of Marian. And he does not seem to have liked it, more's the pity, although he said it was just what he wanted me to do. I don't think that I go any farther than you do. You see Marian's degree from Radcliffe gives her a means of support in case of emergency and one that she could use much more effectually than her medical diploma, for medical practice can hardly, you know, be taken up as a means of livelihood on short notice. And so I quite agree with you that every woman ought to be equipped with some trade or profession which she could use if needed, but then as I believe in every girl having a very thorough education, that comes in with my system. Allen has never said anything about going home this vacation, but you know he probably would not, even if he knew. He has that instinctive secretiveness which seems out of place in the younger generation of our family and that is the only thing about him that really irritates me. I should think Aunt Mary would be wild to have him come. Have they ever been separated so long before? Yet I doubt if it would be a very happy time for either of them.

Your comparison of Miss Amos to Virginia is fine, as far as I remember "Sir Percival." It has set Edith to reading it over and has made me resolve to read it this next vacation. I dare not now, for I am just finishing "Anna Karenina" and though I have been a whole month over it, it has been very upsetting and has eaten up lots of time that should have been spent on other things. As I remember Virginia there was a strain of hardness in her and that Miss Amos has too, but she has besides a curious kind of denseness that is very amusing and I think must be distinctively English. She says such funny, impossible things, naively and unconsciously conceited, so that Edith and I have to keep from looking at each other for fear we shall laugh. We are beginning, since we have heard so much about the Russells, to think her views very mild and to wonder why she has not been carried farther by her friends. But she does not even live up to her own views; she has "shreds of atavism" still clinging to her which she is rather ashamed of and which the Russells consider most deplorable; such as sacrificing herself for her brother, refusing to urge her young girl friends to read Maupassant and the de Goncourts; thinking that a home is a necessary and sacred thing.[a] All of which saves her in spite of herself from unlovableness. I am sure Mrs. Russell cannot be lovable.

What interesting things you are reading and thinking and hearing about. I wish I knew something about sociology. Next Summer I mean to read some of those books of yours up at Mackinaw. I don't read anything outside my work except reviews of things, and that is a dreadfully superficial way of acquiring information. While I was in Wilmington I had some long talks with Margaret's aunt, Miss Madge Hilles.[b] She is the finest woman. She is interested, practically interested, in social purity, temperance and municipal reform and she is immensely respected by the men of Wilmington. It was her social purity work that specially interested me, for I always have a feeling that that is the work I ought to take up. It seems to me that a medical woman ought to be able to handle the question more intelligently than another, don't you? Only it is so hard to make up one's mind what one ought to do. Miss Hilles has looked up the big questions and decided them to her own satisfaction but her decisions would not in the least satisfy me and it seems to me that one has no right to take up a work practically before one has decided the aim to which all work of that sort ought to tend, what is the real solution of the whole matter, even though one has no hope of bringing about that solution oneself. And that means more study than I can do now. Last winter I looked up the matter a little from the German standpoints; the scientific, which advocated the legal admission and restriction of vice and offered as its sole possible eradication a thorough training of boys and girls from childhood up; and the socialistic, which saw no remedy except in a total change of economic conditions, either permitting marriage to all with no restrictions from want of money, or abolishing marriage altogether and bringing in the old Spartan method of free polyandry, with State support for the mothers, and State ownership of the children. Further than this my reading has not gone and you see I am horribly ignorant. I want to begin to read up the matter during this winter and then hope for some chance to get a little practical experience in it next year. You see I have this Dispensary work on my hands this year and though I don't believe it is particularly useful, I am in it and cannot give it up for something hypothetical. I wish that I were, like you, always most interested in the practical side of a question, but instead it is always the theoretical that appeals to me, just as it did in medicine. I cannot care as much for the details of schemes for practical reformation which may have immediate good results but which do not touch the real cause, as I do for big theories of general reform which of course I can never hope to really help in at all. So you see you probably will do more for your poor whites than I ever shall for social purity.[c]

Alice

a. Edmond and Jules de Goncourt, French historians, art critics, and novelists, were among the pioneers of naturalism.

b. Margaret Hilles Shearman, who received a B.A. from Bryn Mawr College and later studied sociology, was active in the National Consumers' League and the woman suffrage movement; she remained a friend of the Hamiltons.

c. AH did not in fact participate in the social purity movement.

ALICE HAMILTON's response to Bonté Amos and the Russells reveals the limits of her own individualism. Like other American women of her class and religious upbringing, she rejected the doctrine of sexual freedom advanced in literary and artistic circles in the 1890s. Her discussion of the social purity movement reflected uncertainty about the best method of elevating relations between the sexes, a preoccupation of many of her contemporaries, as well as about her own reform vocation. The social purity movement took many forms in the late nineteenth century, but American reformers generally shunned both of the solutions Hamilton discussed—socialism and licensed prostitution (the latter favored by many German scientists). Instead, they stressed personal and moral reform and advocated a single sexual standard—purity for men as well as for women.

The demand for trained bacteriologists and pathologists was limited, and Alice Hamilton feared that she might not be able to find a job. She dreaded the prospect of being forced back on her family for support when she ought to be helping her younger sisters complete their professional training. "I am tired of being a failure," she wrote Agnes during the winter. By spring she expected to work as a pathologist at the Utica State Insane Asylum in upstate New York. Then came an offer to teach pathology at the Woman's Medical School of Northwestern University in Chicago, a position for which she had been recommended by her Hopkins professors and which Osler urged her to accept. The post offered ample time to do her own research work, four months' vacation, and a thousand dollars. Living in Chicago would also permit her to take up settlement work, either at Chicago Commons, headed by Graham Taylor, a Congregational clergyman, or at Hull House under Jane Addams; she rather thought it would be the Commons. At the beginning of June, she went to Chicago to investigate the job (which she accepted) and the settlements (neither of which had room for her).[13]

24 To Agnes Hamilton

[Fort Wayne]
June 13*th* [1897]

My dear child
    I have wanted to write to you ever since I came back from Chicago last Saturday, but my four days there tired me very much and when

Edith and Margaret came on Monday they insisted that I had grown thin, so my family have taken the alarm tremendously and since Tuesday have allowed me to do nothing but lie around and eat and drink and take cod-liver oil. I must acknowledge that I am tired, though I have no idea what has made me so, it came upon me all of a sudden. They all seem to think that they can make more impression on me by talking about my looks than in any other way, so they spend their time telling me how old and thin and worn I look. It really does not take much to impress me, however, for now that I have accepted the Chicago position I feel that I have sold in advance my strength and work for next year and I grow panic-stricken when I think that they might possibly fail me. It used to seem impossible to us all that our strength should ever be insufficient, but we seem to have been all made to see that it has its limits. I have had to give up all idea of going to Hull House or Chicago Commons and I am dreadfully disappointed. You see I had supposed that I could go there and simply sit around and imbibe, without doing anything useful myself. But I find that one is expected to do part of the work and I was distinctly told in both places that they had many applicants and must choose those who were of the most value to them. They asked me what classes I could teach, whether I could give entertaining talks to the boys' clubs, or lectures on different scientific subjects, or teach Italian or drawing. And you know I simply could not undertake work outside my college work even if I knew how. Hull House of course was fascinating and Miss Addams, who took us over it, attracted me more than ever, but the whole thing made me feel very small. It is so tremendously cultured, and all the people seem to be specialists in sociology or kindergartening or manual training or art or music or anything else that is taught there. I know I never would be accepted by them. You see they have a dreadful custom. They accept you for a six weeks' probation and then vote whether or not you are valuable enough to remain. Wouldn't it be pleasant to be voted out?

Chicago Commons, as I suppose you know is a large, shabby old dwelling house stranded down among a lot of tenements and stores. A girl who lived there told Kate [Angell] that she had to walk three quarters of an hour to get a sight of a bit of green grass. It looks like a place that simply would shrivel up with heat in summer. Three cable car lines pass by it and near it is the big Haymarket Square where on market day the farm wagons begin to arrive at two o'clock in the morning, so that a light sleeper finds it impossible to sleep from that time on. You see I am warning you what you must expect. About the Commons itself I cannot tell you much for Professor Taylor was in Michigan and everybody says that the Commons without Professor Taylor is Hamlet with Hamlet left out. The only person I saw was a dead-alive-looking, slatternly girl, who sat and looked at me in a spiritless way and answered my questions with yes and no, and was apparently only anxious to get rid of me. I could not get anything at all out of her except that Professor Taylor would not care to have me come unless I could teach or lecture. Yet almost all the people who live there are supporting themselves, and Katherine says that he urged people just to come and live among the poor. But this young woman said that they had many applicants and must choose the most valuable.

I have quantities more to tell you, but I am going out to lie under the trees awhile and after that there are four necessary letters which must be written. It is very blissful having the girls here and I could talk on about them for pages. It is also nice to feel that I am to be in Chicago next year, and earning a thousand dollars. I know this is not even an apology for a letter, but do take the will for the deed and write me all about Allen and Marian, especially Marian—

Very lovingly
Alice

ALICE SPENT the summer in Mackinac preparing for the academic year. In early August she wrote Agnes that although she had finished only two-thirds of her lectures for the coming term, she was "rest-curing completely," a practice the sisters followed whenever they felt fatigued.[14]

Some time that summer, she heard from Jane Addams that a room would be available at Hull House after all.

IV

HULL HOUSE
1897–1907

ALICE HAMILTON ARRIVED in Chicago in the fall of 1897, as she recalled it, "a very inexperienced and ignorant young woman just taking my first job." Her surviving letters are strangely silent about her work at the Woman's Medical School of Northwestern University, where she was professor of pathology and director of the histological and pathological laboratories, but in her autobiography she called it "a lonely life." A medical and scientific backwater, entirely separate from the men's medical college, the school was in its declining years. Before her arrival, pathology had been taught only by lecture, and she initially had to rely on colleagues at nearby Rush Medical College for her specimens. Later she gathered her own teaching materials from autopsies performed in the morgue of Cook County Hospital.[1]

It was settlement life that really engaged her. By choosing to live in a run-down neighborhood on Chicago's west side, "a region of unrelieved ugliness," Alice Hamilton was participating in one of the great social movements of the decade. Hull House, the third American settlement, had opened unceremoniously in 1889 when Jane Addams (1860–1935) and Ellen Gates Starr (1859–1940), former classmates at Rockford Seminary, rented a decaying mansion in one of Chicago's foreign colonies. By 1900 there were more than one hundred settlements, some with university or religious affiliations. They drew scores of educated young men and women, stirred by a thirst for adventure and the call to service. Many remained a year or two; a few settled in for life. Alice Hamilton was one of those who stayed: she lived at Hull House for twenty-two years and thereafter returned for several months each year during Jane Addams' lifetime.[2]

111

Inspired by the example of Toynbee Hall in London, the founders of Hull House hoped, rather vaguely, to bridge the gap between their own privileged class and the socially disadvantaged. By living among the poor and renouncing "the luxury of personal preference," they sought to avoid traditionally patronizing forms of charity. More than anyone of her generation, Jane Addams articulated the reciprocal relation between the classes and the mutual benefits of settlement life. For the poor, the settlement provided access to the larger world as well as to education and sociability that their straitened circumstances otherwise denied them; at Hull House they also found respect for their old-world customs and identities. The settlement offered the well educated a place to employ their skills and knowledge in socially useful ways, and thus to overcome the sense of impotence that, for women especially, often resulted in years of invalidism, as it had in Jane Addams' own case. To such individuals, their immigrant neighbors also brought a steadying realism and a knowledge of "life as it is lived."[3]

Inside, Hull House provided beauty and comfort. Outside, residents encountered the squalor and misery that their neighbors had come to take for granted; years later, many still vividly recalled the garbage and the rats. The settlement district, a densely populated section of Chicago's nineteenth ward, was a microcosm of urban life. Although the ward housed a notorious criminal district and eighty-one saloons, its principal inhabitants were the working poor. English was virtually a foreign language in a community representing some eighteen nationalities, including Italians (who became the dominant ethnic group), Irish, Jews, Bohemians, Poles, and Greeks. Housing, sanitation, and education were all hopelessly inadequate. Families crowded into rear tenements and wooden shanties, some in tiny alleys. Tuberculosis and child mortality rates were high. So was unemployment. In a neighborhood where many families lived on less than five dollars a week, the sweatshop wages of women and children often spelled the difference between starvation and subsistence.[4]

One of the first activities at Hull House, a reading of George Eliot's *Romola,* reflected the founders' eagerness to share a life of culture and refinement with their less privileged neighbors. They never abandoned this goal; Jane Addams insisted on the necessity of beauty and pleasure in everyone's life. But as residents of the settlement responded to emergencies of all kinds, they soon became trusted figures in the neighborhood. (Addams' early acts of compassion included saving an old woman from the poorhouse and delivering a baby when more experienced, but also more moralistic, neighbors shunned the mother, who was not married.) As their perception of need broadened, residents ran an employment bureau, assisted truant youths, collected statistics on unemployment, child labor, and housing, sat on city and state boards, and engaged in overtly political action. They also supported, and sometimes helped organize, labor unions, then an unacceptable form of collectivity to many Americans of the middle and upper classes.[5]

No one did more to turn the efforts of Hull House residents outward than Florence Kelley (1859–1932). A socialist who had translated Friedrich Engels, Kelley arrived in Chicago with her three children late in 1891, shortly after separating from her Russian-born husband. She brought a consciousness of class and a crusader's zeal that galvanized residents into action. With characteristic vigor, she plunged into an investigation of industrial conditions in the neighborhood, setting a pattern of study followed by action that became the settlement's characteristic mode of reform. Her work helped pave the way for passage of a state factory act in 1893 that limited the hours of work for women and girls, prohibited child labor, and regulated sweatshop industries. Kelley became chief factory inspector and appointed several other residents to her staff. They strove to enforce the law until Kelley lost the job in a political shuffle four years later. "It was impossible for the most sluggish to be with her and not catch fire," Alice Hamilton later wrote of Florence Kelley, to whom she attributed much of her own political education and whose "bigness and manliness and warmheartedness" she greatly admired.[6]

Beginning as amateurs, Hull House residents gradually transformed themselves into effective reformers and administrators. One of the most successful was Julia C. Lathrop (1858–1932), a Vassar graduate who had studied law. While serving as an appointed member of the Illinois State Board of Charities, she became a specialist in the problems of public dependents—the aged, the insane, paupers, and, above all, children. In 1912 she became the first head of the Children's Bureau, an innovative federal agency that focused national attention for the first time on the shockingly high infant and maternal mortality rates in the United States and set out to lower them. More self-effacing than Florence Kelley, she was no less effective a reformer; persuasive advocacy rather than fighting was her method, and it became Alice Hamilton's as well. Julia Lathrop also had an infectious sense of humor and an appreciation of the absurd that help to explain why, among

Hull House's major figures, she became Alice Hamilton's closest friend.[7]

One of Jane Addams' greatest strengths was the capacity to enlist the talents and loyalties of such women. It was not just that she drew them to her. Convinced that a settlement must be a place for enthusiasms, she permitted residents to find the sources of their own creativity. Impartially she nurtured specialists in factory inspection, child labor, juvenile delinquency, and birth control. At once visionary and pragmatic, she refused to be bound by traditions, even those she had created. She was remarkably unpossessive and allowed, even urged, her residents to leave when she thought they could do more good— or find greater happiness—elsewhere. Yet she did not spare herself if she thought she could be helpful.[8]

Always Miss Addams or J.A., even to longtime associates, she inspired intense loyalty in the most individualistic. Neither authoritarian nor sentimental, hers was a charismatic leadership compounded of moral authority, intellectual integrity, and an indefinable quality that made others want to live up to her expectations. Her sadness, mirrored in her eyes, haunted those who knew her well or met her for the first time. After hearing her speak in the spring of 1897, Alice Hamilton described Jane Addams as "having looked upon the misery and sin of the world and having accepted them as an inevitable burden which she must bear with no hope of ever reforming them." Addams' spiritual loneliness set her apart from her colleagues, but they found her neither solemn nor unapproachable. She was without illusion or personal pride, and she had none of the love of martyrdom of some farsighted people.[9]

By 1897 Hull House had become a genuine neighborhood center, a channel for social action in city and state, and a model for reformers throughout the nation. Already the most famous American settlement, it accommodated some twenty-five residents, three-fourths of them women, and received several thousand visitors each week. It harbored a kindergarten and nursery, a cooperative boarding house for working women, a gymnasium, theater, music school, and art gallery; it also offered adult education classes ranging from Greek art and Dante to English composition and mechanical drawing. Open to speakers of the most radical stripe, the settlement early earned a reputation for radicalism and for godlessness as well. Some enterprises, like the dances and the Coffee House (which offered cheap and nutritious food as well as sociability) were thought to provide more salutary entertainment than public dance halls and saloons, institutions that most residents found alarming. Others, including the public baths and the playground, proved so useful that the city took them over.

Alice Hamilton's encounters with Jane Addams and Hull House were the most important of her adult years. She later claimed that life at Hull House "satisfied every longing, for companionship, for the excitement of new experiences, for constant intellectual stimulation, and for the sense of being caught up in a big movement which enlisted my enthusiastic loyalty." Accurate as far as it goes, this retrospective appraisal ignores the difficulty she experienced during her early years at Hull House in reconciling conflicting personal imperatives and in finding her own, authentic voice.[10]

25 To Agnes Hamilton

Hull-House
Wednesday October 13*th* 1897

Dearest Agnes,

This is only a beginning, a faithful attempt, and perhaps if I am lucky it will succeed. It is my first free evening since so long ago that I can't remember. Last night Rachel [Yarros] came over and I lay off and rested while she talked to me. It was very nice and I loved to have her but I had planned to write to you. It was very forgiving of you to write to me again. Frankly I am awfully glad you didn't come back with me. Last week was interesting but not as interesting as I had expected, and this week I shouldn't have had any time at all for you. And I am not at all in the swim yet and could not go around and show you things as I could later on. No, January will be much, much better. As soon as I can I will ask Mrs. Kelley about your coming. You simply must take the first two weeks here, you know. Mrs. Kelley I am growing tremendously fond of. She has made me very uncomfortable, however. Sunday evening she was talking to me and asking me if I thought the life here would be bad for me, and she added "I hope not, for I feel partly responsible for your coming. I urged Miss Addams very strongly to accept you." And when I asked why, she said "Oh well. I liked your cousins so much when I met them in Fort Wayne that I knew if you were anything like them we had better have you in Hull-House." Which of course made me quite miserable. For of course I am not at all like you and if, as I think you wrote me, you and Katherine spoke to her about the sweating system and what things not to buy, then I am less like you than ever, for I don't even know whether I believe in not buying sweaters' clothes. So I sat there and felt like a miserable hypocrite and wished you were here to do the things they will expect me to do. Mrs. Kelley I find approachable and I can enjoy talking to her very much, but Miss Addams still rattles me, indeed more so all the time, and I am at my very worst with her. I really am quite school-girly in my relations with her; it is a remnant of youth which surprises me. I know when she comes into the room. I have pangs of idiotic jealousy toward the residents whom she is intimate with. She is—well she is quite perfect and I don't in the least mind raving over her to you, because by January fifth you will be just as bad as I, every bit.

This is a typical evening here. Here in the back parlour I am sitting at the table and opposite to me is Miss Johnson, who is the street-cleaning commissioner.[a] She is having a most killing time interviewing an old Irishman who wants a job from the city. He has brought his wife with him and she is scolding him for saying he is sixty. She says he is only forty-five, although she insists he fought at Balaclava. Miss Johnson wants to know when his birthday comes, and his wife says "Say the fourth of July, it sounds well." Now they are having a fuss about his naturalization papers. He says he never had any, his wife says they were lost in the snow out in Utah when Brigham Young was dedicating the temple. At the other end of the table sits Miss Brockway, the sweet little girl who is engaged to Miss Addams' nephew.[b] Mrs. Kelley is lying off on the sofa. In the front parlour are Mr. Deknatel and Mrs. Valerio.[c] Mrs. Valerio speaks Italian and she and the man with the queer name—he is our mournful widower—are taking the names of the people who are registering for classes. Miss Addams is on the sofa with a very nice North End man, a friend of Miss Anderson's. They are looking over plans for an addition to the coffee-house. Miss Watson, Mr. Swope, Mr. Ball, Mr. Hooker, Miss Pitkin, Miss Gyles and all the others are managing classes and clubs in various rooms.[d] In a few minutes a certain Dr. Blount is coming.[e] It is a she-doctor and a socialist. I met her some time ago and she it is whom Miss Addams destines to help me in some scheme for the amelioration of the condition of the Italian "neighbors." Then at nine o'clock we are to have a residents' meeting, to divide up the duty of tending the door and showing people over the house.

Dr. Blount has just gone. We went up to Miss Addams' room and discussed a scheme which I haven't time to expound to-night. The chief part of it is that I needn't take any part in it till after Christmas. Meantime I am to take a class in Physiology and one in art anatomy. Do please help me with that last. Norah spoke of some lectures of Mr. Cox' which she said she would try to get for me when she was in New York, but I am sure she forgot.[f] Do you suppose I could get them by writing to the Art League? And what book would be good? Norah insisted that all were iniquitous, but that was simply Mr. Cox' opinion. Last night I interviewed a nucleus of the Physiology class. It is not to be started unless six apply, and so far only four have: two nice-looking shop girls, a tired-looking middle-aged woman, and a fat Jew, who looks like an old-clothes man.

Well, here it is five minutes past nine, and I have reached my tenth sheet and haven't said one word about the Medical School. One would think it played a very small part in my life. Next time I'll tell you something about it.

It gives me a big tugging at the heart to think that Norry is on the ocean to-night. Write again soon do. Think of me as a lonely stranded heathen among many elect who scare her.

Very lovingly
Alice.

a. Amanda Johnson, a graduate of the University of Wisconsin, became garbage inspector of the nineteenth ward following passage of a civil service law. The appointment was the culmination of Jane Addams' efforts to improve sanitation in the ward.

b. Wilfreda Brockway, a native of San Antonio, taught at the Hull House kindergarten; at Jane Addams' instigation, she later studied nursing. Jane Addams' nephew was John Linn, a former resident who became an Episcopal clergyman and YMCA secretary.

c. Frederick H. Deknatel, president of a small hardware manufacturing firm, had charge of the Boys' Clubs at Hull House, where he had come following the death of his wife. He remained a loyal supporter of the settlement, later serving as auditor and trustee. Amelie Robinson Valerio taught French and Italian classes; she later worked as an inspector for the Chicago Department of Health.

d. Lulu Watson had charge of the evening clubs. Gerard Swope (1872–1957), a graduate of MIT, lived at Hull House from 1897 to 1899 while working as an engineer for the Western Electric Company. He taught evening classes in science and mathematics and also worked with the Boys' Clubs. Frank H. Ball was director of manual training. George Ellsworth Hooker, who held degrees in law and divinity, was a longtime resident who specialized in civic affairs. A member of numerous city and state commissions, he was for a time an editorialist for the *Chicago Tribune*. May Pitkin (later Wallace) taught evening classes and directed the relief bureau. Rose Marie Gyles, a graduate of Rockford College who lived at Hull House for many years, had charge of the gymnasium; she was also on the faculty of the Chicago Froebel Association.

e. Probably Anna Ellsworth Blount, then an intern at Cook County Hospital, whose husband, Ralph E. Blount, was a Hull House resident.

f. Kenyon Cox was a New York landscape and portrait painter with whom Norah Hamilton had studied.

WITH ITS LONG drawing room, carved white marble mantelpieces, and lofty ceilings, Hull House reminded Alice Hamilton of her grand-parents' house in Fort Wayne. The communal life also recreated a pattern familiar to one who had grown up in a large and clannish family, one that was primarily, but not exclusively, female. But there the similarities ended. For, as her uncertainty about boycotting sweat-shop products suggests, by selecting Hull House as her home Alice Hamilton had chosen a path that led away from the values of her family. These she later described as "right-wing Liberal, Godkin's Nation was exactly suited to my parents' ideas." This meant, in her father's case, a passionate devotion to free trade, individual liberty, and home rule, combined with a distinct distrust of the masses and social solutions like those advanced by Hull House. In 1896, fearing that a victory by William Jennings Bryan would unleash the forces of social disorder, the normally Democratic Hamiltons had supported William McKinley.[11]

Alice Hamilton recalled that on her very first evening at Hull House she had been bewildered to meet John Peter Altgeld, who as governor of Illinois had earlier pardoned three anarchists imprisoned for a bomb explosion during a demonstration in support of the eight-hour day. The "Haymarket affair" had taken place in 1886, but it still divided Americans a decade later; like many propertied families, the Hamiltons considered Altgeld a "dangerous Radical." The exposure to new ideas was unsettling. Distressed that her ignorance of economics left her "at the mercy of every wind of doctrine," the following summer Alice wrote Agnes: "I am holding on to myself to keep from toppling off the fence which is my only safe and comfortable place."[12]

Intending to concentrate on her work at the medical school, Alice Hamilton initially seemed to view herself more as an observer at Hull House than as a participant. But, eager to do her share and to prove herself, she volunteered for many assignments. In addition to her classes in art anatomy and physiology, she tended the door and "toted" visitors around the settlement one or two evenings a week. She became known as a "woman-of-all-work," and in the next few years variously

118

taught English (first to Russian Jews and later to Greeks), visited sick children in their homes, directed the men's fencing and athletic club, and worked with the boys' clubs. Some years later, she even engaged in the "rather idiotic task" of teaching artistic basket weaving to Italian women, idiotic because she did not know how to do it herself. After 1902, she had charge of the Sunday evening lectures, which often drew an audience of 750.[13]

The work that was most her own was an informal well-baby clinic. On Tuesday and Saturday mornings she installed herself in the shower room in the basement of the gymnasium, where she presided over the baths that constituted the clinic's chief activity. (She tried to have the sick babies come on Tuesdays, and either treated them herself or took them to dispensaries or hospitals.) Her real goal was to offer health hints to the mothers, mainly Italian, whose children suffered disproportionately from rickets. She learned to overcome the mothers' resistance to bathing by anointing the babies with olive oil, but her admonitions to restrict the babies' diet to milk until their teeth came went unheeded. When she discovered that fried eggs and cupcakes did them no apparent harm, she laughed at her efforts to go by the book in the face of those more experienced than she. She was far less sanguine about the mothers' fatalism toward contagious disease, against which she made equally little headway.[14]

The big event during Alice Hamilton's first year at Hull House was the campaign to unseat Johnny Powers, alderman of the nineteenth ward. As chairman of both the Finance Committee of the Chicago City Council and the Cook County Democratic Committee, Powers was one of the city's most powerful politicians. He was also one of the most corrupt: he received bribes from powerful businessmen in exchange for valuable franchises while neglecting the sanitary and educational needs of his own neighborhood. In addition, Powers owned saloons and a gambling establishment, suspect institutions in the eyes of the settlement.

On two prior occasions Hull House residents had tried, unsuccessfully, to challenge Powers' control of the ward. They moved again after Powers eliminated Amanda Johnson's job as garbage inspector early in 1898. The reform candidate was Simeon Armstrong, a long-time dweller in the ward, an Irish Catholic, and, though running on the Republican ticket, a Democrat. Because women could not vote, the Hull House Men's Club spearheaded the campaign. But everyone at the settlement joined in the effort. Powers, who took more notice of this campaign than he had of the others, claimed that Hull House residents were jealous of his charitable work and would soon be driven from the ward. Some priests in the neighborhood openly supported him and charged the settlement with being anti-Catholic. (Alice Hamilton reported that "the most idiotic stories" were circulating, among them "that we keep a Presbyterian clergyman.") Powers' followers also tried to capitalize on hostility to a female presence in politics. Powers easily trounced his opponent. A master of the uses of patronage and of ministering to his constituents in times of need, he retained his post as alderman until his retirement in 1927.[15]

26 To Agnes Hamilton

Hull-House
Sunday April 3rd [1898]

My dearest Agnes,

I have such a delicious relieved feeling—Rachel Yarros has just telephoned me that they are invited to Dr. Carey's [?] to dinner and so they cannot have me there to supper this evening. It is horrid to say that I feel relieved but I do. One needs good bracing December weather to make one feel up to Mr. Yarros, and in Spring time like this one longs for a little more vagueness and ramblingness and one gets cross and weary when a person insists on settling all questions at once in the hardest, most decided and positive way.[a] I must go there to-morrow evening but to-morrow is a long way off and anyway we are going to see Frau Sorma in the "Doll's House." Perhaps by to-morrow I shall be rested. I have been dead tired for three days now. That campaign literature business was lengthy and dreadful. We began on Thursday in the afternoon, as soon as the new registration lists came in. I found Miss Gernon, Miss Brockway and Miss Bartlett at work when I came back, and the rest of us went at it right after dinner.[b] All the first evening it was addressing envelopes. I found that I did two hundred and eighty in three hours. That kept up until late Friday night when the last of the twelve thousand was done. Then we adjourned to the dining-room and folded campaign literature. Mrs. Kelley and the men joined us at half past ten and we worked on till all hours of the night. Then Saturday morning we went at it again. A good many volunteers came in and of course Miss Addams insisted on working, though she is still on the sick list. We gave up the dining room to it and had lunch in the coffee-house and by half past four we really had finished and were the raggiest, flabbiest looking lot of people you ever saw. I had a Thomas concert on hand but I sent the ticket to Mary Hill and lay around on sofas and the longer I lay the tireder I grew.[c] The poor men, however, really looked even more weary than we. Mr. Hill is on the last edge and Mr. Bruce looks like a wreck.[d] While we were sitting around in the evening the big Powers' parade came by. It was Powers' last, supreme effort and was very imposing indeed. We all ran out to see it as it came down Polk street and turned north on Halsted. Of course we had run out with nothing around us, so the men had to lend us things and we stood all wrapped up on the corner and with our bare heads showed very plainly, I am afraid, that we came from Hull-House. First came the Father Mathew's cadets from the Jesuit church, the ones whom Powers uniformed two years ago.[e] They say that he paid a thousand dollars to the church to have them march. He wanted Father Lambert to march at their head, but the reverend gentleman replied haughtily that he might hire the band, but he couldn't hire him.[f] After them came the Cook County Democracy with Powers among them, all in Prince Alberts and silk hats and canes. Then a lot of carriages and then crowds on foot with banners and transparencies. One had "No petticoat government for us," another had a picture of the House with the legend "God bless the Hull-House." But the worst was a great long transparency with a picture of Mr. Hill at a window of the House dangling Mr. Armstrong at the end of a string. Mr. Murray and several others said that on the other side was a picture of Miss Addams tearing out Dr. Valerio's hair, but we didn't see it and I can't see the point of it anyhow.[g] Except for cheering loudly for Powers as they passed the house nobody did anything much, indeed I think it was only small boys who yelled "Down with Hull-House." Mr. Murray says that a good many mothers are very angry with Father Lambert for letting their boys who are Father Mathew's temperance boys march in a saloon-keeper's parade. To-day things seem very quiet. The men are all away, Miss Addams is lying down, Mrs. Stevens type-writing and politics seem to have melted away for a little while.[h]

A dreadful piece of news was broken to us to-day. Miss Gernon is going away for six months, going to Europe on five days' notice. What we are to do nobody knows. You can imagine what her going means. Somehow the house seems all upset. Mr. Hill goes away, and the Moores; Mr. Valerio comes in to stay in Miss Gernon's room, Miss Thomas moves down to one of the Moore's rooms, Miss Gyles to the other, Miss Howe into Miss Gyles' room, and Miss Watson comes back to take her room again.[i] But you see, except for Miss Watson no worker comes into the house and we are left without Miss Gernon. I just wish you could come back. Miss Addams told me this morning that she was afraid they would have to keep a permanent toter in the house and pay her twenty-five a month until the Fall, but of course you couldn't stay in the city. Mrs. Stevens says that it is very trying having the Valerios in the house together, for they jar dreadfully and it makes a very strained, uncomfortable time for everybody. I feel forlorn about the changes. I wish things would stay as they have been.

It is Sunday afternoon and I haven't been out all day and I think I will put on my things and go for a ride on the grip, as my back doesn't feel up to walking.[j]

Miss Strong[?] is ill and Mrs. Stevens has announced that she really cannot tote any more if she is to do any outside work and now with Miss Gernon going we really feel pretty despairing. Miss Addams is going away for a week after the campaign to rest, in Cedarville, which will be forlorn too.

Well I must go or I shall have no fresh air to-day. I send you our campaign literature and an amusing catechism[?] which Miss Johnson brought in.

lovingly
Alice.

a. Victor S. Yarros, a journalist, came to the United States as a young man to escape arrest for radical activities in his native Ukraine. Initially a writer for *Liberty,* a journal of philosophical anarchism, he later became an editor of the *Literary Digest* and an editorialist for the *Chicago Daily News.* He also held a law degree and was associated with Clarence Darrow.

b. Maud Gernon (later Yeomans) lived at Hull House while working as a visitor for the Chicago Board of Charities. Miss Bartlett (probably Jessie Bartlett) taught sloyd, a Swedish manual training system that involved woodcarving.

c. Theodore Thomas was conductor of the Chicago Orchestra. Mary Dayton Hill was studying with John Dewey at the University of Chicago and taught at his Laboratory School; she roomed with AH during part of her residence at Hull House. In later years, she was active in the Henry Street and Greenwich House settlements in New York City.

d. William Hill, who taught economics at the University of Chicago and was active in the Municipal Voters' League, chaired the campaign against Powers. Andrew Alexander Bruce, a former resident who had been attorney to the Illinois State Board of Factory Inspectors, later became professor of law at Northwestern University.

e. Father Mathew was a nineteenth-century temperance advocate whose name was attached to temperance organizations long after his death.

f. Rev. Aloysius A. Lambert, a Jesuit, was assistant pastor of the Holy Family Church, Powers' home church.

g. George Murray, the neighborhood policeman, was friendly with Hull House residents and assisted them with their reform activities. Alessandro Mastro-Valerio, in an effort to reduce Powers' influence with Italians in the ward, had launched *La Tribuna Italiana* in 1898; he edited the weekly and later became its sole owner.

h. Alzina Parsons Stevens, a union organizer and former assistant factory inspector under Florence Kelley, was one of the few residents who had personally experienced the hardships of factory labor; in 1899 she became the first probation officer of the Cook County Juvenile Court.

i. Ernest Carroll Moore lived at Hull House while working on his Ph.D. at the University of Chicago. He later taught philosophy and education and was instrumental in founding the University of California at Los Angeles, where he served as vice president and provost. Dorothea Lummis Moore, his wife, taught physiology at Hull House and was Jane Addams' liaison with the police. Elizabeth H. Thomas, a socialist, was secretary of the Dorcas Federal Labor Union, which met at Hull House. Gertrude Howe (later Britton) was director of the Hull House kindergarten from 1897 to 1907; she later worked for the Juvenile Protective Association.

j. "Grip" was colloquial for grip-car, a cable car.

AT THE END of her first year, Alice Hamilton remained in Chicago to work up an experiment. She read extensively for it (by early June she had completed twenty of the eighty references she had collected) and did some lab work as well. In addition to her own work, she spent two days a week tending to Hull House duties, which included collecting milk samples for laboratory analysis. She also found new avenues for enjoyment and friendship.

27 To Agnes Hamilton

[Hull House]
July 3rd [1898]

Dearest Agnes,

I wonder if you are still in Fort Wayne or if the hot weather has driven you and Jessie away. It has been very, very hot here since Wednesday, but to-day the wind comes from the lake and it is quite endurable again. It is Sunday morning. Mrs. Ball is sitting in the next room, all the rest are scattered. I am in the octagon and the breeze is coming through the window delightfully. This is my last Sunday here for I go to Mackinaw, next Friday or Saturday. Think how much more time I shall have for writing when I am up there and have nothing interesting to write about, while here I have hundreds of things to tell you and never any time. The only reason why I am writing now is because I needn't send my weekly letter home this time. Margaret and Edith got to Mackinaw on Thursday and they are very cross because they have to wait a week for me. I think they are getting jealous of the House and suspect that my arbeit is but a pretext. Which, however, is not true. I cannot love even Hull-House in hot weather and I would have left day before yesterday if only I could have finished my literature. But I simply cannot do hard work when the thermometer is over ninety. Yesterday it was ninety-four and my dispensary in the morning swarmed with children and Oh how hot they were! I soaked myself in a tub of cold water after it was over and got partly cooled off, but when I got up to the Newberry [Library] in the afternoon I was so hot again that my brains melted away and after three hours of it I just gave up and came home and got into the tub again. I think I could not live through these days if it were not for the thought of a bicycle ride in the evening. Mary Hill and Mr. Swope and I purchased three wheels on the same day—all Monarch '97 and all the same price, twenty-five dollars. Last Tuesday I went for the first time on mine and I have been every evening since. It is simply delightful. We leave the nineteenth ward steaming and choking and melting and in fifteen minutes we are on the lake-shore drive spinning along with the air fresh on our faces and the lake before us and the moon just coming up. We usually go to one of the most distant beer-gardens, leave our wheels and stay for an hour or so listening to the music and drinking delicious cold Bavarian beer from stone mugs, then we mount again and reach home between eleven and twelve. And after a cold bath one goes to bed feeling deliciously instead of all melted and miserable as I usually do in such weather. One night it was so dreamy and still that we kept on far, far out along the lake shore and when we felt we had left the world behind us we got off our wheels and went down to the water and took off our shoes and stockings and sat on the embankment paddling in the cold water. The lake was as still as a mirror and the moon was up and there was not a sound to tell us that we were near a big city. We sat there until midnight and did not reach home until one o'clock. We have so infected the house with enthusiasm that Mrs. Kelley has begun to take lessons and last night Miss Addams went out with us on a tandem with Mr. Swope, all the way to Oak Park and back. I think it will be the best thing in the world for her. This afternoon we three bicyclers, Mr. Ball, Miss Addams and Mrs. Kelley are going to Winnetka by train, taking our bicycles and the tandem with us. Then Miss Addams and Mrs. Kelley are going to the Smiths' for supper while we are going to the lake side where there is a farm house and a lovely Scotchwoman who gives us bread and butter and honey and eggs and lets us eat it on the bluff over-looking the lake. Then we shall come home by moon-light on our wheels. Mary and Mr. Swope are planning another trip for the fourth but I think I shall take my Italian children off for the afternoon. Mr. Deknatel says he will go with me if I do and that will be a great help. I am going to take thirty Italian children over to the Deweys.[a]

Speaking of Mr. Deknatel brings up Miss Brockway. She came back here last Wednesday to get ready for the hospital, for she is going there next Wednesday. Thursday night when I had just got into bed she came in and told me that John [Linn] had come back and it was all right again. Oh how angry and disappointed I was. I couldn't say anything against him of course, but I did make her promise not to be engaged to him again, simply good friends until there is some prospect of their marrying, and to go on with her plans all the same. You see, John went to Cedarville—Miss Addams sent him—and stayed with his uncle and his conscience got stirred up and he didn't see his country-girl and he came back with a long tale of how she had utterly mis-understood him and he had only meant that the beauty and purity of their relation had been polluted by people's gossip and had never meant that he loved Louise and all. I think that that evening the relief was so great that she forgave him fully, but the next morning she kept her promise to me and told him that they were only friends now, and she would not give in to all his pleadings. That day she told me that she could not feel in the same way toward him as she had before. Meantime she and Mr. Deknatel are getting very near to each other. Really it is striding on with marvellous rapidity. Last night they sat out together on the porch and he told her all about his wife. I know she still thinks that she loves only John but what she needs is somebody to love, not any particular person, just somebody.

You mustn't think that all the residents I wrote you about are going for good. Mrs. Kelley's plan has fallen through—and the rest—all but Miss Thomas, are going only for the vacation.[b] Except of course Miss

Brockway, but those two places will be taken by Miss [Ellen Gates] Starr and Miss Benedict.^c So I think we shall not have anybody new at all.

It is almost dinner time and this is a very long letter. Write me soon.

Very lovingly
Alice

a. John Dewey and Alice Chipman Dewey, founders of the influential Laboratory School of the University of Chicago, were closely associated with Hull House. John Dewey was a trustee of the settlement.

b. In June AH had informed Agnes that Florence Kelley hoped to secure a position investigating garment workers in Boston.

c. Enella Benedict, head of the Hull House art studio from 1892 to 1943, also taught at the Art Institute of Chicago.

DESPITE ALICE HAMILTON's initial apprehensions and an attack of lice—a common penalty of "slumming"—the first year went well. During the summer she acknowledged her ability to "teach respectably and do some independent work as well." Returning to Hull House in the fall, she felt a comforting sense of belonging: it was rather like being an "old girl" at Farmington. Alice hoped that Agnes, who had visited Hull House the previous winter, would return. But Agnes and Jessie Hamilton spent the next two years in Philadelphia studying at the Pennsylvania Academy of the Fine Arts.[16]

28 To Agnes Hamilton

Fort Wayne
November 26th [18]98

Dearest Agnes,

I came home with a resolve to write all my overdue letters before I went back, beginning with Miss Thomas and Allen Williams and instead I am writing to you who are not necessary at all. I did want, however, to thank you for your letter about Miss Addams and to apologize again for not having asked Miss Benedict about the scenery painting. I keep thinking of that and mentally kicking myself for my forgetfulness, for I feel sure my letter reached you too late. Then another stupid thing I did. Miss Addams gave me two invitations for that Saturday evening, and I sent one to Edith and then the other I did not dare send home to be forwarded for it was Thursday then and I was afraid it would take too long. And I thought that 3716 Chestnut was your old address, not your new one, so I addressed it to the Drexel Art Institute, which Aunt Phoebe says is not the place you are working in at all. Anyway it would not have reached you before you went to see Miss Davies.[a] Miss Addams made me promise that I would read her all you and Edith wrote about the evening, but of course I couldn't read more than extracts, the letters were both too full of adjectives. Mrs. Kelley was immensely entertained at the story of the woman who taught the deaf and dumb and was doing work just like Miss Addams'. "She was quite right" Mrs. Kelley said "that is just what the Lady Jane had been doing all that evening." Miss Addams herself was much depressed over her paper and really seemed cheered to hear what you and Edith said about it, for she said she had never had a more unresponsive audience. She said "Your cousin was looking so pretty, prettier than I ever saw her, in the daintiest waist."

She is not coming to Philadelphia again this year, as I thought she was, she is going only to Boston. I shall make Allen go and hear her. He amazed me the other day by sending me Mrs. Stetson's last book on "Women and Economics."[b] It is the first sign I have had from him since I was in Baltimore. I suppose he sent it because it takes the same view on women being self-supporting that he and Marian do. It is very cleverly written and I agree with it much more than I did when I talked the matter over with Marian. Mrs. Stetson is in Chicago now and I met her the other day. She is very attractive. She is the woman whose story I told you, do you remember? who couldn't stand being married and so agreed with her husband to have a divorce and now is on excellent terms with her former husband and his present wife who takes care of Mrs. Stetson's little girl, an arrangement which I fancy I should approve of more than you. I was glad to get a sign from Allen, and have an excuse to write to him again. Chreighton says he is still at Pinckney street.[c]

Miss Addams says she saw Miss Thomas in New York and found her enthusiastic in her work. She is living in Miss McDowell's part of Miss Wald's settlement and has lots of Jewish pupils.[d] She told Miss Addams that Miss Wald's was the only settlement in New York which had the least bit of settlement atmosphere about it. The University settlement and Rivington street she found bleak and business-like. I think Hull-House must be even more unusual a place than we thought.

127

Miss Addams says that Miss Wald has never got over Mr. Bruce's performance that night at the House. Miss Dudley was speaking to her about Miss Wald's visit and asked "When was it that she was so dreadfully hurt by a savage attack upon the Jews?"ᵉ You can imagine how badly Miss Addams felt about it. It is a lesson to me never to bring up the subject of Jews in any company whatever. Miss Addams says she is going to tell Mr. Bruce.

Mr. Hooker is not back yet, neither is Miss Starr, but the latter comes in about a week's time. Mr. Deknatel confided to me that he dreads her coming and he believes he will move down to our end of the table. I am so glad I am safely there already.ᶠ We are all worried about Miss Brockway. We have had to brace her up most vigourously all the time, but this last month it has really been almost impossible to keep her from throwing the whole thing up and breaking her contract. To be sure it was her month of night duty and she was worn out, but worse than that was John Linn. He writes to her constantly and has been twice to see her and she has not enough strength of character to send back his letters and refuse to see him, so the affair is half off and half on and she cannot devote herself exclusively to her work and lose herself in it, because her mind is full of him. The result is she is a very poor nurse and she does not take the least interest in her patients or in learning things. It is a pretty serious outlook for the poor child. Aren't things badly mixed in this world? Here is a girl who hates independence and longs to be shielded and protected and managed and has no cravings after latch-keys and money of her own earning, while hundreds of much-fathered and mothered girls would gladly change places with her.

Miss Johnson has gone over to Catholicism as far as she can without actually being confirmed. She told Mrs. Kelley that she had promised her mother not to do that during the latter's life-time. But she goes to confession—can you imagine that independent, hard-headed girl confessing?—and she goes regularly to mass. She has very little to do with anyone in the house except the Valerios, indeed she seems very much changed. Mrs. Kelley feels very badly over it.

I cannot think of any new doings in the house since I wrote last. I have organized a club of little Italian girls, who meet Thursday afternoon after school and sew and play games, but it is not under my charge—two kindergarteners have it.ᵍ My own work goes on just the same. Sometimes I get discouraged over it, for it ought to increase every year, but I cannot give more time to it than I do. I am trying

128

to do some work of my own at the laboratory this year, you see.

Well I have not written a word about Fort Wayne but I know you would rather hear about the house, for you get the rest from your home people.

Very lovingly,
Alice—

a. Anna Freeman Davies was headworker of the College Settlement in Philadelphia.

b. Charlotte Perkins Stetson (1860–1935), better known as Charlotte Perkins Gilman, argued in *Women and Economics* (1898) that to be genuinely independent a woman must be self-supporting. She had suffered a severe breakdown following the birth of her daughter and recovered only after she left her husband.

c. Creighton Williams (1874–1958) was Allen Hamilton Williams' younger brother. He became a lawyer and later practiced in Fort Wayne. AH usually misspelled his name; her original spelling has been retained.

d. Lillian D. Wald (1867–1940) founded the Henry Street Settlement (originally the Nurses' Settlement) on New York's lower east side in 1893. She was a leading figure in the establishment of public health nursing and in many social reforms, particularly those having to do with children. Helen McDowell, a member of the Henry Street circle, opened her house (which was in back of the settlement) for many events.

e. Helena Stuart Dudley headed Denison House, a college settlement in Boston.

f. Although no longer as close to Jane Addams or as central to the management of Hull House as she had once been, Ellen Gates Starr continued to spend time at the settlement until the late 1920s. She was considered difficult, but AH liked her better than she had expected.

g. The *Hull-House Bulletin* lists this as the Alice Hamilton Club. It had twenty-five members.

ALICE REGULARLY kept Agnes informed of romances between residents. She was not infallible in her predictions: in the fall of 1898 she thought that Gerard Swope and Mary Hill were not "progressing," but three years later they were married at Mackinac Island by Jane Addams and a Congregational minister. The romance between Frederick Deknatel and Wilfreda Brockway, which Alice did much to encourage (even while hinting that Agnes might be "a dangerous rival" for Deknatel's affection) also led to marriage some years later.[17]

Of romances of her own the letters are silent. Perhaps only in her attachments to children is there even a hint of regret at her decision to give up a life that included children of her own. She liked working with babies and children and sometimes took as many as thirty Italian youngsters on a weekend expedition. Although she once claimed that she had not influenced a single person, even a child, during her years

129

at Hull House, she made a distinct impression on some of her charges. With cries of "Doctor Hammel," they trooped down the street after her and vied for her attention. She was dismayed that a girl to whom she had given a doll for Christmas had put the gift away, unable to imagine playing with anything so lovely. When she became god-mother to twelve-year-old Alice Wilson, she wished she could take the girl away from the alcoholic grandparents with whom she lived. Her attachment to children is also apparent in her sadness at the de-parture of Florence Kelley's teen-aged son Nicholas (Ko), who was leaving Hull House to join his mother in New York, where she had recently taken up her new job as secretary of the National Consumers' League.[18]

29 To Florence Kelley

Hull-House
May 31st [18]99

Dear Mrs. Kelley,

It is just about the time you used to come home from the library and people are sitting around looking as if they were waiting for you. Mr. Deknatel is in here with me, Miss Addams and Miss Starr and Mr. Twose are in the next room, talking over motor industrial mu-seums and Dewey schools for Italian children.[a] It would be pretty nice if you should ring the bell just now and bring your milk bottle in and join in the talk. Then I would tell you about our trip to Rockford and assure you that Ko went to bed early this evening and will be all rested by to-morrow. Our trip ended much sooner than we expected for we had hoped to have four days, but here we are all home again at the end of the third. Perhaps it is just as well for Ko and me—we are not up to quite as much as the other three are, but they were very charitable to us and put us on the tandems and let us stop for rests and treated us as the babies of the party.

[Fort Wayne]
June 2nd

My letter stopped there though I cannot remember what interrupted it, but I know I had not another minute for writing before I left. It would have been easy to leave in a quiet, dignified, leisurely way if I had not been coaxed and bullied into spending the whole of my last day bicycling on the North Shore. Ko says that he has written you about this last expedition and his unlucky encounter with Mrs. Vance's dog. Don't worry about that at all. His leg was pretty much bruised but the skin was broken only in a place the size of a large pin-head and I squeezed out the blood immediately and cauterized with pure carbolic so that I am sure no germs escaped. And the dog is a perfectly healthy fox-terrier who has a way of snapping at anybody he does not know. They usually have him tied, so I suppose that is why we have not encountered him before. Mary Hill wrote me that Ko had had the nurse inspect his leg and there was absolutely no inflammation. He stood the cauterizing in a way that made me feel like crying. I can't tell you how I felt when I said good-bye to him. One can say good-bye to grown people for a few years and feel that one will see them again, but Ko won't be Ko when I see him again. He stayed for my last lunch and risked being late to school, which pleased me immensely, and I think he really hated to have me go. But children can't be as fond of you as you are of them, do you think they can?

The House is going to be very empty soon and Miss Addams will have a lonely time of it until you come. Miss Starr must have left by now. Miss Smith went yesterday and the less important ones will be dropping off one by one.[b] We are all hoping that Mr. Deknatel will go to Germany with Mr. Hooker, for he is looking very badly and needs a thorough change.

I have come home to be a domestic character for the next four months. My mother has already left me and sails on the seventh to join the two younger girls in Italy.[c] So I am devoting my mind to my grandmother and Quint and the linen closet and camphor chest and sweetmeats and a new waitress and other unusual things, and in a little while I shall forget that I was ever a scientist or a settler. Already Chicago and everybody in it seems very far away.

Don't think you must answer this. I know you have more to do than you have time to do it in and I wrote to you just as I do to Miss Addams, trusting to you not to feel that I expected any answer.

Affectionately yours
Alice Hamilton

a. George Twose, an Englishman who lived at Hull House for many years, taught manual training in a high school. Something of an eccentric, he could be counted on to enliven theatrical productions and to make himself useful even while expressing his cynicism about the possibility of real reform.

b. Probably Eleanor Smith, longtime director of the Hull House Music School.

c. Margaret and Norah Hamilton were in Europe studying biology and art respectively.

THE PLAINTIVE tone of the letter to Florence Kelley hinted at some underlying distress. In June Alice wrote Agnes that the year had been "just a series of failures to look back on, and not only individual failures but one great fundamental one." Away from the confusion and excitement of Hull House, she realized what "an unthinking, restlessly active unquiet life" she led there, existing "in a condition of what I suppose the physiologists would call 'reflex activity,' that is, responding always to outward stimuli, not originating one's springs of activity from within." Yet she felt grateful for the experience of Hull House: "To know Miss Addams and Miss Lathrop is gain enough to make the two years seem worth while." Though vowing to do better, she distrusted her power to "get hold of myself next year any better than I have this." Five days later, in response to a concerned inquiry from Agnes, she elaborated.[19]

30 To Agnes Hamilton

Fort Wayne
June 23rd [18]99

My dearest child

I ought not to be writing to you the least bit in the world. I ought to be making a bathing-suit and answering four letters and writing to Europe and not thinking of spending this time on you, but I am going to all the same. I was so glad to get your letter this morning. It was very, very dear of you to write me at once, that is why I am doing the same, that and because I want to tell you things. I knew I had not been writing much to you, but I did not realize that I had kept up so badly with you as I seem to have. Let me go back and tell you what the winter has been like, and then you will see what I mean by my failure. You see, of course, I was to work among the Italians. In some way I was to try to teach them more sensible and hygienic ways of doing things. Of course I made something of a start last year, but you know how small it was. Well, this year it has just been a fizzle. I had any number of children come to me, but the mothers got more and more into the habit of sending the babies with their older sisters

132

and brothers, so that sometimes I would have over thirty children and only one or two mothers. Now I don't mean that it was valueless. I could teach the chil[dren] at any rate to care for cleanliness, but you can see that it was not what I meant it to be. And to make it worth anything I should have had to do neighbourhood visiting and I could not. All winter long I kept feeling what a farce it all was and how Miss Addams was classing me with the people she is always talking about, who have had scientific or literary training but are utterly unable to put their knowledge into a form useful to simple people. I would worry over it and wonder how pathology could ever be applied to Polk street, and whether I could go in scientifically for rickets although it does not come in with my work at all. And then somehow the people in the house got into the habit of referring to me everything which had the remotest connection with medicine so that in a way I had a good many demands on my time, though it was all so scrappy that I had the feeling of being pulled about and tired and yet never doing anything definite. About January I began to get very tired and I think my work at the school would have been just about all I could manage, but you know how it is about refusing to do things at Hull-House. You simply cannot. And so my work at school was done lifelessly and my work at the House, with weariness and irritation, and along with it all I had a bitter feeling inside that my own work was going to the winds, that I never could be a scientist and was making spasmodic efforts to get up a paper for the Pathological Society, a paper which was so poor that I blush whenever I think of it.[a] You can see how it was, can't you? how I was trying to do three things at once and failing in all of them. In March I got so tired and irritable that I used to dread to go back to the House, I couldn't bear to meet the people. The place seemed so crowded and they were all talking and the bell kept ringing all the time. And Rachel [Yarros] would get at me and tell me that I was ruining not only my health but my career. And Miss Addams would talk about the work of a settlement being to "translate knowledge into terms of life" and I would go upstairs and wish I were in Kamchatka. When it got quite bad and I had to go and hide myself for fear I would cry if somebody spoke to me, Rachel came and carried me off for three days to her house. She was very busy and Mr. Yarros was away so I just lay in bed for three days and nobody talked to me. When I came back I found that Rachel had been talking to Miss Addams and Mrs. Kelley and telling them that if ever I was to do anything in science I must

133

leave Hull-House. Well, I was feeling that pretty strongly myself and Mrs. Kelley thought me a goose to hesitate for a minute. I did hesitate, though, for a whole month and it wasn't a happy month. You see if I could make up my mind just to be professor of pathology in the Woman's Medical School and do what work I could at Hull-House I think I could manage it all right. But I don't think that would be right, even if it were possible. I think I can do good work in science, the men I have worked for have all told me so, and if one can do good work one must, don't you think so? Then in justice to my family too I ought to aspire to a better place some time than this. So I must do scientific work. The only nice thing about all that time was that Miss Addams seemed so genuinely distressed at the thought of my going away. She said that I must stay and do nothing at all, that she was sure it would be better for me in a place where I was happy and had people I liked around me than in a lonely boarding-house. Well of course there is no need for me to tell you that I am going back. I simply must, you know you would too. But as for doing nothing, one might as well resolve to go to a small-pox hospital and not catch small-pox. Only I am going to be wise and quiet and slow and self-controlled. That is really the whole thing. If I could only get myself in hand I could rule my life so that it should not be scrappy. Anyway I will try next year. I am going out to the University to do an arbeit under Dr. Donaldson—the neurologist.[b] That sounds as if I meant to do more than last year but it isn't really so. It will put my mind at rest about my own work if I take two days a week or parts of two days and go out there, away from students and interruptions and with somebody over me to direct and keep me up. My work at the school I am going to crowd into the other three days. As for Hull-House— well Miss Nancrede is coming and I am to direct her work among the Italians.[c] Miss Addams says that is all I am to do. Of course it will not be, but never mind. I must try it one more year anyway. If only I were strong! But I am not. I tire so easily and then I cannot sleep and my work goes to pieces and then I think I shall have to give up everything.

This is the most egotistic and almost morbid letter I ever wrote, but I cannot write about last winter without having all the old feeling of failure and impotence come back. One thing I have resolved upon. As soon as I feel myself growing tired I will go to the Y.W.C.A. for a week and not see anybody for all that time, anybody outside of the school I mean. You see there are many chances that next winter may be my last at Hull-House. Now please don't talk about this to your home people, for I have not told anybody all of it except you. Mother and the girls would worry so, and really nobody can decide but myself.

Edith is home and I am realizing how lonely I was before. One must have somebody to talk one's own language to, you know. She is quite well rested at last and we are doing nothing here so the resting goes on. I had the nicest letter from Hallie Campbell. She says that Maud and Lina were her bridesmaids.[d] She seems very happy, which is natural, and very much like her old self.

This letter sounds worse when I read it over than I thought it would, but I am going to send it all the same.

Very lovingly
Alice—

a. On March 13 AH reported to the Chicago Pathological Society a case of tumor of the brain that showed a peculiar form of degeneration. It was published later the same year in the *Journal of Experimental Medicine*.

b. Henry Herbert Donaldson was professor of neurology at the University of Chicago.

c. Edith de Nancrede, daughter of AH's professor of surgery at Michigan, taught dance and worked with the young people's theater groups. She remained at Hull House until her death in 1936.

d. Farmington friends.

By HER OWN ACCOUNT, Alice Hamilton had worn herself out trying to do too many things, none of them, she thought, successfully. Underlying her inability to control her life was a fundamental uncertainty about what she wanted to do or be. Years later she admonished the graduating class of a girls' prep school that each of them must decide what kind of person she wanted to be. She hoped they would decide to be magnanimous, for the great-hearted would be "incapable of smallness, pettiness, of envy, jealousy, narrowness" and were certain to find their place and especially their work. While acknowledging the triteness of the sentiment, she claimed that "to acquire the joy of work is to have a certain assurance of satisfaction in life, because it grows with use, it is not dependent on other people and even when it tires it does not leave one with the sense of futility that follows pleasure for its own sake."[20]

She spoke from personal experience. Her own struggles to find satisfying work (and perhaps also to transcend some of the attributes of character she later deplored) persisted for a decade. Intense doubt about her own abilities and goals, combined with exceptionally high standards and a deep need to prove herself, made her best efforts seem inadequate. She defined her dilemma as an inability to integrate her scientific work with the Hull House ideal of service; not only did they seem to be hopelessly incompatible ideals, but she found it difficult to define her relationship to either. Career crises are not unique to women, then or now, but the fact of her sex was at the core of Alice Hamilton's dilemma. Science and service represent the classic polarities separating men's and women's spheres. It was a division represented as well in her own family by the contrasting temperaments and values of her parents: she depicted her father as a somewhat remote intellectual, more at home in the library than in life, and her mother as a woman of "generous enthusiasms and indignations."[21]

Neither domain, science or service, satisfied her fully. The laboratory, a world of men, gave her an opportunity to develop a special competence and, she hoped, to prove her worth as an individual. But in the end she found it rather sterile and removed from life. In the early days she insisted that theory interested her more than practice, but her penchant for the absurd incongruity and the emotionally compelling incident, revealed in her letters, belies this self-assessment. There was little of the stereotyped image of the scientist, abstracted and self-absorbed, in Alice Hamilton.

By contrast, Hull House afforded her a glimpse into a world that seemed more real than either the laboratory or the protected environment of her childhood. The compositions of a class of Russian Jews, all socialists, she once wrote Agnes, "interest me more than anything I read during the week." She welcomed the spontaneity and honesty of working-class people; by involving herself with them, Alice Hamilton—like Jane Addams—was able to escape the constrictions of her own sheltered background. But there must also have been something about the claims of the settlement that reminded her of the self-sacrificing nature of most women's lives, a pattern she was determined to avoid. She proclaimed to Agnes that she lacked the true settlement spirit, and she felt impelled to apologize for her selfishness. After she had held out for the career she wanted, despite her family's initial disapproval and subsequent financial pressures (unlike Edith, who had given up the Ph.D. for an assured income), her failure to find complete satisfaction in her work must have been all the more distressing. So was her conclusion, some years later, that she would "never be more than a fourth-rate scientist." But she held her ground: her need to find clear and definite work in which she excelled was too compelling.[22]

Alice Hamilton registered as a neurology student at the University of Chicago in the fall of 1899, and for the next two years she did research there with H. H. Donaldson. She later claimed she had undertaken this work in order not to "lose the desire for research for its own sake." At the time, she wrote Agnes that she "must get up a bit of a reputation to fall back upon" because she expected her medical school to be absorbed into the men's school in the next few years and her job to be eliminated. For Donaldson, whose specialty was the quantitative study of the brain and nervous system, she investigated whether the cells of the gray matter in the brains of newborn rats, unlike those in humans, regenerated after injury. She also continued her own research on the pathology of the human brain.[23]

During her third year at Hull House, a devastating family tragedy overshadowed purely personal concerns. In early October Norah suffered a severe breakdown in Europe, where she had been studying art. Gertrude Hamilton sailed immediately to join her daughter, who had been found delirious in a Lucerne hotel. In November Gertrude thought they might be home by Christmas, but mother and daughter remained in Europe until the summer. Norah spent most of the intervening time in two Zurich institutions, one the famed Burghölzli Mental Hospital, headed by Eugen Bleuler, who later developed the concept of schizophrenia. Between hospital stays, when Norah had seemed better, there was an abortive trip to the Italian Riviera. In letters to her "at home people," Gertrude wrote in minute detail about her daughter's fluctuating condition, always worse during her menstrual periods. At such times, Norah heard voices and complained of "an electrical current that pervaded the house, and affected people's minds." Her symptoms also included extreme restlessness alternating with dreaminess and listlessness, an inability to concentrate, a disinclination to eat, drawings and writings that were sometimes childlike in form or content, and alternating periods of rejecting and clinging to her mother. During Norah's bad times she could scarcely see her mother at all. On one occasion, Gertrude reported: "Each action and word when I am with her mostly seems almost or quite sane, yet when I look back on the time, I find as a whole she was not rational."[24]

Alice's state of mind mirrored the reported changes in Norah's condition. In early November, when the first good news had come, she rejoiced: "At least the terrible part is over. We need not think of her as in the uttermost wretchedness and despair with no possibility of comfort." In late April, after a period of extreme discouragement, she wrote: "When I look back on this winter it seems all Norah . . . everything I have done has had her for a background. For almost two months until now I have had no hope, but now it has come again. If a relapse comes like this last I think I can never let myself hope again." Three weeks later the news was again "very, very bad." Norah's mental confusion and rapid shifts in mood persisted the following summer at Mackinac.[25]

31 To Agnes Hamilton

Mackinaw
August 8*th* 1900

Dearest Agnes,

I have an insecure and unsatisfactory feeling about writing to you in Baddeck because two letters from Edith to Jessie have gone astray and who can tell but what this one will? Papa says that we ought to address them "Nova Scotia," but we never used to and I should hardly think the government would have forgotten where Baddeck is in one year's time.

Of course I am sorry about Hull-House, but not really disappointed for I had never expected it. It was one of the things that are too good to be true. I wonder if you could come for a month or so next winter. I know I could always squeeze you in somewhere.

Norah has just gone to bed. I live so much in her these days that it seems the only thing I have to write about, except "Cell-Division in the central nervous system of the Newborn Rat," which would not interest you.[a] It is my latest arbeit which should have been finished in May, but my head got tired and I could not do anything with it so I kept it until I should feel fresh up here.

Well, as to Norah. I have been with her almost constantly for the three weeks since she came and in that time she has slowly gone through many changes, but all, I am almost sure, are for the better. The first one was to wake up to great life and energy. She took swinging, rapid walks and she was eager to do housework. She was

138

interested in discussions but not really in us. She did not warm to any of us for the first part of the time. Gradually her excessive energy died down, her arm began to trouble her again and we gave up house-work, except on Monday and Tuesday, and had Frances come over. Then she began to grow very considerate and gentle and thoughtful of us. She began to ask questions about last winter and she seemed really to care. I had grown very hopeful until one day when Mother came over, for the first time since the day after they arrived. Norah showed a simply intolerant hatred of her, and after she was gone she was restless and finally told us that she could not help it, she must tell us of Mother's perfidy lest we too fall victims to it and suffer as she did. There was a long story of how Mother and the doctors and nurses were leagued together to keep her in the hospital and use her for psychological experiments. It discouraged us very much, for she told us the feeling had been growing on her and we despaired when we thought of next Fall. She has never spoken in that way since then and I begin to hope that she may be changing in that as she is in other things. She is quite gay now sometimes but by far the best symptom is that she has begun to talk of last winter and tell me all that she thought and felt in the hospital. And as she goes on she lets me stop and ask her if this thing actually happened or only "mentally" and she begins to distinguish between the real and the unreal. That, you know, is a great stride ahead. She says frankly that the dream hap-penings are more real than anything in her life but that it is a different reality from the every day one. It is a strange story, all this mental life of hers last year. When she tells it to me I cannot help thinking of what mine would be if my judgment were suspended and my fancy ran riot. I am sure it would never be anything as sweet and pure and fanciful and high-minded as hers has been. It would be things I would be eternally ashamed of, I am sure. I let her talk of it all for it seems better to try to help her unravel it, yet sometimes I am afraid too. Today she has seemed languid and sunk in thought and she wants to speak of it all the time when we are alone. But indeed on the whole I am hopeful.

Margaret has gone to bed and I must go too. I had a note from Marian Walker to-day. She is in Boston and is seeing Allen over a week, and in September they are to have two whole weeks together.

I have not had a word from Hull-House except this letter of Mr. Deknatel's. Miss Lathrop and Miss Addams landed last week, but I think Miss Addams is in Chautauqua. This Miss Collson whom Mr.

Deknatel speaks of is the one who is doing part of Mrs. Stevens' work.[b] She is a Unitarian minister and a Socialist, not very attractive but most zealous. She is there only for the summer, so are Miss Bartlett and Miss Bowman.[c] You see there was a great exodus from the house this year and they came to look after the really necessary things. Miss Howe has gone to Denison House for the summer and is taking charge of Kindergarten work in the Vacation Schools in Boston. Mr. Deknatel got very down on the House last Winter, and I am sorry he still feels in the same way. I think it is true that there is a pose in the house, which is copied a little from Mrs. Kelley and perhaps Mr. Twose but it is only a pose. We are desperately afraid of thinking ourselves great reformers, of treating our work with an enthusiasm which would betray freshness and lack of real experience and we get into the habit which I think all large, clannish households do, of laughing at outsiders. But even with Miss Addams gone, people like Miss Gernon and Miss Howe, the Valerios, Miss Starr, Mr. Twose and Mr. Deknatel himself, would go on spending their lives in much the same way, just from love of the people and the work. Don't you think so?

It is in the morning now and Norah is sketching me as I write. All the time I feel with her how little we any of us come into her inner life. She has decided that last winter was a time of crucial test for her, that she failed and now she never can be an artist, but must go in for some trade, some form of designing. Yet she speaks of it very quietly, as one would to a bare acquaintance. She is just beginning to realize that last year was one of trouble to us, and she cannot understand it, it puzzles and worries her. She seems to have thought we were all happily indifferent to the whole thing.

I do hope this letter will reach you.

Lovingly
Alice.

a. The article was published in the *Journal of Comparative Neurology* in 1901.

b. Mary Collson took over the work with the juvenile court following Alzina Stevens' death in June. She later became a Christian Science practitioner in Boston.

c. Ella Bowman assisted in the Hull House kindergarten.

WITH NORAH's recovery still in doubt, Alice learned that Allen Williams might die of blood poisoning: "All the feeling that such things are too terrible to happen is gone." She also wrote Agnes that she "would give anything to go to bed and to sleep and never wake up again." For the next few years, Alice's health—sturdier as a rule than that of the other Hamiltons—became a matter of concern to her sisters and cousins. She lost weight, looked gaunt, and took to resting in the summers. Two years later she observed: "How I should have despised myself if I could have seen myself, some years back, sitting on the porch and working cross-stitch—yet that is what I do a good deal of the time."[26]

In the fall of 1900 Alice rented an apartment in Chicago so that Norah could be with her. It was her sister's preference, and the plan had the approval of the Chicago neurologist Alice had consulted. Norah took well to Hull House, to which Alice introduced her, and later became a resident there. But recovery was slow, and for the rest of her life she continued to experience periods of depression and incapacity. Switching from painting to etching and drawing, she continued working as an artist. She illustrated several of Jane Addams' books (and later Alice's autobiography) and after 1921 headed the children's art program at Hull House. Supported at least in part by her sisters, Norah led a roving life, moving often between Hull House and Greenwich Village and living in Europe during the late 1930s. Alice took special responsibility for her sister's health and kept in touch with her physicians. In later years she believed she had often disappointed Norah and blamed herself for Norah's disinclination to share a home with her and Margaret.[27]

When Alice returned to Hull House to live in the fall of 1901, she still hoped that Agnes would join her there. But the following January her cousin went instead to the Lighthouse, a small settlement founded by Esther Kelly in Kensington, the center of Philadelphia's textile industry. During her remaining years at Hull House Alice roomed with Clara Landsberg (1873–1966), a Bryn Mawr College classmate and close friend of Margaret Hamilton. The daughter of a prominent Reform rabbi from Rochester, New York, Landsberg had charge of the evening classes at Hull House, where she lived from 1899 to 1920. For most of that period she also taught German at the University School for Girls. She remained close to the Hamilton family for the rest of her life.

In the winter and spring of 1902, Alice Hamilton faced the possibility of unemployment. As she had anticipated, the Woman's Medical School of Northwestern closed, the victim of inferior facilities, inadequate funding, and the growing appeal of coeducation to women medical students. The school's closing was also a sign of the deteriorating position of women in medicine. By the following year only three women's medical colleges remained, and subsequently the proportion of female students at coeducational schools declined from the peak years of the 1890s, as did their proportion—and even their absolute number—in the profession.

Threatened with the loss of work, Alice Hamilton labored hard at her own research. She prepared several articles for Albert Buck's *Reference Handbook of the Medical Sciences* on topics ranging from lung tumors to acute and chronic spinal meningitis; although she wrote "slowly and laboriously," the work brought in extra money. In March she delivered a paper on poliencephalomyelitis at the second annual meeting of the American Association of Pathologists and Bacteriologists. This meeting, she wrote Agnes, brought her "again into close touch with scientific men as I had not been for five years and made me feel so strongly that I must do my own work well and not half way." By April she thought she would take a research position that would pay only twenty-five dollars a month but would permit her to teach one class at Rush Medical College. She subsequently considered two other positions: the deanship of the women's department at the University of Michigan and an administrative post in the burgeoning campaign against tuberculosis.[28]

32 To Agnes Hamilton

[Fort Wayne?]
[mid-June? 1902]

Dearest Agnes,

Edith is here writing letters beside me and we have been talking so much about you that I feel like writing to you. Edith says that you are going back to Miss Kelly next year which fills me with mingled feelings, for after what Miss Addams said about your looks and what Margaret said still later I am wondering if next year will be your last year and after that you will have to go back to Fort Wayne with the feeling that you will never be strong enough for any other sort of life. I wish I could persuade you to take up some outside work, not to spend your whole time in Settlement work. I do not believe anybody can do it and keep perfectly healthy and normal. Clara tried it this year and will never do it again, yet Hull-House life cannot be nearly so intense and emotional as that in your settlement. Couldn't you do regular work somewhere else every day, teaching drawing or something? Really in these five years of my life here, I have never seen anybody who could stand undiluted settlement work and I don't believe it is right to. I think what one ought to do is to live one's life ordinarily among the people of the neighbourhood one chooses, doing one's avocations and taking one's spare time for one's neighbours. That is really the sort of thing one can settle down to for life, but do you think yours is? Could you spend winter after winter as you have this last one? I don't mean only from the physical point of view, though that is the chief thing, but even mentally it must be exhausting. I do so dread your breaking down and giving it all up forever and yet that is what I am afraid you will do, if you do not spread it all out more instead of intensifying it all in one short space of time.

Edith and I go up to Mackinaw on Thursday. She came up here sooner than she expected because I telegraphed her to come and help me decide about a most seductive offer which had just been made me and which had to be decided at once. She helped me to refuse it and I think it was wise but I have doubts which are very disquieting. For it is a big fine work and somebody is going to have a chance to do something well worth-while and perhaps I should have tried to be that somebody. But I did not feel capable of it and it is not my specialty. It was to go to New York and be secretary of a committee on tuberculosis, taking charge first of the investigation of conditions, then later of the application of remedies. The details of the plan were very well worked out and if they are followed it will be a fine piece of work, but it needs someone with executive ability, power with people, power to organize, to plan work and see that others do it, in short the very qualities which I do not possess. Yet it was very hard to refuse it, especially a salary of $1800 a year for the next five years. I hope I did right.

The work I have for next year is really not remote and useless. I believe at last I shall be able to bring together my scientific life and my settlement life. I shall be doing research into the causation of scarlet fever and I think it will not be hard to extend it and include investigation of scarlet fever in our neighbourhood and stir up the Board of Health on the subject of isolation and disinfection. It seems like a real practical use at last of my knowledge which has always seemed so remote and academic.

These two jobs have given me such a very pleasantly secure feeling. You will never know what it feels like to face probable idleness and a return to dependence upon one's sisters, yet that is what I faced between January and May. Now I feel that, come what will, I can always earn my living and up to this time I have never felt sure of it.

Margaret is so much better. I rejoiced whenever I looked at her, there was such a contrast compared to last year. She is in Mackinaw now with Clara.

What do you think of my passing the New York job on to Marian Walker?[a] I wrote to her about it. The money would tide them over beautifully and I think she is cut out for the work.

Lovingly
Alice—

a. Marian Walker Williams was not interested in the position because she considered it too scientific. This amused AH, who had turned it down because it was not scientific enough.

THE POSITION Alice Hamilton turned down was probably the secretaryship of the Committee on the Prevention of Tuberculosis of the Charity Organization Society of New York, a pioneering voluntary health association that held its first meeting in June. Relieved that she would be returning to Hull House, Alice disagreed with Agnes that "being happier in a place has nothing to do with it. One is happiest in a place into which one fits easily and in which one feels successful. And it is only the exceptionally strong who can stand the process of moral 'hardening,' who grow finer and sweeter through unhappiness and struggle. Most of us can only maintain an equilibrium of growth and waste in pretty favorable surroundings and grow narrow and sour when we are in square holes instead of round ones. Of course I may some day be obliged to go away but I know I shall feel as if I had cut off part of myself."[29]

The position she accepted was as bacteriologist at the Memorial Institute for Infectious Diseases, which opened that year. The director, Ludvig Hektoen, a pathologist and bacteriologist, taught at Rush Medical College and the University of Chicago; he was also pathologist at Cook County Hospital. A pivotal figure in Chicago medical circles, with a national reputation, Hektoen did much to encourage research and to advance standards in his specialties. Although she worked with

144

many remarkable teachers, Alice Hamilton claimed that no one did more for her than Hektoen. Initially one of a staff of four, she found the institute a companionable place. (She had hoped for an academic appointment at Rush, where two of her colleagues taught, but did not receive one.) Her new position left her time to pursue her Hull House interests and to study at the Pasteur Institute in Paris during the summer of 1903. It also placed her at a leading center of pathological and bacteriological investigation. Memorial Institute, one of the first research centers of its kind in the United States, combined theoretical and practical goals: it soon established the *Journal of Infectious Diseases*, edited by Hektoen, and also produced and distributed diphtheria antitoxin. In the next few years Hamilton published articles on the pseudodiphtheria bacillus, scarlet fever, and gonorrheal vulvovaginitis in children. She also wrote on the opsonins (a type of antibody) and the opsonic index (a measure used in vaccine therapy), subjects then much in vogue. (The Institute's greatest success came in the 1920s when George and Gladys Dick established the cause of scarlet fever and developed a skin test for susceptibility to the disease.)[30]

Returning to Chicago in the fall of 1902 in the midst of a severe typhoid epidemic, Alice Hamilton seemed to have found the perfect opportunity to integrate her research with settlement life. The nineteenth ward, with less than 3 percent of the city's population, had suffered more than 14 percent of the casualties. Since the district drew its water and milk from the same sources as neighboring wards, Hamilton believed that some local condition must be responsible for the high death rate. Poor sewage disposal seemed the most likely explanation. Two Hull House residents visited 2002 dwellings and found that only 48 percent had modern sanitary plumbing; their careful maps and charts graphically illustrated that the incidence of typhoid was "greatest in those streets where removal of sewage is most imperfect." Hamilton suspected that flies might be the principal agent in spreading typhoid in the ward, where open privies (though illegal) abounded and sewage arrangements were unusually antiquated. She and her coworkers captured quantities of flies near two undrained privies. Bacteriological tests at Memorial Institute revealed the presence of the bacillus of typhoid fever, apparently confirming Hamilton's hypothesis. She presented her findings to the Chicago Medical Society in January 1903, and the following month her paper appeared in the *Journal of the American Medical Association*.[31]

It was probably the most acclaimed discovery of her career. Certainly none had more immediate impact. Hull House residents launched an attack on the Chicago Board of Health for its lax enforcement of existing sanitary regulations. A Civil Service Commission inquiry followed and found evidence to support the charge. Many of the sanitary inspectors were subsequently dismissed; five were also indicted for bribery. But the residents also met with resistance. Hamilton reported in the fall of 1903 that *The Chicago Chronicle*, a Democratic newspaper, "accused us of godlessness, then they went at the Commons for being religious and then they attacked us both for socialistic ideas." The mayor refused to remove the Commissioner of Health, but following the appointment of a new chief sanitary inspector, conditions in the neighborhood improved as landlords were forced to comply with the law.[32]

Alice Hamilton's own triumph turned out to have been premature. The real cause of the severity of the epidemic was a broken water main that had allowed sewage to seep into the drinking water, an even more shocking indictment of the Board of Health, which had covered up the incident. Never again did she achieve such dramatic results.

After 1902 Alice Hamilton assumed fewer internal responsibilities at Hull House but took a more active part in neighborhood and city-wide reform efforts. As with her baby clinic and the typhoid investigation, most of these forays touched on one or another dimension of public health, a rapidly developing field of settlement work.

Her first major public venture, an investigation of cocaine, began in 1904 when Hull House residents became aware that a number of neighborhood boys were addicted. Aided by police officer George Murray, Hamilton and two other residents rounded up suspected samples of the drug. She learned the toxicological tests, initially performing them on laboratory animals at Memorial Institute but later, after a lawyer had impressed a jury by dwelling on scientists' cruelty toward animals, testing the samples on her own eyes. With other residents, she lobbied for legislation that would increase penalties for sellers of cocaine, mainly druggists. In 1907 a new state law, providing for a thousand-dollar fine or a jail sentence or both for selling the drug, incorporated many of her suggestions. But a clever attorney, by managing to get the cases thrown out on a technicality, effectively nullified the statute. These and other experiences in the courts engendered a distrust of the American judicial system that Hamilton never outgrew.[33]

Alice Hamilton also played an important role in the first phase of the antituberculosis movement in Chicago. Although she did not even mention them in her autobiography, these efforts provided the context for her earliest exploration of the relation between occupation and disease. In January 1903 she helped to organize the Committee on the Prevention of Tuberculosis of the Visiting Nurse Association; she also served as secretary. By the end of the year she had resigned the latter post but was still a member of the organization's medical committee and its committee on statistics and literature. A full-scale investigation of the distribution of the disease proved impossible, but she wrote up what information she had in *A Study of Tuberculosis in Chicago*, a pamphlet published in 1905 by the City Homes' Association, based in part on a house-to-house investigation of a block in the Jewish district near Hull House. The following year, in a report for *Charities and The Commons*, she identified unsanitary conditions, low wages, excessive fatigue, and long and irregular hours as the principal occupational conditions that fostered tuberculosis. She also took charge of a Hull House investigation that tested, by means of an ergograph, signs of fatigue in young women factory workers. Though the findings were negative, Jane Addams and Alice Hamilton maintained, in a paper presented at the International Congress of Tuberculosis in September 1908, that more accurate tests would demonstrate that the excessive fatigue caused by the piecework system was a predisposing factor in tuberculosis.[34]

In addition to the campaigns against tuberculosis and cocaine, Alice Hamilton also served on a committee on midwives, jointly sponsored by Hull House and the Chicago Medical Society. Accepting the fact that midwives were "socially inevitable" for the foreseeable future, the committee recommended that standards of training be raised and that practitioners (many of them suspected of performing abortions) be more closely regulated. Hamilton also conducted a pioneering study of working-class families that correlated "excessive child-bearing" with high infant mortality rates. If such efforts did not bring her the satisfaction in work that she craved, they did mark her entry into reform circles. She began to write articles on such subjects as "The Settlement and Public Health" for the national social work press, and she attained public recognition for some of her activities, most notably her investigation of cocaine.[35]

Alice had been extremely disappointed when Agnes had decided to take up her work at the Lighthouse rather than at Hull House. But the contrasting nature of the settlements they chose reflected the cousins' diverging interests. The Lighthouse had opened in 1895 when Esther Kelly, daughter of a prosperous sugar broker, established a restaurant and social rooms for men to counteract the influence of the numerous saloons in the neighborhood. Its scope broadened in the next few years, but the atmosphere remained intensely evangelical. Where Hull House residents earned the disapproval of the orthodox by eschewing activities of a religious nature and refusing to preach morality to their neighbors, the Lighthouse held Bible classes and distributed total abstainers' cards. (Agnes took the pledge in January 1903.) It was probably such differences, and possibly a lingering tendency to idealize her cousin, that prompted Alice to take an apologetic tone about the less sober doings at Hull House, of which she feared that Agnes, and especially Esther Kelly, would disapprove. Jane Addams had always rejected the view that settlement life was to be a sacrifice, and Hull House residents made a point of not taking themselves, or at least their reform vocations, too seriously. They also enjoyed such worldly entertainments as concerts and theaters, the latter evidently frowned on by Kelly. The differences in the cousins' religious and social views, long apparent, increased over the years, but their friendship and mutual respect persisted.

33 To Agnes Hamilton

Hull-House
January 28th 1904

Dear Agnes,

Of course Miss Kelly and Miss Wilson can come for a week. Miss Addams says she will be very glad to have them and we can surely get them in, though perhaps we may have to put them into one room. How would that be? Would they be uncomfortable? It is possible that we may have two little rooms, but I can't be sure. What do you think about it? If Miss Kelly were a really High Church Episcopalian and belonged to the Holy Cross and went to confession and liked rosaries and prie-dieus, I could put her in Miss Starr's flat, but as it is I cannot be sure Miss Starr would be nice to her.

If you think I am not quailing inside you are mistaken. Miss Kelly paralyzes me. She makes me feel uncouth and ill-mannered. And I am so afraid of the impression the House will make on her. You never can tell how they will act. Only this very evening Mr. Twose and Mr. Yeomans and Mr. Riddle began to scrap in the reception room and actually got to wrestling, and it would give a stranger such a queer impression.[a] And Miss Lathrop won't be here and Miss Addams may be away part of the time and Miss Starr may act up and everything may be queer. All the same please tell her we are most pleased and happy and all the rest.

Last Sunday morning Miss Addams toted some Gospel Settlement people into the Lecture Hall where Miss Nancrede was folding theatrical costumes and Miss Goodrich was doing cross-stitch and Mr. Yeomans was popping corn.[b] What am I to do if that happens while she is here?

I don't know about the money. It is so hard to manage breakfasts and lunches because we have them in the Coffee-House and pay for them. Dinner is punched off our tickets and that meal is easy to manage, but we have a lot of trouble over the others, guests are always going up and paying for themselves. It would amount to seventy or seventy-five cents a day apiece.

Honestly I do hope Miss Kelly will see something nice about the house and I will do my best and take her to the Commons and Association House, which is the dearest little settlement I know, and do my best, but it scares me to think of it.

I thought you would write me something about Marian. I have not had one bit of news since Sunday afternoon, which means that Aunt Mary has had none, for she promised that Chreighton would bring it over at once. Of course there is comfort in that, fatal news would come at once.[c] Poor Allen.

Lovingly
A.H.

This does not sound terribly cordial, but really I mean it to be.

a. Charles and Edward Yeomans, both Hull House residents, had a family business that manufactured pumping machinery. Both men worked with the boys' clubs, and Charles directed the glee club. The Yeomans in question was probably Charles, who was twenty-six at the time, eleven years younger than his brother. Joseph B. Riddle, probation officer of the Juvenile Court of Cook County, taught classes for boys and was also treasurer of the Hull House Men's Clubs.
b. Helen Goodrich assisted with the Hull House music program.
c. Marian Williams was seriously ill; by early February she was out of danger.

In her early thirties, Alice Hamilton had yet to find her vocation. Shortly after Agnes moved to the Lighthouse, Alice contrasted her cousin's new mission, with its "close personal work," to her own life, which seemed "cold and formal and remote," adding: "I had to tell myself that after all Hull-House was not my work, that I was a professional woman and had only a little time for this part of my life and so I must not expect too much." A few years later she still worried about how adequately she was fulfilling her family responsibilities. As she wrote Agnes: "I do hope that in that after-life in which you have such confidence we shall at least have a little more light on right and wrong. Whoever invented the saying 'Be sure you are right and then go ahead' was surely a very simple and naïve creature. As if the last part were not a lot easier than the first."[36]

34 To Agnes Hamilton

Hull-House
December 1st 1906

Dear Agnes,

When I was dating my slides this morning I suddenly remembered that it was your birthday and so I am going to write you this very evening, unless somebody comes along and interrupts me. I am so glad Edith went to stay with you and that you went to Baltimore. They all, and Edith especially, needed to be taken out of hospital air and thoughts and made to see other people and think about less personal and worrisome things.[a] That is the great trouble of family life— one cannot get away from one's worries, even at meal time. I think people ought to eat their breakfast alone, their lunch perhaps with the family, and for dinner they should always have some outsiders so that family subjects cannot be discussed. Then there is a period of real rest and change.

Edith wrote me a letter full of Kensington and you and Mrs. Bradford and the other Kellys.[b] She did enjoy it ever so much. Your last letter filled me with envy. I have been so irritatingly busy with laboratory work this Fall that I have done nothing else except a few things squeezed in here and there. And every now and then it comes over me that the whole thing is so little worth while, the results so utterly insignificant, for of course I shall never be more than a fourth-

150

rate scientist. I do sometimes want to give it up and work in the same way over social questions. Frances Kellor was here not long ago, and told about her study of immigration and made me wildly envious.[c] And Mrs. Van der Vaart does investigations of women in industry, which I should like too.[d] However, of course if I were not held down to a paid job, probably I should sleep late and take afternoons off and go to the theater, for I doubt if I should have conscience enough to keep me steadily at work.

Norah wrote me that you had met Bob Colarossi. I wish you could cultivate him a little.[e] He needs it and is a very lovable boy when one gets to know him, but dreadfully tangled up in his own temperament. He would love to know you, I am sure.

I am afraid I am getting dull and unobservant. Sue was here with me for two weeks and was constantly asking about and exclaiming over things which I have come to look on as a matter of course. Perhaps I don't see things as interestingly as I used—certainly I don't find them as exciting. Margaret and Edith say my H.H. letters are not nearly as full of things as they used to be. When I try to think what we have been doing and talking about this Fall it seems to come down to truancy and immigration. We have a really good and interesting investigation of truancy which I must send you as soon as it comes out, next week. Then Miss Addams is much wrought up over the bill for restricting immigration and we have had lots of people here talking about it.[f] She is going on to Washington next week to see the President. Cocaine came to a complete standstill for five months, while we waited for a judge's decision which proved unfavorable. I believe now there is no use doing anything till we have a new law, and the legislature does not meet till January. Meantime I fuss a little with Italians and run my Sunday evening lectures and that is about all.

Norah writes that Margaret is coming home in ten days. I am afraid she dreads it. I wish I could help, it is ever so hard being way off here.

Lovingly
Alice

Please remember me to Mrs. Lothrop [?] and Mrs. Bradford.

a. Margaret Hamilton had recently had surgery.
b. Esther Kelly had married Robert Bradford, a lawyer and Lighthouse resident.

151

c. Frances Kellor, a social reformer, specialized in the problems of immigrants.

d. Harriet Van der Vaart was secretary of the Illinois Consumers' League.

e. Bob Colarossi was related to the head of the metalwork shop at Hull House.

f. The U.S. Senate had endorsed a literacy test for entering immigrants. The measure, which stirred up heated opposition among immigrant groups and their allies, was not included in the new immigration act of 1907.

IF SHE HAD not yet found the work she wanted, Alice Hamilton had begun to climb down from the political fence. She still disliked controversy, but after the Board of Health episode in which she had figured so prominently, she observed of the "revulsion against the house" by some wealthy patrons: "Of course it means less money but I can't see how people can possibly do work that is worth while without getting some people down on them." As her experiences at Hull House drew out a latent empathy for the underprivileged, facile references to "slumming" gave way to deeper feelings. Even Edith, whose "natural bias" was on the other side, could understand after a visit why the residents talked so much of socialism. Alice Hamilton never became a doctrinaire socialist, but she was radicalized by police brutality and by the intolerance of Chicago's establishment toward those who were poor and foreign-born. In her early years at Hull House, a policeman shot two neighborhood men—killing one of them—for sitting on garbage boxes in front of their homes. She also lived through several witchhunts, including the attack on Hull House for allegedly harboring anarchists that followed the assassination of President William McKinley in 1901. Repelled by intolerance and violence, Alice Hamilton rejoiced that Hull House remained open to all points of view. Living in a settlement, she later noted, made it impossible for anyone to "divide people into sections and file them away in labelled cubby-holes, and then think you know all about them."[37]

V

EXPLORING THE DANGEROUS TRADES
1908–1914

ALICE HAMILTON ATTRIBUTED her interest in industrial diseases to her long experience of living in a working-class neighborhood. Her general concern took a more precise turn after she read in close succession a muckraking article by a young Chicago journalist and Thomas Oliver's exhaustive compendium, *Dangerous Trades*. The year was probably 1907. Setting out to learn everything she could on the subject, she discovered a rich European literature on potters' consumption, on lead and mercury poisoning, among other illnesses, and on the efforts of several nations to control them. But in the United States she was met with "an almost complete silence." Everyone assured her that the superiority of American working and living conditions made such diseases impossible. She was skeptical, and in her first article on the subject, published in *Charities and The Commons* in September 1908, maintained that it was "very improbable that all American employers would voluntarily take the precautions which stringent legislation has had to force upon the foreign employers." Still, in the absence of reliable information, she could only describe some of the ills that afflicted European workers and urge Americans to take up the subject and join "every civilized country" in proving that industry could be carried on without the "sacrifice of life and health."[1]

Alice Hamilton's article was one of the first on the subject to appear in the United States. During the next few years interest grew, as the campaign to improve the conditions of life and labor for all Americans gained momentum. Seeking to define and implement minimum standards for all, settlement workers, university professors, and other reformers pressed for immediate improvements—minimum wages, shorter working hours, safer and healthier working conditions—and for a long-range system of social insurance, like those in Germany and England. State and federal agencies as well as voluntary associations commissioned investigations of every aspect of industrial life, the first step, so it was believed, to effective social change. The subject of industrial accidents received extensive and dramatic coverage, leading to the creation of the United States Bureau of Mines in 1910 and to the passage in the next few years of a number of state workmen's compensation laws for accidents on the job. Interest in industrial diseases lagged behind the more newsworthy subject of accidents, but these too gradually came under the glare of publicity.

No organization did more to promote the goal of industrial hygiene than the American Association for Labor Legislation (AALL), founded in 1906. Although its objectives were wide-ranging, the AALL chose industrial diseases as one of its first points of attack. Between 1908 and 1912 it established a national commission on industrial hygiene, sponsored two conferences on industrial diseases (the second included a joint meeting with the American Medical Association, marking the first session devoted by that body to the subject), and mounted an exhibit, a favorite device of the era for conveying information to the public. The AALL also drew up a memorial requesting President William Howard Taft to appoint a national commission to investigate the incidence of occupational diseases and methods of preventing them.[2]

Early in 1908 Alice Hamilton was among the handful of physicians approached by John R. Commons, professor of economics at the University of Wisconsin and then secretary of the AALL, for advice on how to proceed in the matter of industrial diseases. Later that year Commons, through Irene Osgood, assistant secretary of the organization, asked her to prepare a short piece on "laws for the prevention of diseases and of the use of poisons." Although her first article on industrial diseases had already appeared, Hamilton felt unequal to the task, claiming: "Mine is not the pen of a ready writer." Osgood subsequently sought her advice on how to conduct an investigation of phosphorus necrosis (popularly known as "phossy jaw") among matchworkers.[3]

154

35 To Irene Osgood

My dear Miss Osgood

I am sorry that I cannot tell you much about the match industry in this country. Undoubtedly white phosphorus is used in the Diamond Match factories, but phossy jaw is said to be unknown in this country. I have been told that their factories are models, but again I have heard that there are badly managed factories, here and there. No investigation has ever been made in this country I am sure.

I am sorry, but I cannot write your paper for you and I do not know who could except Mr. Henderson.[a] It would be utterly foolish for me to advise as to how investigations into industrial poisons [should be made], whether by one organization or another, for I have no information on this subject at all, nor could I do more than to guess how extensive such an investigation should be.

You must turn to somebody who has had experience in such work, if not Mr. Henderson then somebody in the Pittsburgh Survey who knows how long and how thorough such inquiries must be.[b]

Sincerely yours
Alice Hamilton

a. Charles R. Henderson, professor of sociology at the University of Chicago and a leading advocate of social insurance, had been interested in industrial diseases for several years.

b. The Pittsburgh Survey, financed largely by the Russell Sage Foundation, was the first major study of an entire community by a team of social researchers. The model study for reformers of the era, it appeared in six volumes published between 1909 and 1914.

THE INVESTIGATION of phosphorus poisoning in the match industry, conducted by the new executive secretary of the AALL, John B. Andrews, was the organization's most important early contribution to industrial hygiene. Although many European nations had recently prohibited the use of white phosphorus in the manufacture of matches, American match manufacturers denied the existence of phosphorus necrosis, a painful and disfiguring disease that might cause the loss of an eye or require removal of a lower or upper jawbone. But Andrews' study, which covered fifteen of the sixteen match factories in the country, documented 150 cases of the disease, four of them fatal. His report, published early in 1910 by the U.S. Bureau of Labor, was one of the most effective reforming documents of all time, certainly in the field of industrial diseases. It prompted passage two years later of the White Phosphorus Match Act, which established a prohibitive tax on matches made of white phosphorus and banned their importation and exportation, thus effectively eliminating the poison from the industry. The triumph over phossy jaw dramatically demonstrated how successful a well-planned investigative and lobbying campaign could be. But, as Alice Hamilton observed, the circumstances were unusual: phosphorus was a recognized poison used in only one industry; a safe and reasonably inexpensive substitute existed; and its extremely disfiguring effects aroused immediate sympathy for its victims. Few cases were so easily resolved.[4]

When Andrews conducted his investigation, Alice Hamilton was already engaged in an enterprise that was to change the direction of her life. In December 1908 Charles S. Deneen, the reform governor of Illinois, appointed her, along with eight men, to the Illinois Commission on Occupational Diseases. The group included Charles R. Henderson, a veteran of numerous city and state commissions who had recommended Hamilton's appointment, and Ludvig Hektoen. Lacking any hard data, the commission could do no more in its 1909 report than list a number of industries in the state that might be exposing thousands of men and women to serious diseases. Citing the vastness and complexity of the problem and the need to study the European literature, the commission asked the legislature to fund a two-year investigation of industrial diseases to be conducted by medical experts with a knowledge of bacteriology, chemistry, and pathology. After a delay of nearly a year, the legislature acted favorably on the proposal, although it provided funding for only nine months.[5]

In March 1910 Alice Hamilton accepted the position of medical investigator for the survey, resigning from the commission to do so. Awed by the dimension of the task, the commission decided to concentrate on the poisonous trades, where the connection between the disease and the occupation was "direct and demonstrable." Hamilton was relieved that Hektoen was to have general oversight of her work: "I do feel pretty much lost, for it is starting out into a great unknown and nobody seems to know the first step." In April she observed: "This industrial diseases work is like trying to make one's way through

156

a jungle and not even being able to find an opening. I go to the factories, white lead and lead pipe and paint so far—and I see conditions which make for lead poisoning and then it is the most desperate work finding any. Of course the foremen deny everything and the men will not talk and they live in all parts of the city and employ any number of physicians . . . I shall never be able to get more than an approximate statement about any place."[6]

In addition to supervising the entire project, Alice Hamilton conducted the survey of lead poisoning. Lead, which enters the system gradually and is slow to be eliminated, was the most widely used industrial poison, and the most insidious. The toll in illness and death was high, for repeated small doses, which caused no immediate discomfort, accumulated in amounts sufficient to cause severe poisoning. In acute cases, victims suffered from excruciating attacks of colic or from encephalopathy, characterized by convulsions and even temporary blindness. The poison also injured the nervous system, causing paralysis—the most common form was wrist drop—and premature senility; in extreme cases, men of forty looked and acted twice their age. Far more common were the generally unrecognized symptoms of chronic lead poisoning—extreme pallor, loss of weight and appetite, indigestion, constipation, and rheumatic or gouty pains. At the time, apparent variations in the ability to tolerate lead were attributed to differences in individual susceptibility, a point often made by Alice Hamilton.

She began by trying to determine which Illinois industries used lead. Starting with the known users, she picked up gossip about other likely sources by talking to labor leaders and factory inspectors as well as physicians and apothecaries. It was painstaking and often laborious work: she had to consult thirteen commercial chemists before determining that there was no lead in Illinois pottery glazes. Hamilton and her assistants visited 304 establishments and discovered more than seventy industrial processes that exposed workers to lead poisoning. Among the most unexpected were making car and can seals, polishing cut glass, wrapping cigars in "tinfoil," and laying electric cables.[7]

She was proudest of her discovery that the enamel paint used to make bathtubs contained lead, for nothing in the European literature suggested it. The hospital record of a Pole with colic and double wrist drop provided her only clue. The managers of the sanitary-ware factory where he had been employed assured her that no lead was used in the enamel coatings, and allowed her to observe the work. Finally

157

she tracked down the victim, who informed her that she had witnessed only the final touching-up process; the real enameling consisted of sprinkling a finely ground powder over a red-hot tub. Securing a specimen from a worker whose wife used the powder to scour her pans, she found that it contained as much as 20 percent soluble lead.[8]

In addition to Hamilton's investigation of lead, the Illinois survey included reports on arsenic, brass manufacturing, zinc smelting, carbon monoxide, cyanide, and turpentine, as well as on caisson (compressed air) disease, occupational deafness, and miners' nystagmus (a rhythmic oscillation of the eyes). Twenty-three individuals—physicians, medical students, and social workers—officially participated in the investigation, while many others assisted. There had been several earlier studies of occupational diseases, but none on such a scale. Moreover, with the exception of Andrews' phosphorus study, these had been mainly summary statements of likely dangers rather than contributions of new information. The Illinois survey, and particularly Alice Hamilton's report on lead, was the first to make extensive use of hospital and dispensary records and to correlate medically diagnosed cases of an illness with specific occupations. By following up all leads and visiting workers in their homes, Hamilton was able to document 578 cases of lead poisoning, which she acknowledged to be a conservative estimate.

The commission presented its sobering report to the governor in January 1911. Having demonstrated beyond a doubt that thousands of workers were casually exposed to disease and even death, the commission asked for legislation to secure healthful conditions in all workplaces in the state and special protection for those employed in the most dangerous industries. Several months later Illinois passed an occupational disease law, one of six states to do so in 1911. The law required employers to provide safety measures and monthly medical examinations for employees who worked with specified types of lead and arsenic, in the manufacture of brass, or in the smelting of lead and zinc. All cases of illness were to be reported to the Department of Factory Inspection, which had the power to prosecute violations. Like most of the early occupational disease laws, the Illinois statute did not provide for compensation.

The apparent effectiveness of the Illinois survey and Andrews' study together suggested that the principal bar to industrial hygiene was lack of information, an assumption congenial to reformers of the period. Leadership in this regard came not from physicians, who demonstrated little interest in the problem, but from individuals and agencies interested in working conditions, among them the United States Bureau of Labor. The Bureau had commissioned its first study of industrial diseases in 1903 and in the next few years had published several bulletins on the subject. Early in 1911 Alice Hamilton accepted an invitation from Commissioner of Labor Charles P. Neill to investigate the lead industries. She had met Neill the previous year in Brussels at the Second International Congress on Occupational Accidents and Diseases, where she delivered a paper entitled "The White Lead Industry in the United States." Both Americans had been profoundly embarrassed by the unequivocal declaration of a Belgian physician: "It is well known that there is no industrial hygiene in the United States." They hoped that her investigation would constitute an important first step.[9]

Alice Hamilton began in 1911 by studying the white-lead and lead-oxide industries. The "Old Dutch" process employed by the vast majority of white lead plants to change metallic lead into basic carbonate (white lead) entailed the handling of large quantities of dry lead. This method of "dry separation" produced considerably more dust than did British and German processes, which effected the separation under water. In the absence of effective dust-collecting systems, the conditions Hamilton found were appalling: exterior walls encrusted with white lead that blew into the workrooms when the wind was right; dusty scrap piles of lead on floors; men eating lead-smeared sandwiches in their workrooms; dusty processes carried on in the same room with clean ones, thus needlessly exposing workers to the poison.

One of the white-lead factories that Alice Hamilton visited was Wetherill and Brother, an old Philadelphia establishment. She found conditions in several rooms to be exceptionally dangerous and tracked down twenty-seven "Wetherill Cases," mainly in local hospitals. Soon after her visit, at the request of Webster King Wetherill, secretary and treasurer of the company, she submitted detailed recommendations for making the plant safer. Among these were strict control of dry white lead as well as required use of respirators and overalls for all men who came in contact with it, and weekly medical examinations. Noting that Wetherill might find her suggestions "sweeping and radical," she observed that each of them was in force in some white-lead factory in the country and all were compulsory under the new Illinois law. Wetherill answered the next day that he intended to put her recommendations into effect immediately, except for her suggestions that canvas covers be provided for white-lead cars and that workers be given milk in the morning (rather than at noon). A few days later, the physician Wetherill had asked to examine the workers wrote Hamilton for suggestions.[10]

The day she sent her assessment to Wetherill, she also wrote the foreman, Mr. Foster, who had been her guide. In her notes she observed that Foster, who had been with the company thirty-eight years and had experienced even more primitive conditions, "naturally considers the present arrangements as almost ideal and is unable to suggest any way of improvement."[11]

36 To Mr. Foster

Philadelphia, Pa.
May 22, 1911

My Dear Mr. Foster:

I have just been writing to Mr. Wetherill about our conversation of last Friday, as he asked me to put my recommendations down on paper. Now I want to write you also, because I feel that you are really the one upon whom the reforms depend. The factory which is safe and clean, is the factory which has a foreman who wishes it to be safe and clean. He is the most important factor, for he is there all the time. That is why I am sending you, as well as Mr. Wetherill, a list of the things that I think ought to be done.

1. There ought to be a rule against any white lead which is dry, being exposed to draughts of air, especially on the floor in rooms where men are working, except of course the dry pan room where it cannot be helped. As long as your roller room has piles of white lead on the floor and in open trucks, you will always be having lead poisoning. You see you will never be able to make your men careful under those circumstances, for they get so used to dust and untidiness, that they do not know it when they see it. Make a rule that the floor must be kept clean and all white lead covered up. I know two white lead factories, old ones at that, where you would not know what was

being made in the place, where the floor is so clean you could eat your dinner off it, and no white lead is to be seen anywhere. This should be the rule for every white lead factory.

Keep at the men all the time about dust. Teach them to watch each other, and when you see a man raising dust, tell the other men that he is poisoning them and they must watch him. Make them wear respirators and overalls which does not mean only trousers. It will be slow work, but it can be done. They have done it at John T. Lewis. I have been able to find only three cases from John T. Lewis for 1910, and twenty-seven from Wetherill, yet both factories employ about the same number of men, both use the old Dutch process, both have an oxide department.

There is one matter that I want to speak about: the widow of one of your fatal cases told me that her husband contracted lead poisoning very quickly because he was doing an unusually dangerous piece of work. He was packing barrels and he was induced by the promise of extra pay to ram down the white lead in the barrel so as to be able to pack in an extra quantity. I hope this is no longer done at Wetherill's. There could not possibly be a more dangerous piece of work.

It is slow work and difficult to train a lot of foreigners to take care of themselves, but I know you can do it and I am sure there is no man in the white lead business who is more genuinely anxious to protect his men than you are.

I shall be very grateful if you will let me know later on, whether any changes have been made in the factory.

Yours sincerely,[a]

a. This is an unsigned carbon copy.

ALICE HAMILTON continued her investigations of white-lead factories in the east during the spring of 1911. In June she attended a meeting of the National Conference of Charities and Correction (NCCC) in Boston, where she delivered a paper on occupational diseases for a session arranged by the Committee on Standards of Living and Labor. She became a member of this committee, which included many prominent reformers and which the following year set forth the minimum standards required by every American for a life of decency, among them a living wage, an eight-hour day, and the right to a home. (Hamilton was one of two members who submitted a minority report that suggested a method for financing the higher standards; although Henry George was not mentioned, their proposal called for the untaxing of buildings and increasing the tax on land values.)[12]

37 To Julia C. Lathrop

Staten Island
June 21st 1911

Dear Julia

I am afraid a pencil won't improve my writing but I am waiting in the bleak reception room of a Staten Island Hospital with nothing to look at but portraits of benefactors and nothing to occupy my hands but a note-book and pencil. So you will have to put up with the results.

I am back at work after two weeks wandering which took in Chicago, Boston, Bar Harbor and Farmington. Boston was not much of a success because I felt ill most of the time and the last two days could do nothing at all, except my own paper. Miss Addams went to the last evening meeting, the "Drunkenness" section, and said it was very good, only that R. A. Woods made his usual mistake of talking an hour instead of fifteen minutes.[a] Dr. Stelzle was fine, she said.[b] Everybody was talking about Mr. Fox's effort to set the Trades Unionists to talk against the regular speakers.[c] The story had leaked out and they say Mr. Fox's days in Annual Conferences are over.

It is a night's journey up to Bar Harbor so, as I stayed only over one night and took a night train down, I had just two days there. Mrs. Bowen was most kind and I saw a good deal of the island in that short time.[d] Miss Addams' cottage is charming, her coloured maid most competent and there is every reason why she should have a very pleasant summer. The view from her porch is lovely, like one of the English Lakes, and parts of the shore drive are very beautiful, but on the whole Bar Harbor disappointed me a little. There is very little real forest left, it is almost all recent growth of tiny birches and firs like Michigan on the way up to Mackinac.

Mrs. Bowen's place, too, looks so recent, as if it would be very

162

beautiful in ten years' time. However very few of her flowers were out and no doubt in August her garden is glorious. The house is nice on the outside, but frankly ugly indoors.

Miss Sears and Miss Vittum were fellow guests and ever so nice.[c]

From there I went to Farmington, primarily to see Allen Williams, my twin cousin of whom I am so fond, and who has had so much trouble lately. His little boys are in a cottage there. Allen practices in Hartford and lives on the trolley line and Marian, his wife, is in Saranac, with tuberculosis. I am going up to see her on my way home.

In Farmington too is an old school friend who has built a beautiful Colonial house on a hillside, with meadows stretching on all sides and the valley spread below, the loveliest spot I have seen in this country to live in.[f] She motored me into New York the next day, a seven hour ride through Connecticut and Westchester County and I wondered why anyone need go to England for beauty of scenery. We spent the night at the St. Regis which made the glories of the Touraine fade away. Then I came back to Henry St. and felt like myself again.

Work has started up finely. I spent all yesterday dictating to a stenographer in the National Lead offices all of the criticisms that could be made of their factories. Think how nice of them to wish me to do it. When I had finished, one of the Vice Presidents, a kindly old man, said to me "There has been a good deal of discussion among us as to the propriety of our allowing you to do all this for us without any compensation but I told the other officers that I would not be the man to offer you any pay." Of course I assured him that I was only too glad of the chance to do it, and on my way home I suddenly thought what an awful thing it would have been to take anything from them and then have it said that I had sold Government material to the Trust.

I found two letters waiting for me from independent manufacturers both telling me that they meant to send me the blueprints of their new wash-rooms and bath-rooms and one of them saying that he had already installed a physician as regular inspector. So of course I am more than ever eager to make out my report wisely and cautiously.

Mother writes me that you and Mrs. Case may come up to the cottage in August.[g] That will be delightful and will more than make up for Boston.

Margaret is in Plainfield waiting for news from Clara. It was all arranged that Mrs. Landsberg was to be taken to Baltimore to see Dr.

Kelly and Margaret was to meet them there, but a telegram from Clara said her mother's temperature had suddenly gone up and she could not be moved.[h] The poor girl is having a most anxious time.[i]

a. Robert A. Woods, one of the most prominent male settlement leaders, was head resident of South End House in Boston.

b. Rev. Charles F. Stelzle, superintendent of the department of church and labor of the Presbyterian Church in the U.S.A., disclaimed being a prohibitionist, but worked to enlist trade union support against the saloons.

c. Hugh F. Fox, a civic leader in New Jersey who helped to shape the state's welfare legislation, was active in the NCCC. He was also secretary of the United States Brewers Association at a time when most social workers favored temperance if not outright prohibition.

d. Louise deKoven Bowen (1859–1953), a prominent Chicago civic leader and philanthropist, was a financial mainstay of Hull House, which she also served as trustee, as treasurer, and as an officer of its Woman's Club.

e. Amelia Sears worked for the United Charities of Chicago. Harriet Vittum was head resident of the Northwestern University Settlement in Chicago.

f. The old school friend was Theodate Pope Riddle.

g. Anna Lathrop Case was Julia Lathrop's sister.

h. Howard Atwood Kelly, Esther Kelly Bradford's brother, was an influential professor of gynecology at the Johns Hopkins Medical School whom the Hamiltons and their friends often consulted.

i. This letter is unsigned.

IN THE SUMMER of 1911 Jane Addams sent Alice Hamilton a copy of a manuscript on prostitution, a subject then attracting considerable national attention. Published in 1912 as *A New Conscience and an Ancient Evil*, the book drew heavily on thousands of case records of the Juvenile Protective Association of Chicago, some of them known personally to Hull House residents. Addams emphasized inadequate wages as the major reason women turned to prostitution, a standard view among reformers of the era. She was concerned as well with the living arrangements and recreational opportunities available to young women at work in large cities without the protection of traditional family controls. Though compassionate and less alarmist than some works on the subject, the book is more moralistic and less illuminating than most of Addams' writings.

38 To Jane Addams

Mackinac Island
August 14*th* 1911

Dearest Lady

We spent yesterday evening reading the manuscript and every body is interested immensely, some of them deeply thrilled. We shall know to-day whether or not Edith Wyatt is coming.[a]

Now please tell me if you want anything more than a general comment. Do you want the English picked to pieces, and do you care for any specific criticisms. Because of course if it is practically finished that is quite different. I rather hope it isn't and that you will let yourself take this book a little less breathlessly at the very end. This doesn't mean that I would wish to change much, but I should like a few things to be different.

It is splendid that you have taken up this theme and my sisters and cousins to whom all your material is absolutely new are much moved by it, so I know the public will be. But to me it is you only in spots. Every now and then I catch my breath as I listen and think "There now, that is really she," but I do wish there were more of those passages. Only perhaps for the general public the book may tell better as it is.

Please write just what you want me to do with this manuscript. If no other change is made, please change the word "officer." I found that all the girls thought you meant police officer and did not know what "own" [?] meant.

The stories are awfully good and illuminating, especially Olga's.

Is this what started from Edith Wyatt's article, I wonder. I recognized her Marie [?] Druot [?].

Thank you ever so much for sending it.

Yours ever
A. Hamilton

I haven't said how much I like certain things about it, the reserve and frankness together. I think you have quite wonderfully escaped using ugly words and using uglier euphemisms. But all that is what one knows you would do.

a. Edith Franklin Wyatt, a Chicago writer, was co-author of *Making Both Ends Meet* (1911), a study of self-supporting women who lived on their own in New York City. The inquiry had been initiated by the National Consumers' League.

HAMILTON'S STUDY of white lead, published by the Bureau of Labor in July 1911, conclusively demonstrated the prevalence of lead poisoning in the industry. Visiting all but three of the twenty-five known American white-lead factories, she documented 358 cases of poisoning, including sixteen deaths, that had occurred in a sixteen-month period. Even by this early date, she had developed the techniques that would become her trademark: thorough investigation of factories, correlation of illness with specific industrial processes, and compilation of medically diagnosed cases of lead poisoning. Though models of "shoe-leather epidemiology," her studies did not employ sophisticated statistical techniques, a limitation that no doubt reflected her own weakness in mathematics—she once observed that one "can get on very easily through life without arithmetic"—as well as her primary interest in prophylaxis. Her ability to achieve practical results was striking. Even before publication of the white-lead report, eleven of the factories had significantly improved conditions and five more had indicated their intention of doing so; only three seemed likely to continue entirely as before.[13]

In the next three years Hamilton went on to study lead poisoning in the pottery, tile, and porcelain enameled sanitary ware industries, the painters' trade, smelting and refining, and the making of storage batteries. By 1915 she had moved on to the rubber industry. With the creation of a full-fledged Department of Labor in 1913, her home base shifted to the new Bureau of Labor Statistics there. By then she was irrevocably launched on a career that was unusual not only because as a woman she worked in a male world but because she successfully combined the unlikely roles of scientific investigator, skilled negotiator, advocate for the helpless, and crusader for the public health.[14]

Before heading for the field, Hamilton studied the technical side of an industry with scrupulous care. She once claimed that the most pleasing compliment she had ever received was the declaration of an expert on smelting: "Here is a woman writing on the metallurgy of lead who knows her job perfectly." Since her post carried no automatic right of entry, she had to talk her way into every plant. She resorted to any technique she could—she even worked through a Farmington connection whose father owned a factory—and was rarely turned away.[15]

Once inside, she refused to take anything for granted. After inspecting a factory for lead dust and fumes, she undertook the more difficult task of establishing the extent of morbidity. Sometimes she encountered overt subterfuge—as when the managers of a smelter in a dismal community in Joplin, Missouri, known as Smelter Hill ordered the men with lead poisoning to stay out of sight while "the lady from Washington" visited. To forestall such occurrences, she made it a practice to visit workers in their homes and on at least one occasion examined striking union members in a saloon. But since labor turnover in the lead industries was high, even a careful examination of current employees would reveal little about the incidence of lead poisoning in the industry. Insurance companies and unions—the best sources for such information in England and Germany—were of little help: the principle of workmen's compensation had yet to be established and, though union sickness and death-benefit lists helped in some cases, most of the lead industries were unorganized. Hospital and dispensary records proved by far her best sources, but even these were inadequate, for the majority of workers did not seek treatment from such institutions. Moreover, given general medical ignorance about the symptoms of chronic lead poisoning, many cases went unrecorded.[16]

Since there were as yet no reliable diagnostic tests for lead poisoning, Hamilton early adopted a strict standard for reporting cases: the presence of a clear lead line (a bluish line along the gums caused by a deposit of black lead sulfide) accompanied by a typical case history, or a diagnosis by a physician during an acute attack. Her results were thus deliberately understated. Eager to defend herself against the suspicion of sentimentality, she wrote up her findings in dryly technical documents: "Nothing can be more cold blooded than a hospital history," she observed.[17]

The collective impact of Hamilton's work was dramatic, underscored as it was by her essential caution. By demonstrating that American factories had higher morbidity and mortality rates than their European counterparts, her studies dispelled the myth of the superiority of American industrial conditions. In the white-lead industry, for example, an American plant employing 142 men had twenty-five cases of poisoning in one year, while a German factory with 150 men had only two cases and an English plant with 182 employees had none. Not only were American rates higher, but American workers suffered from the most severe forms of lead poisoning, which had virtually disappeared in England and Germany. Alice Hamilton attributed the difference to the effective European factory inspection systems—workers were transferred to other jobs at the first hint of trouble—and to such

167

improved techniques as the English wetting-down system that minimized the hazards of dust.[18]

If she had done no more than reveal the deplorable conditions in the lead industries, Alice Hamilton would have performed an invaluable national service. She was unwilling, however, to stop with a "cold, printed report" and went far beyond her assigned investigative task. At a time when the federal government lacked enforcement power, she assumed personal responsibility for persuading owners to reform the conditions she had brought to light. Despite her dread of conflict, she made it a rule to present the person in charge with her findings, no matter how unpleasant. Many were initially suspicious of her motives, fearing that she intended to hold them up as public examples. But, temperamentally predisposed to work by persuasion rather than confrontation and convinced as well that the evils resulted from genuine ignorance, she disavowed the techniques of exposure and protest favored by her muckraking contemporaries.[19]

While avoiding sensationalism, she used every other means available to her. During the Illinois survey, for example, she enlisted the aid of Louise deKoven Bowen, a major stockholder in the Pullman Company, after discovering that workers who sandpapered and painted the interiors of Pullman cars suffered from acute lead poisoning. The company responded to Bowen's formal protest by installing a medical department: the number of cases declined dramatically from 109 during the physician's first six months on the job to three per year a short time later. More typically, Alice Hamilton had to fall back on her own ingenuity. In the early days, owners usually denied her claims, or else blamed the illness on the worker's fondness for alcohol or his wife's abysmal cooking. Even those employers who accepted responsibility for protecting their employees thought they had handsomely discharged their obligation by providing washrooms and showers and by requiring workers to wear gloves, while doing nothing about atmospheric dust. In fact, rapid labor turnover was the most accepted mode of prophylaxis.[20]

Alice Hamilton was extraordinarily effective in her self-assigned mission. Her greatest assets were her evident sincerity and her ability to meet the employer on his own ground. Convinced that any man of goodwill would do the right thing once he knew the truth, she had an uncanny ability to appeal to the best instincts of others. Her persistence and persuasiveness, combined with her carefully assembled evidence, helped to overcome the doubts and suspicions of many owners. One whom she came particularly to admire, Edward J. Cornish, vice-president and later president of the National Lead Company, which controlled a large proportion of the white lead industry, initially told her: "I don't believe you are right, but I can see you do." When she presented him with twenty-two cases severe enough to require hospital care, he instituted the reforms she recommended, including medical inspections. Over the years, he frequently invited her back to evaluate current conditions.[21]

In addition to her efforts to influence the practices of individual manufacturers and to secure legislation, Alice Hamilton waged a quiet but insistent campaign to publicize industrial diseases, a subject she considered "a neglected branch of medical and social science." Placing the blame squarely at the door of her own profession, she declared: "One would find it difficult to point to any disease equally prevalent, equally serious, equally controllable, which has been so neglected by the medical profession in America as industrial lead-poisoning." In her view lead poisoning was, next to tuberculosis, the great destroyer of the working class; she considered it endemic to industrial districts as malaria was to swamps. But in contrast to Europe, where industrial diseases engaged the best intellects in the profession, American physicians had dismissed the subject as "tainted with Socialism or with feminine sentimentality for the poor." In print and on the platform, she urged the profession to overcome its ostrich-like attitude and to treat industrial diseases as it would any other public health problem: by taking the trouble to learn about them, by teaching them in medical schools so that every physician would learn the essentials of diagnosis, and by insisting on mandatory reporting of cases.[22]

Alice Hamilton's report *Lead Poisoning in Potteries, Tile Works, and Porcelain Enameled Sanitary Ware Factories* appeared as Bulletin 104 of the Bureau of Labor in August 1912. The following January Dr. H. T. Sutton, a member of the Ohio State Board of Health who was also employed by the American Encaustic Tile Works of Zanesville, Ohio, attacked the report of "this woman" as "a striking example of exaggeration, either a false and apparently a malicious and slanderous report, or an erroneous one." In her study of sixty-eight potteries and factories in nine states, Hamilton had emphasized the high incidence of lead poisoning in the art and utility pottery industry that centered

169

in Zanesville, a consequence, she believed, of the low wages received by workers in the area, who were not unionized, and of the large amount of lead used in the glazes. Sutton, who claimed that little lead was used in Zanesville industries, disputed her conclusions about the high incidence of lead poisoning and the prevalence of low wages.[23]

After learning of Sutton's attack, Hamilton defended herself in a letter to Charles H. Verrill, an economist who was her superior at the Bureau of Labor. Her reply reveals her disillusionment—rarely expressed—with the Illinois occupational disease law, which she believed was being undermined by the loyalty of physicians to the companies that employed them. Elsewhere she noted that since the law required only company doctors to report cases of lead poisoning, individuals seen by private doctors would not be included in the official records.

39 To Charles H. Verrill

Hull–House
February 12, 1913

My dear Mr. Verrill:

It is very distressing to be told that one has issued a false and malicious and slanderous report and I suppose that the Bureau of Labor does not enjoy its share of the unpleasantness.

I hardly know what to say as to Dr. Sutton's accusations. He is the physician employed by the American Encaustic Tile Works of Zanesville whenever they need emergency care and this fact may perhaps color his opinions. I have been going over the notes I made in Zanesville and I find the following jotted down during my talk with Dr. Sutton. "Member of the State Board of Health. He has seen a good deal of lead colic and is often called to the American Encaustic to treat cases of hysterical convulsions in girls, but he does not know whether these girls come from the glaze room or not. He never asked, thinking it was only hysteria. He has had some men with lead colic but cannot give their names. It would not be worth my while to go to the Samaritan Hospital, for there have been no cases there in the last two years, although there must be plenty really, only it is not rightly diagnosed. Dr. Sutton sees about a dozen cases a year and he thinks that a good deal of lead poisoning is not recognized as such. He recalls one case he treated recently, a man working in glaze in Crooksville, an obscure case in which the diagnosis was made largely by elimination and on the ground of the man's occupation." You will see that he made no denial of the existence of lead poisoning in Zanesville.

The three physicians I quote on Page 50 were Drs. Sellers, Bateman and Bainter. Other men who had much to say about low wages and lead-poisoning were Drs. Hanna, McDowell, McCormick, McBride, Rambo, Gorrill, Klemm, and Shinnick. The actual statements as to the amount of wages paid came chiefly from work people, but also from Dr. Sellers and Dr. Hanna.

As to the amount of lead in the glazes used in Zanesville, some official in each establishment was responsible for each statement made on pages 25 and 26, except in the case of No. 1, and there I was careful to say that I had only hearsay information.

Do you wish me to do anything about it? Would there be any use in issuing a counter statement? Of course I could give the names of my informants, but I could not guard against their repudiating their statements.

I tremble when I think what the bath-tub people will say, for that is really the worst part of the report.

Perhaps laws requiring reports from physicians may do some good, but I doubt if there is any very noticeable effect at first. I have been very much discouraged by the results of the Illinois law. Recently there was a meeting held in New York of all the physicians employed by the National Lead Company and four independent companies to discuss what was a reportable case of lead-poisoning. Four of these men were from Illinois plants. They sent me the minutes of the meeting and it is very evident that the policy they have adopted is to first protect the company which employs them. They refuse to admit that any but the most extreme cases are to be called lead-poisoning. This explains why the physician in charge of the National Lead Company's smelter in Collinsville, Illinois has had but four cases to report in the last six months out of a force of four to five hundred men. I discovered a hundred and five cases for 1912 when I was down there. The doctor had treated twenty-seven in six months in his private hospital, cases of colic in lead workers, but according to him not cases of lead colic. I do not know how such a situation can be remedied. I really found my work much harder than when I was there in 1911, before the passage of the Illinois law, for the fact that there is a company doctor

171

who treats the men without charge makes the men go to him, and the other doctors now treat only a minority, yet they are the only ones who are willing to give any information.

I have finished eight smelters and have found this much the best lead industry thus far studied. Not that there is not plenty of lead-poisoning but there is far less than formerly and the men in charge, especially those in the Trust smelters are quite alive to the dangers and welcome suggestions. It will be a comfort to be able to get out one favorable bulletin.

Some time ago you wrote me about lead-poisoning in zinc smelters. This industry is of course very closely related to lead smelting and might perhaps be taken up in connection with it. When in Pueblo I went out to the big zinc smelter at Blende because it seemed an economy to do so while I was there. They told me then that Illinois, Kansas and Oklahoma were the only states with large zinc smelters. Do you think it would be best for me to delay the lead smelter report until I could finish zinc, or might we perhaps combine zinc and copper smelting, for copper is also mixed with some lead.

When I was in Denver I succeeded in speaking to the International Congress of Master Painters as an employee of the Bureau of Labor, not as a friend of the National Lead Company, and as I spoke against the use of white lead in interior work, I think no one connected me with them in any way. When I had finished, a master painter from Washington asked why the government did not begin at home, and went on to say that the interior work in the federal building is all done with lead paint and sandpapered. This is like the statement I wrote you before as to the dangerous methods used in the naval yards.

The White Lead people have been much agitated over that Cincinnati inquiry and I am hearing from them on all sides. You will let me have the chemist's report when it is ready, will you not? I should like to write the thing up for the Journal of the American Medical Association.

I enclose a list of establishments visited in connection with Bulletin 104. It is very mortifying to have to confess that I can find notes of but 38 white-ware potteries instead of 40. I have cases from two more but cannot find the record of any visit made to them.

Yours sincerely.[a]

a. This is an unsigned carbon copy.

IN 1914 THE MOVEMENT to attain woman suffrage was in high gear. The major suffrage organization, the National American Woman Suffrage Association (NAWSA), had concentrated its efforts on state-by-state campaigns and had suffered many defeats over the years. In 1913 Alice Paul, an ardent young suffragist, founded the Congressional Union to work for passage of a federal amendment. Initially a committee of NAWSA, the Congressional Union split off from the parent body in February 1914. Alice Hamilton, though a member of the Chicago Equal Suffrage League, was not an active campaigner.

40 To Agnes Hamilton

Baltimore
March 1st 1914

Dear Agnes,

You ought not to send me birthday presents but this one is so nice I can't help being glad you did. It has introduced efficiency into my method of making engagements and now I shall never let it lapse, for it will always be in my bag and I can't miss. It was very, very thoughtful of you. I got back here for my birthday though I had quite forgotten it was my birthday till the family produced new slippers and candy. In Washington it was ever so nice. Julia Lathrop was as lovely as possible to Kitty [Ludington], but I am afraid she found her a bit exhausting. Kitty is such a tremendous talker and so eager to do what Julia calls, "social philosophizing." However she does most of it herself.

I worked daytimes mostly but one morning I went sightseeing. Kitty's friend Felix Frankfurter, a brilliant young lawyer in the War Department, took us to a session of the Supreme Court and then to hear Mr. Brandeis argue on interlocking directorates before the Judiciary Committee.[a] Then we lunched with the Brandeis and found Mrs. Brandeis very attractive. You know she is the Goldmarks' sister.[b]

Also I met three friends of Allen and Marian, a doctor and his wife, and a Mrs. Croly. I am going to write Marian about them. At that same dinner I happened to say that I did not know any antisuffragists whereupon I was invited to lunch to meet two. They were very grand. One was a sister of Mabel Boardman and of Senator Crane's wife.

She fixed a coldly sweet eye on me and said "You see *I* trust the men." Also as we were leaving the table she said "Miss Hamilton, you know poor people. Do tell me if married life among the poor is always unhappy. One reads such shocking stories, but I suppose there are exceptions." I should greatly like to study that lady's pedigree and see how many generations she is from a dinner pail. The other was Senator Lodge's daughter.[d] She was distractingly dressed in a rough tweed of mustard green and peacock, and a long lavender sweater, and she smoked cigarettes between courses. She thinks only a small body of educated men should vote. Really antis are worse than I imagined.

There is a terrible mix up between the Congressional Union and the National Suffrage Board and everybody says the latter bungled most stupidly. I was perfectly amazed to hear people speak of the Federal Amendment as a possibility. I had supposed it was only a shrewd advertizing scheme.

I saw my chief, Royal Meeker, and had a talk with him.[e] He is a nice boy, but I am glad Mr. Verrill still runs my part of the Bureau. They are so poor they cannot make a contract with me for an investigation of rubber, but I mean to do it anyway and trust to their making it in July, the new fiscal year.

Please tell Miss Shipley the School has found an investigator for me and I am very much obliged to her.

Also remember me to Mrs. Bradford.

Yours ever
Alice.

Clara writes me that the "New World," the Archbishop's organ in Chicago, sent a man to Hull-House to say that our Italian men's club must not be allowed to give a play called "Galileo," on pain of being attacked by the New World. Miss Addams said she would have the play read carefully and if it were an attack on religion she would forbid it, but not if it were only anticlerical. The man said the Church recognized no distinction between the two. Well the play was read and proved to be strongly anticlerical but nothing more, so it was given and now we are waiting to see what the Archbishop will do.

Clara writes too that there is a big strike of waitresses in a downtown restaurant and that Miss Starr is picketing and passionately longing to be arrested. I do hope it will be over when I get there, Miss Starr is so difficult when she is striking.

a. In 1914 Felix Frankfurter (1882–1965) was working in the Bureau of Insular Affairs in the War Department. Louis Dembitz Brandeis was at this time a reform lawyer and a close adviser to President Wilson.

b. Alice Goldmark Brandeis was the sister of Josephine and Pauline Goldmark, two prominent social reformers and associates of Florence Kelley at the National Consumers' League.

c. Florence Sheffield Boardman of Cleveland was the sister of Mabel Boardman, a prominent leader of the American Red Cross, and Josephine Boardman Crane, a civic leader and wife of Senator Murray Crane of Massachusetts.

d. Constance Davis Lodge was the only daughter of Senator Henry Cabot Lodge.

e. Royal Meeker, an economist and statistician, served as Commissioner of Labor Statistics from 1913 to 1920. He was four years younger than AH.

IN AN EFFORT to promote and improve occupational disease legislation, the AALL drew up a "standard bill," with special reference to lead poisoning. Although there existed an estimated 150 trades in which lead was used, the bill included only those workers engaged in industries that had already been investigated in the United States (mostly by Alice Hamilton). It also specified safety precautions in minute detail. Hamilton, a member of the AALL's General Advisory Council, favored a more general statement that did not name specific lead industries or exact safety requirements.

41 To John B. Andrews

1312 Park Avenue
Baltimore
December 4th 1914

Dear Dr. Andrews:

As I read over your Standard Bill, I do not see anything in it which would prevent your including under its provisions such industries as the making of storage batteries and the smelting and refining of lead, to say nothing of the making and grinding of paint, which last I have not yet investigated.

I wonder if you are not willing to consider recasting this Bill and making it general instead of specific. Of course, I know that it is the result of much thought on your part and on the part of others in the Association. Nevertheless, I am sure you will not mind my putting my own views on the matter before you. There are, to my mind,

175

two objections to this Bill. The first is that by mentioning certain specific industries you at once arouse their antagonism, not only because they do not wish to be regulated, but also because they resent being classed as specially dangerous and in need of regulation. You will remember, perhaps, that there was absolutely no opposition to the passage of our Illinois law, which is couched in general terms, and I am sure there would have been had we made a list of the industries to which the law would apply.

The second objection I have is that industries in America change their methods with great rapidity, and that a law which seeks to regulate the processes becomes obsolete sooner or later. For instance, I am sure that all of the details as to the making of white lead which are to be found under "A" of Section 3, were worked out with great care. At that time, Miss Erskine's rule about the cloth dust collectors and their area with relation to the amount of air passing through was doubtless as good as could be made at that time, but I have recently seen in Illinois a new form of cloth dust collector which does not at all conform to this standard, and which yet is, I believe, a great improvement over the old.[a]

The same thing is likely to be true of others of these details. It seems probable that the white lead industry will eventually be carried on without the necessity for any drying system, and the methods substituted for those now in use will require a different sort of safeguard. In the smelting industry, though I have just finished a careful study of most of the plants in the country, I should feel quite unable to lay down special rules for its control, because here, too, methods and styles of furnaces change from month to month. In other words, a law like your Standard Bill will have to be modified from year to year, not only to cover additional industries, but to provide for new developments in those already covered. In a state like Wisconsin with its Industrial Commission, the difficulty would not be very great, but in the other states, I think it would be very great.

After all, our Illinois law, poor as it is, muddles along very well. Any industry in which lead is used can be brought under its provisions and these last are broad and vague enough to fit every case. I really believe that the situation is better controlled in Illinois than in any state, except, perhaps, New Jersey, where the Department of Labor seems to have the power to draft its own special rules.

Please write me what you think about this. I am writing to Mr.

Ahrens, but I feel confident that his Company will oppose a law in which their industry is specifically mentioned, while they might be willing to come in under a more general law.

Yours sincerely,
Alice Hamilton

a. Lillian Erskine was a special investigator of industrial diseases for the AALL and for the New Jersey Department of Labor. She had a long career as an industrial hygienist and helped AH investigate poisons in the munitions industry during World War I and in the viscose rayon industry in the 1930s.

THEODORE AHRENS, president of the Standard Sanitary Manufacturing Company in Pittsburgh, objected to many of the provisions of the occupational disease law (based on the AALL's "standard bill") that had been passed by the Pennsylvania legislature in July 1913. Hamilton, who hoped that the existing law would be expanded to cover all lead industries, answered his criticisms in detail. A short time later, Ahrens wrote that he would be "very glad to co-operate with you and your friends" in preparing another bill for the next legislative session. Nothing came of the matter at the time.[24]

42 To Theodore Ahrens

Hull House, Chicago
December 4*th* 1914

Dear Mr. Ahrens:

I have been studying carefully the letter you sent me at Harrisburg with the objections of your Company to the so-called Lead Bill, which was passed in 1913. I noted, of course, that you are willing to accept most of the requirements called for under the Act, and that, in fact, you believed the greater part were already being complied with voluntarily by your Company, but as I read over the law, it does not seem to me that if the sections you object to were omitted there would be very much of importance left, and I feel quite sure that the [American] Association for Labor Legislation would not be willing to back a law which failed to cover so many points of the present law.

Let me go a little into details. Your first objection is to the requirement in regard to overalls and other necessary clothing. We have had this provision in our Illinois law since 1910, and it has been found to be entirely practical by large numbers of manufacturers. For instance, the Pullman Company furnishes overalls and caps not only for the workmen exposed to lead, but for their brass workers and for the men who handle acids and alkalies and cyanides and other poisonous substances. I am sure that they would tell you that they have no desire to give this up and go back to the old way, that they find it really pays to give special care to these men.

The objection to the respirator could be overcome if you would provide the English form of respirator—a simple muslin bag with two tapes to tie it over the mouth and nose. Nor do I believe that the men would find it impossible to wear, even while enamelling. For the mill hands any kind of respirator could be used as they are not exposed to heat and great exertion.

Your third objection I cannot make out, because the paging in the copy of the bill which I was able to obtain is evidently different from yours. I should suppose, however, that it referred to the provision of the lunch room for the mill hands and enamellers and forbidding them to eat their food without first washing up. I can understand that this would be a difficult matter in case of the enamellers and their helpers, for their work is supposed to be continuous during their six or eight hour shifts and they are not supposed to stop while the furnaces are working. Still, I am sure you will recognize the danger of a system like this which gives the man only two alternatives, to work for six or eight hours without food, or to eat his lunch with lead-smeared hands. Nothing favors lead poisoning more than an empty stomach.

As for the monthly examination of the men who are engaged in this kind of work, I agree with your objection to that Section of the Pennsylvania law, but I find that the objection would not have arisen if the law had been left in its original form. As it stands now, the examining physician is required practically to insist on the discharge of any workman who has symptoms of lead poisoning from the job which exposes him to lead. This would mean, in your industry, that a skilled and well paid enameller would be degraded to the ranks of unskilled labor—the greatest hardship in his eyes which could happen to him. It would undoubtedly lead to such a man's concealing his symptoms for fear of losing his job. But if the law had been left in its original state, the Section in question would have read as follows:

"The employer shall not continue the said employe in any work or process where he will be exposed to lead dusts, lead fumes or lead solutions, *nor return the said employe to such work or process without a written permit from a licensed physician.*" That last clause, you see, makes it possible to put the man back at work as soon as he is fit to go.

I am writing you thus fully about it, because I feel sure that you are our best hope in this situation. The Pennsylvania Lead Law should be made to cover all the lead using industries as does the Illinois law, but as long as companies as powerful as your own oppose this, I do not believe there would be any use in our attempting any further legislation. Will you not think it over and see whether it is really true that your Company cannot accept the regulations which are accepted by the big industries of Illinois? When you next come to Chicago, I wish you would make a visit to Pullman and see for yourself not only how admirable is their system, but how enthusiastic they are over its results.

I am venturing to send you the pages from my much criticized government report which apply to the sanitary ware industry. I wish you would read them. There are not very many, and please believe that they were written from no prejudiced point of view, that I began my study of the trade without the least idea of what I should find. It is not at all a notoriously dangerous industry and nothing I have read in English and German factory inspection reports had led me to think that I should find it dangerous, yet this is what I did find, and I wrote it without exaggeration.

I feel sure that once you are convinced of the necessity for greater safeguards in the making and use of enamel you will insist on their being introduced into the plants controlled by your Company, and that is why I am turning all my efforts upon you. The Survey is anxious to have me go at the matter in another way and by writing up the question for that magazine to try to arouse public opinion and a general demand for these reforms, but that does not seem to me the right way to go about the matter.[a] You may be sure that nothing you write me will be published or used in any way. I feel that I know you well enough to be confident that we shall in the end attain our ends through you.

Sincerely yours,
Alice Hamilton[b]

Not till 1938, 24 years later, did Pennsylvania pass such a law.[c]

179

a. *The Survey* was an influential weekly for social workers and reformers.
b. This is a signed draft.
c. This handwritten note was added many years later.

By 1915 ALICE HAMILTON had become the foremost American authority on lead poisoning and one of a handful of prominent specialists on industrial diseases. She was not, as some admirers later asserted, the only person in the field; nor was she responsible for all progress, although she has even been credited with eliminating phossy jaw. But she concentrated her efforts to an unusual degree and may have been the only person doing the work full time. She also brought to it a unique single-mindedness.

In her forties, she had finally discovered her true vocation. Her initial forays into the field had been marked by characteristic tentativeness and self-doubt. But it soon became apparent, even to her, that no one knew more than she did. The pioneering nature of her work gave her real pleasure, and she made much of it, both at the time and later. Never given to self-glorification, she could claim, without seeming to boast, that almost everything she learned in the early days was new, and much of it "really valuable." To one who had always emphasized her own limitations, this hard-won confidence was profoundly liberating.[25]

The nature of the field was central to its appeal. Industrial toxicology has been described as particularly gratifying work because of the one-to-one relationship between cause and effect so rare in medicine or any other endeavor. By establishing which were the poisonous trades and by securing relatively modest changes—respirators, exhaust and sprinkling systems, medical examinations—Hamilton could feel certain that she was preventing illness, perhaps even death. Here was work, "scientific only in part, but human and practical in greater measure," that provided the perfect resolution to the conflict that had plagued her for a decade. It also permitted her to achieve the goal, articulated sixteen years before, of leaving behind some "definite achievement, something really lasting . . . to make the world better."[26]

To a woman for whom doing good work was so important, it was doubly unconscionable to permit men and women to be poisoned while doing the work for which they had been hired—particularly when modest changes could eliminate much of the danger. Alice Hamilton attributed this cavalier form of exploitation to the distance separating workers in the dangerous trades, most of them foreign born, from their employers, a gulf in her view no less than that between European peasants and nobles. Amazed at the casual attitude toward lead poisoning, she observed: "One would almost think I was inquiring about mosquito bites." When she demurred at the claim of an apothecary in Salt Lake City that there was no lead poisoning in the vicinity of a smelter, he explained: "Oh, maybe you are thinking of the Wops and Hunkies. I guess there's plenty among them. I thought you meant white men."[27]

Alice Hamilton thus found her own voice by championing a despised and neglected social group. Just as occupational illnesses had been designated diseases of slaves in ancient times, so, she maintained, contemporary workers in these trades belonged to "a class which is not really free." As she early observed: "For an employer to say to his work-people, 'If you don't like the job, get out,' may in many instances be like a captain at sea saying to his sailors, 'If you don't like the ship, get overboard.' " Even if they knew the dangers to which their work exposed them—which often they did not—many workers chose to risk their health rather than possible unemployment. Always sensitive to the claims of the helpless, Hamilton had felt too inexperienced in the past to be of much use. By acquiring a specific skill and subjecting her emotions to the discipline of objective fact, she overcame her own feelings of uncertainty even as she worked on behalf of the oppressed and powerless.[28]

The work also permitted her to escape the constraints customarily imposed on women of her class without violating her own notion of femininity. Chafing at what she later called "the tiresome tradition that a woman is something different and must be treated differently," she delighted in her newfound ability to go anywhere and do anything a man could do; she found the dangers—and her ability to face them without revealing her fear—exhilarating. Like most women of her generation, she believed that men and women differed in essential temperament, and she once claimed that when she wanted advice on an intellectual problem she took it to a man, but that for wisdom on a personal problem she would always consult a woman. Although she chose to work for the most part in a world of men, by emphasizing the human side of science and by relying on personal persuasion as an instrument of change, she was adopting a classically "feminine" strategy.[29]

Alice Hamilton's father died in June 1909, as she was on the threshold of her new career. His death may have freed her to give up the laboratory for work of which, in view of his politics and his contempt for sentimentality, he might not have entirely approved. It is also possible that by reclaiming the victims of lead poisoning—then often called lead intoxication—or by preventing the disease altogether, Alice Hamilton may have been doing for others what no one had been able to do for her father, who had wasted his talents, first in an unsuitable occupation and then in drink.

Hull House remained her home. Norah often joined her there, and so at times after her husband's death did Gertrude Hamilton, who helped with theatrical productions and taught English to foreign visitors. Despite her frequent protestations that she lacked the true settlement spirit, Alice Hamilton had quickly made her way into the inner circle that included Jane Addams, Julia Lathrop, and Florence Kelley. With maturity, her schoolgirl awe of Jane Addams gave way to a more realistic—though still deeply admiring—appraisal of her mentor, and even, on occasion, to gentle criticism. She early assumed special responsibility for guarding Jane Addams' often precarious health, a role she filled for the rest of Addams' life.

Alice Hamilton's enthusiasm for her new life was immediately apparent. Soon after she had begun her first federal survey, Norah observed: "Alice seems at rest about her work lately—I think it may be refreshing for her to get away often." Her work for the federal government permitted her to indulge a lifelong love of travel. On a paid trip to Europe in the summer of 1912 she studied leadless glazes and investigated factories, and she and Norah spent a week in northern Ireland, where she made inquiries about her forebears. An informant near Londonderry, she reported, told her that the Hamiltons were "upstarts" and "had nothing to do with the country outside their own estate."[30]

The letters of the period reveal a new contentment as well as confidence in herself and her ability to direct her life. Alice Hamilton enjoyed working hard and using her powers to the fullest: "I am having a lovely time," she wrote on one of her trips. "Really I never worked so hard before, except in my dispensary service in Boston, when I was young and strong. It has been awfully good for me, though, I am bursting with fatness." Delighted that no one would "keep tabs" on her, she also liked being responsible for her own work. Since she received no regular salary—she was paid for each investigation—her income was small and precarious, and she supplemented it with fees for articles that appeared with some regularity in *The Survey*. But despite the financial uncertainty, she turned down a job in Boston that carried a large salary but did not otherwise appeal.[31]

Alice Hamilton's inner confidence was matched by recognition from the outside world. In 1910, a few months after she had begun work on the Illinois survey, the University of Michigan awarded her an honorary master's degree the same day it gave her brother Quint his earned one. She served as president of the Chicago Pathological Society (1911–12) and in 1914 became vice-president of the newly organized industrial hygiene section of the American Public Health Association; two years later she chaired the section.

While settling into her life's work, Alice Hamilton retained an appealing remnant of youthful romanticism. This was reflected in her dealings with a prostitute who had written Jane Addams following publication of her book *A New Conscience and an Ancient Evil*. Taking up the correspondence in Addams' absence, Hamilton was impressed by the woman's frankness and by her professed eagerness to change her life. Despite the provoking skepticism of Jane Addams and Julia Lathrop, she stopped off in Toledo, hoping to effect a rescue. (She checked in first with the local Charity Organization Society because Gertrude Hamilton was "frankly nervous" at the thought of her daughter's visiting a brothel.) The meeting disappointed both parties. Alice Hamilton, who had envisioned a caged and "pitiful little bird" found instead a woman of mature years, at once cold and passionate, who had taken up her work voluntarily and had no intention of leaving until she found a man to set her up in a flat, preferably with a servant. For her part, "Adelaide" found the occupation of her would-be rescuer disgusting ("That is not the sort of thing I could possibly do") and her youthfulness distressing. "Lavender and Old Lace, that is what I thought," she told the physician, who was then forty-four.[32]

VI

THE WAR YEARS
1915–1919

IN APRIL 1915 ALICE HAMILTON joined a delegation of nearly fifty Americans, led by Jane Addams, who attended the International Congress of Women at The Hague. The Congress had been called by European leaders of the suffrage movement to protest the war and to work for reconciliation. Although the United States was still neutral, both the Congress and the women's decision to attend met with ridicule and criticism. Ex-president Theodore Roosevelt considered the affair "silly and base," and the more conservative leaders of the American suffrage and peace movements also condemned it.[1]

43 To Agnes Hamilton

> 1312 Park Ave.
> Baltimore
> April 5*th* [1915]

Dear Agnes,

Will you think me utterly mad when I tell you that I am going over to The Hague with Miss Addams and the rest of the Peace delegation. I made up my mind quite suddenly and I fear not on very noble grounds, but of course I could not decide really till I knew just how mother would take it, so I came on here Saturday, ready either to go over or to settle down to a printing investigation. But she seems not to worry about it at all, she takes it as serenely as possible. Only Edith is worrying and she does not count. I shall stay here all the rest of the week and on Sunday I shall go on up to New York, Margaret with me. Now cannot you come up too and stay at the Brevoort with Margaret and me and see me off Tuesday noon. Do come. I can't stop in Philadelphia for Mother wants me till the end and on Monday I have to go to the Survey for several hours.

We go to The Hague for a week, then probably to England for two weeks more and then home again, getting back before the first of June.

Write and tell me you will come.

Yours ever
Alice—

THE *NOORDAM*, a Dutch vessel, carried most of the American delegation across the Atlantic and through the war zone. Alice Hamilton, who thought of herself as Jane Addams' "confidante in white linen," initially viewed the peace venture with ironic detachment and even outright skepticism. Nevertheless she followed the shipboard discussions of war and peace closely and delivered an account of the medical aspects of the war that a fellow passenger found "quiet but almost intolerable."[2]

44 To Mary Rozet Smith[a]

Holland-American Line
[S. S. *Noordam*]
Thursday April 22*nd* [1915]

Dear Mary,

This will be a rocky letter, I am afraid, for we have had a following wind and a heavy roll for the last two days and my head is most uncertain. Indeed it never has been smooth enough to make either Miss Addams or me quite comfortable but then it has not been rough enough to make us sick.

It is a most novel trip. It is like a perpetual meeting of the Woman's City Club, or the Federation of Settlements, or something like that. Really it makes the day go amazingly quickly. I never was so little bored on a trip. Always one is going to meetings or discussing the last meeting or reading up something which somebody has said is peculiarly illuminating on the underlying causes of the war. It is interesting too to see the party evolve from a chaotic lot of half-informed people, and muddled enthusiasts, and sentimentalists, with a few really informed ones, into a docile, teachable, coherent body, only too glad to let itself be led by those few. We have long passed the stage of poems and impassioned appeals and "messages from womankind" and willingness to die in the cause, and now we are discussing whether it is more dangerous to insist on democratic control of diplomacy than it is to insist on the neutralization of the seas. There are still some five or six whom we regard with a little mistrust and who may possibly disgrace us at the last moment but most of us are very quiet and tractable.

I find the discussions ever so interesting and get quite absorbed in them, and then all of a sudden the whole thing looks absurdly futile. I suppose I shall always be a doubting Thomas and a pessimist. Miss Breckinridge has been a great help.[b] She and Grace Abbott and Miss Balch of Wellesley and Mrs. Post are acknowledgedly the leaders.[c] Miss Addams is really having a good time. She has made every woman on board feel that she is an intimate friend and they all adore her. And she has the pleasant conviction that she has done a good job with not very promising material.

We are right next to Miss Breckinridge and Miss Abbott and have the door open between our cabins so we are very chummy. I like Grace Abbott better all the time but my special crony is Miss Kittredge, Norah's friend and landlady.[d] I hope she will stick to us, for I foresee a Cinderella kind of time if we go to England. Everybody on board wants to go to England with Miss Addams and though she is lying low and not saying whether or not she will go, I know that when she does she will have annexed a lot of them and I shall be lost in the shuffle. In that case I shall break loose with Miss Kittredge.

Thursday

We were off the Scilly Islands when we came up from breakfast this morning, then came all sorts of wild excitements, a big ship westward bound, very high out of the water and marked in great letters "Belgium Relief," evidently empty and going back for more supplies. Then a fleet of fishing vessels which roared angrily at us and made us go out of our way in a big curve to avoid their nets. And then Eddystone Light and the Lizard, which would be so beautiful if

186

only that foggy island were not at its foggiest. It is disappointing. I had looked forward to this day along the English coast and now it is only a dim outline.

We know absolutely nothing of what has happened during these nine days. The wireless clicks away and the Captain has sheafs of despatches brought to him at the table, but no word do we hear. He announced that he would allow no news to be posted because on one of the trips the passengers got into such a fierce controversy that he was afraid of trouble. It seems a little provoking that forty elderly female Peace delegates cannot be trusted to keep from fighting over war reports.

They say we are to reach Dover tomorrow morning, though other reports are that we shall not steam at all tonight but just lie and wait for daylight. At Dover the excitement is supposed to begin, for there the British warship is supposed to meet up with us and inspect us. Two wretched stowaways, Germans I suppose, came out from the coal hold a few days ago, having exhausted their food supply and I suppose the English will take them off. Then there is a pale, melancholy, pinched-looking woman, a German who was caught in America by the war and has only just succeeded in escaping. She has some dreadful American disease which she would never have caught in Germany and the American doctors could not cure her because they have no good medicines now that they cannot get them from Germany. She is in terror of the British and when we try to comfort her she only sniffs in a melancholy and skeptical way.

The Pethick Lawrences are much in evidence and I can see why Rachel Yarros loved them so passionately for they would rather discuss than do anything else.ᵉ They are nice and warm and likeable, but not very intelligent.

I have not said a word about all the wealth of things which you and Mrs. Bowen sent us. It is just as if I were taking them for granted—and perhaps we really all did act a little as if we knew of course we should be showered with luxuries. Anyway we enjoyed them immensely, all of us, for none of us has ever been beyond the possibility of liking to eat between meals and when going to bed. Also we have taken boxes up on deck from time to time and been ostentatiously generous with them. Miss Addams has been saving a box of peppermints and the cake and biscuit box for Holland and now they say that cake and candy will be confiscated at the customs house.

It is really amusing to think that we are actually in the danger zone now, in this gray, still, monotonous ocean, with everything as ordinary and reassuring as can be. One cannot have a single thrill.[f]

a. Mary Rozet Smith (1868–1932) was Jane Addams' closest friend and companion. Daughter of a wealthy family, she contributed financially to Hull House and joined in its activities, but remained in the background. AH considered her "the most universally beloved person" she had ever known.

b. Sophonisba Breckinridge (1866–1948), a lawyer and political scientist, was dean of the Chicago School of Civics and Philanthropy (later absorbed by the University of Chicago as the Graduate School of Social Service Administration). A resident of Hull House, she wrote extensively about women, children, and the family.

c. Grace Abbott (1878–1939), another Hull House resident, was head of the Immigrants' Protective League, an organization of social workers that defended immigrants against exploitation; she also succeeded Julia Lathrop as head of the Children's Bureau. Emily Greene Balch (1867–1961), professor of economics and sociology at Wellesley College, devoted herself after the war to the Women's International League for Peace and Freedom, the permanent organization that evolved from the women's congress; she received the Nobel Peace Prize in 1946. Alice Thacher Post, an editor, was vice-president of the American Anti-Imperialist League and of the Woman's Peace Party.

d. Mabel Hyde Kittredge (1867–1955), an associate of Henry Street Settlement, was the founder and president of the Association of Practical Housekeeping Centers. She was particularly interested in school lunch programs, which she frequently inspected on her travels. She and AH became good friends.

e. Emmeline Pethick-Lawrence was a militant British suffragette and a leader of the international women's peace movement. Frederick William Pethick-Lawrence, also a supporter of suffrage, became a prominent member of the British Labour Party after the war.

f. The letter is unsigned.

AFTER AN ENGLISH boat removed the two German stowaways, the *Noordam* proceeded a short way, only to be detained for four days, without explanation, in the English Channel near Dover. The women reached The Hague just in time for the opening session of the Congress.

It was an extraordinary event, this bid by more than 1100 women from twelve belligerent and neutral nations to influence the course of international affairs. Few of them possessed the right to vote, let alone more tangible signs of political power, and those from the warring nations risked censure as traitors by their very presence. The Congress, over which Jane Addams presided, declared the special interest of women in opposing war and linked the success of peace efforts to the enfranchisement of women in all nations. It also endorsed measures for international cooperation, including a permanent international court, a Society of Nations, general disarmament, freedom of the seas, and, more daringly, national self-determination and democratic control of foreign policy. In their most dramatic move, the women endorsed an American-backed plan calling for the creation of a conference of neutral nations that would offer continuous mediation to end the war. They also instructed two groups of delegates—one of which included Jane Addams—to carry the proposal in person to the heads of belligerent and neutral nations.[3]

45 To Mary Rozet Smith

Hotel Wittebrug
Den Haag
May 5*th* [1915]

Dear Mary,

I am sitting in the parlor of this very pleasant hotel, surrounded by a crowd of Dutch people, whose language is guttural but whose voices are full and agreeable. J.A. is lecturing in Amsterdam, with Mme. Schwimmer and Mrs. [Pethick-]Lawrence and Fräulein Heymann.[a] I could not go to hear her because Miss Kittredge and I are making a desperate and probably quite futile effort to get into Belgium, and we had to stop over in Rotterdam to see the German consul instead of going on to Amsterdam. We have already wasted three precious days in the pursuit of this wild plan and though today the consul was very sanguine we do not really believe we can get in. Even the wives of German officers may not, and our consul in Brussels who was returning to his post was held up eight days before they let him enter.

The Congress is over and since Sunday Miss Addams has been in sessions of the Resolutions Committee, making the final draft. I wonder how much has been reported to you in the American papers. The Dutch papers are mostly contemptuous, the English sometimes quite nasty. To me it was intensely interesting and sometimes very moving. People are saying now that the German note predominated, that it was a pro-German Congress, but that is true only in the sense that the German women were there in goodly numbers and were an unusually fine lot of women, so able and so fair and so full of warmth and generosity. I wish Miss Hannig could hear them talk, not only

189

the real Germans, but the Hungarians and Austrians.[b] The English were only three, and not even a united three, for Mrs. Pethick Lawrence was ignored by the two legitimate suffragists, Miss Courtney and Miss Macmillan.[c] There was a fine Canadian girl there, a niece of Sam Hughes, the Major-General of the Canadian forces. We expected the English delegation to welcome her with joy as an addition to their small numbers, but they were very thoroughly English and evinced no enthusiasm over a Colonial. The Norwegian and Swedish women impressed one very well, but they were the most cautious of all, being in fear all the time lest they do something to violate the neutrality of their countries. Finally there were the Poles and Belgians who were very moving and yet seldom over-emotional. Indeed what I felt all the time was the deep undercurrent of emotion, but an admirable self-control. Only Madame Schwimmer could sweep the Congress off its feet and she did it several times, notably at the end when she succeeded in having them pass the resolution which filled most of us with dismay and which you will have seen in the papers, that the resolutions passed by the Congress be presented by a committee to the various Powers. As you will have seen, J.A. is one of the delegates to visit all the countries except Russia and Scandinavia. She wants me to go with her and of course I will. To me it seems a singularly fool performance, but I realize that the world is not all Anglo-Saxon and that other people feel very differently.

J.A. was simply wonderful as president. She could not have been better. And Grace Abbott and Miss Breckinridge helped her as nobody else could have. I was really lost in admiration of their ability, their clearness and quickness. They are with her in Amsterdam tonight.

I have quantities more to talk about, but I must go to bed, after a long day. Love to Mrs. Bowen.

Yours ever
A.H.

J.A. had two splendid letters from Mrs. Bowen but I have not heard of one from you.

a. Rosika Schwimmer, a Hungarian feminist and pacifist, represented the radical wing of the women's peace movement. Lida Gustava Heymann of Munich, a leader of the German woman suffrage movement, had helped to plan the Congress.

b. Amalie Hannig taught piano and needlework at Hull House.

c. Kathleen D. Courtney was honorary secretary of the National Union of Women's Suffrage Societies. Chrystal Macmillan, a Scottish lawyer who was recording secretary of the International Woman Suffrage Alliance, had been among the leaders calling for the Congress; she became one of its official delegates to the Scandinavian countries and Russia. Emmeline Pethick-Lawrence had broken with Emmeline Pankhurst, leader of the British suffrage movement, on a matter of tactics. The other British delegates had been unable to attend the Congress because all traffic between England and Holland had ceased for a time.

WHILE JANE ADDAMS was in England on the first leg of her mission to the war capitals, Alice Hamilton visited printing establishments and a Belgian refugee camp in Holland and then spent a week in German-occupied Belgium. Her experiences there made an indelible impression. "It is not a question of present horror, nor of starvation, nor pillage, nor any kind of violence, for all is decent and quiet and orderly," she wrote her family. "It is simply a conquered country under the foot of the conqueror and what that means is almost indescribable." It made her feel like a conspirator: spied upon, frightened, even duplicitous. There were the ubiquitous gray-coated German soldiers; the holiday crowds, "silent and submissive as if at a funeral"; the parents who hurried their children out of her presence before they could commit some indiscretion; and three searches of her person to make certain that she was not smuggling forbidden information in or out of the country— the Marshall Field label was removed from her hat and her Baedeker confiscated because it had maps. Her initial amused curiosity at being searched gave way to anger and she was certain that "it cannot be good for people's characters to feel that way." Reaching Holland again "was like escaping from a dark and stifling room into fresh air and sunshine." For the rest of her life Belgium was to be a symbol of oppression.[4]

46 To Louise deKoven Bowen

Amsterdam
May 16*th* [1915]

Dear Mrs. Bowen:

I am sitting in the headquarters of the Woman's Peace Party waiting while Miss Addams goes over minutes and reports with a very meticulous English lady. Downstairs a taxi is ticking away and the thought of it gives me indigestion but Miss Addams keeps saying she is coming.

191

I am wondering if at this rate my money—Mary [Rozet Smith]'s—will hold out, or whether I shall have to fall back on hers. She seems to have plenty.

I was very startled to hear of your operation and of course the letters telling that you were planning it came long after the cable telling you had had it. I know it was dreadfully hard stopping everything and dropping back into invalidism, but I know that in the end you are glad to have it really settled and I am thankful you sent for Dr. Abbe. We seem so far away here that a cable or two saying you are doing all right is a great comfort.

So far things have gone very well and I only hope they will keep on so. Miss Addams got to England when she planned and got out when she planned. I went into Belgium and though they kept me on tenterhooks for days, they let me out in time to meet her. We are to go to Germany now, for people seem to think that if we go fast and come out fast we can manage it before our dear country decides to break off diplomatic relations with Germany. The party will consist of Dr. Jacobs, an elderly woman, very decided, fairly irritable and quite able to see that her own comfort is attended to; her friend Frau Palthe, the wife of a man who has a plantation in Java and is very rich but exceedingly careful about pennies, also elderly; and J.A. and myself.[a] It will not be bad, for a party of two couples breaks up easily into twos.

Miss Addams must have had a wonderful time in England, but you will have heard of it from Miss Breckinridge by now. People here really do take her mission seriously. When it was proposed in the Congress by Madame Schwimmer we all thought it hopelessly melodramatic and absurd and we said Miss Addams would never consent to go about from court to court presenting resolutions. Then as she talked to the foreigners she saw that in their eyes it was both dignified and important and she consented to take the warring countries. For me, of course, it is only interesting, for I have no responsibility, I only trail along as a lady's maid.

Belgium was a wonderful experience. I never knew before what the life of a revolutionist in Russia must be like. It is a choking and an embittering atmosphere. One feels oneself spied upon, one is afraid to be frank ever, one is afraid and at the same time one comes to hate those gray coated men of whom one is so mysteriously afraid. And the Belgians one meets in the relief work are so fine: intelligent, capable, high-bred people, living as simply as possible and giving all

192

their time and strength and money in the care of their poor. And they are so sweetly courteous about it too. I think one can be more polite in French than in English anyway.

Miss Addams is starting and I must go.

With much love to Mary, and to yourself.

Affectionately ever,
Alice.

a. Aletta H. Jacobs (1854–1929), an official delegate to the war capitals, was president of the Dutch suffrage society and had been instrumental in organizing the Congress; she was also a physician and a pioneer in the birth control movement. Mevrow Palthe was Aletta Jacobs' traveling companion.

ON MAY 19 Alice Hamilton left Amsterdam with Jane Addams, who was continuing her mission to the war capitals. The timing could not have been worse: on May 7 German submarines had torpedoed the British passenger liner *Lusitania*, taking the lives of nearly 1200 passengers and crew members, among them 128 Americans. The incident not only fanned anti-German sentiment in Britain but also made many Americans eager to enter the war. To Hamilton's distress, even the most cosmopolitan Germans seemed to approve the deed. Still, as political leaders in Berlin, Vienna, and Budapest received the women courteously—some even hinted that they would welcome a peace initiative—she, like Jane Addams, began to take the mission seriously. Since Alice Hamilton was not an official delegate, she joined Addams only for the interviews with Count Stephen Tisza, prime minister of Hungary, and—the high point of her trip—with Pope Benedict XV. But she talked unofficially with politicians and journalists as well as with members of women's organizations and peace groups. Despite press censorship and one-sided presentation of the news, she discerned significant peace sentiment in each country, even in France and Italy (which had just entered the war). In articles published in *The Survey* after her return, Hamilton maintained that despite the German outrages in Belgium she could not view the war, which still found tsarist Russia on the side of Britain and France, in absolute moral terms. Declaring that "feeling is not everything," she urged her compatriots to keep their heads so they might find a way to shorten the war and help bring about a liberal peace settlement.[5]

The success of the mediation plan depended on persuading Woodrow Wilson to place himself at the head of the neutral nations. Jane Addams and other delegates of the International Congress of Women met with Wilson and his advisers several times that summer and fall. They found the president sympathetic to mediation, but he did not consider the time right for such an intervention and, in any case, preferred to act alone rather than with other nations. By February 1916 he had reached a secret understanding with Britain and France that he would make a peace initiative only with the prior agreement of both nations.

The American peace delegates continued to place their faith in the president; many, among them Alice Hamilton, supported his reelection in 1916. But they were soon disabused of their faith in the American public. On July 9, 1915, four days after she returned from Europe, Jane Addams summed up her impressions of the trip at a mass meeting at Carnegie Hall in New York City. Concentrating on the desire for peace she had witnessed in each nation, she observed toward the end of her talk, almost incidentally, that European governments had to administer stimulants before their troops could be induced to take part in bayonet charges, a particularly distasteful form of combat. Outraged members of the press, misconstruing her remark as an attack on the honor of the soldiers, castigated Jane Addams as "a silly, vain, impertinent old maid who . . . is now meddling with matters far beyond her capacity." The fierce controversy, which lasted for weeks, wounded her deeply and foreshadowed the bitter attacks against her when she later refused to endorse the American entry into the war.[6]

47 To Jane Addams

Mackinac Island.
July 20th [1915]

Dearest lady,

Do you think you have a really good photograph to send Miss Sheepshanks?[a] She certainly earned it by that friendly little dinner.

I wonder how things stand now as regards the President. It really seems too absurd that he should do himself up in cottonwool for fear of any influence except his own.

It looks very much as if the two parties in Germany were at loggerheads—the civil trying to patch up an agreement with us, the von

Tirpitz party going their own sweet way and acting like the devil's own.[b]

Do you know, the more I think over that time with you in Europe the more the wonder of it grows on me and it seems so impossible to even begin to thank you for giving it to me. And you did it so without reservation, good measure pressed down and running over. As I look back I cannot see how you could have been lovelier. If ever I got on your nerves, and of course I must have sometimes, you never let me suspect it, and I believe that is the test of real, good comradeship. I wish I ever could have a chance to [do] something in return.

Margaret is not doing quite so well. It was too much to hope for a recovery without any setback. But she is letting Clara come and I think will like to have her about. We expect her in a couple of days.

I wish so I were going to hear you speak in Chicago. I hope you are not letting yourself be intimidated by creatures like Davis and Everett Wheeler.[c] We certainly were told that regular rations of rum are served on the English side and that before a bayonet charge the Germans give a mixture containing sulphuric ether and the French, absinthe. And you never said they were drunk.

Affectionately ever
A.H.

I am afraid you will have to look over my Survey article or I may say something imprudent.[d] I am sending it to you.

a. Mary Sheepshanks, a British pacifist and suffragette, was also vice-principal of an evening institute for working men and women in South London. She had arranged a dinner with Bertrand Russell.

b. Admiral Alfred von Tirpitz, the German navy minister, was head of the war party and an advocate of unrestricted submarine warfare.

c. Richard Harding Davis, the novelist and war correspondent, and Everett P. Wheeler, a lawyer and founder of a New York settlement house, were among those who attacked Jane Addams most bitterly for her Carnegie Hall address.

d. AH's article, "At the War Capitals," a factual account of the trip, appeared in the August 7, 1915, issue of *The Survey*, and, with modifications, in *Women at The Hague*, a collected work by Jane Addams, Emily Greene Balch, and AH published the same year.

IN DECEMBER 1916 the Hamilton sisters purchased a house in Hadlyme, Connecticut, a tiny community that Alice later described as "a loose collection of groups of houses." Their early-nineteenth-century federal house sat fairly high on the east bank of the Connecticut River, next to the ferry landing. (A ship chandlery came with the property and was later converted into living quarters and Norah's studio.) Alice had known and loved the area since she had begun attending Katharine Ludington's fall house parties in nearby Old Lyme several years before. The sisters initially spent only their summers in Hadlyme, but from the start Alice intended it as a place for retirement. During their first year there they were joined by Gertrude Hamilton, who died in December 1917.[7]

48 To Margaret Hamilton

Hull-House
March 19th [1917]

Dear Margaret

It is Monday morning just after breakfast and my typist will be here in a minute and I have no business anyway to be writing to you, but I can't let this morning's letter go unanswered till I do have time. I don't see how Edith could talk that way to you. If she and Clara are temperamentally antagonistic, and I suppose they are, surely the one thing to do is to try to ignore it as much as possible. She knows that Clara is almost more necessary to you than anybody in the world and that if you are to be happy, it can only be together with Clara, and with Clara happy. And she really cares more that you should be happy than anybody else except perhaps Lucy.[a] So it does seem pretty illogical to distress you about something that is not to be done away with and that sensible people simply have to accept and then ignore. Of course when I thought out Hadlyme I realized that we were getting together a lot of people with strong likes and dislikes and all sorts of temperaments and that you and I and Mother were the only ones who could be trusted to like everybody and get on with them. But that must always be true, even in families. One can't expect entire harmony in any group, and it seemed to me that with the extra houses we could manage especially if we made up our minds that it did not matter so dreadfully

This means that Miss Oliva has come and I am using Miss Addams' Corona while she does my business letters.[b]

Truly you must not worry about Hadlyme and mixed relations. Norah and Clara and Lucy will always be rather ticklish problems—

196

I do not mean to put up with Edith's being—but we are old enough not to take that sort of thing very deeply. And Clara loves you too deeply to do anything for any great length of time that would make you unhappy. As for Norah, she is quite as likely to be down on one of us as on Clara. You certainly need not add me to the list of difficulties. Clara and I have lived together, in the same room, for eighteen years more or less, and I have the tenderest affection for her, and so close a knowledge of her that nothing she could do would antagonize me for long. I could not think of a life in which Clara did not have a great part, she has become part of my life almost as if she were one of us. And I do have a great deal of faith that as years go on things will go more easily between her and the family. It was bad last summer, and again this Fall for awhile, but somehow I do not believe it will [be] again. And if it did, I can always wait quite philosophically for the clouds to roll away provided I do not find that it is making you unhappy. That is the only impossible thing about it. If you can bring yourself to feel that this is on the whole no more mixed a houschold than most households of grown-ups—and indeed I don't know of any less mixed one—and that small jars will come now and then but need not come often if we scatter ourselves through the different houses judiciously, then you can look forward quite serenely to life in Hadlyme. Just don't expect all the people to be as fond of each other as we are of them, and realize, what I am only now beginning to learn, that the people who make the jars cannot mind them as you and I do or they would not make them.

Now I must go and read up on compressed air disease for my last lecture.[c] Remember that the one thing I cannot stand is having you worried over Hadlyme. I got it for you and it simply must make you happy or I shall not know what to do.

Yours ever
Alice—

a. Lucy Donnelly, professor of English literature at Bryn Mawr College, was a close friend of Edith Hamilton's. They had met as undergraduates at Bryn Mawr.

b. AH was explaining the change of typewriters in mid-letter.

c. AH taught a course on industrial hygiene at the University of Chicago from 1916 to 1921.

WITH THE AMERICAN declaration of war against Germany in early April, many prominent progressives, including some who had been active in the peace movement, went over to the war cause, some reluctantly, others with enthusiasm. Jane Addams and Alice Hamilton belonged to the minority who remained opposed to war as a means of solving international problems. In June Jane Addams delivered a speech entitled "Patriotism and Pacifists in War Time" at the First Congregational Church in Evanston, Illinois, in which she denied that pacifists were mere passive bystanders. Conceding that they must abandon antiwar propaganda now that the United States had joined the battle, she insisted that pacifists had a continued obligation to speak out on moral issues and, especially, to promote a postwar international organization. When she concluded, not only was there no applause, but Orrin N. Carter, associate judge of the Illinois Supreme Court and a longtime supporter of Hull House, announced that he was breaking with Addams because he could not condone anything that cast doubt on the justice of the American cause.

49 To Jane Addams

62 Washington Square
[New York City]
June 13*th* 1917

Dearest lady,

Well it was detestable of Judge Carter, wasn't it, but men are going to be that way from now on, as long as the war lasts, and please don't let yourself mind it. I could slay anyone who hurts you, but I do hope you are not letting it hurt. Read John Bright's life and see how people treated him for opposing a war not one hundredth so terrible as this.[a] I feel so sure that blows like this are coming pretty often to you that I long to have you develop a sort of protective covering so that they will not hurt you. I want you to keep on saying things even more positively, no matter what you are called, for in the end it will count.

I did not do anything about Mr. Gavit's letter because I did not know what could be done.[b] It is clear that he and Mr. Lippmann both are willing to take stuff if it is what they want, but they are not willing to pledge themselves to take anything from unknown people.[c] And of course you cannot guarantee the Allens, you have no idea how able they are and you do know that they are quite inexperienced.

I have finished my New York part of the aeroplane inquiry and this morning Clara and I are starting for Hartford to buy household supplies and then to go on down the river to Hadlyme Landing.[d] Clara came to me after the Bryn Mawr reunion and as Mabel Kittredge had to go down to her country place to get it in order for her tenants, Clara could stay here. We have had very good times, she and Norah and I, in the intervals of work. If only the weather would change. Mother and Margaret and Edith are to motor up and reach us Sunday evening, so we shall have time to get the house in some sort of shape, and I have had our nice German waitress come on to help. I do want to give Mother a good impression of it all.

I lunched with Billy Hard and his wife the other day, and heard the most sickening war talk I have ever heard in this country.[e] It sounded like the Daily Mail at its worst. Really he is too bad and she is just a little worse. And yet they are just as affectionate and he spoke with the nicest feeling about you. Then I dined with the Crolys and Jesse Williams.[f] They were all three what I should call pacifistically minded but under the obsession of the fatality of this war. Herbert Croly was the most optimistically inclined man I have talked to. He feels sure the war will not last the winter. Jesse's oldest boy, just eighteen, ran away to join the Navy and is now in Newport. Every friend I have here has sons or nephews in the service, and the lack of enthusiasm is striking. Mrs. Croly told me she thought there was a very decided reaction against the war already.

Please give much love and sympathy to Mrs. Bowen. I am so very sorry she has had a return of her trouble. Please don't bother over militaristic idiots, even when they are federal judges.

Affectionately always
Alice

a. John Bright, a leading British advocate of free trade and a hero of AH's father, had opposed his nation's participation in the Crimean War.

b. John Palmer Gavit, a former resident of Chicago Commons, was managing editor of the *New York Evening Post*.

c. Walter Lippmann, associate editor of *The New Republic* from its founding in 1914 to May 1917, had just become an assistant to the Secretary of War.

d. AH was investigating industrial poisoning caused by the dope used in the manufacture of airplane wings.

e. William Hard, a former Hull House resident, was the muckraking journalist whose article AH credited with arousing her interest in industrial diseases. Anne Hard was a free-lance writer.

f. Herbert Croly, an influential progressive intellectual, was editor of *The New Republic*. His wife was Louise Emory Croly.

ALICE HAMILTON continued her field investigations for the Bureau of Labor Statistics throughout the war. In the spring of 1916 she had turned her attention to the poisons used or produced in the manufacture of explosives. Before the war, the United States had imported most of its explosives from Germany, but as the nation became "the arsenal of democracy," factories producing TNT, picric acid, smokeless powder, gun cotton powders, and mercury fulminate sprang up almost overnight. The manufacture of high explosives carried considerable risk of poisoning as well as accident, but under the pressures of wartime demand, the safety of workers received even less attention than usual. Because of the great secrecy surrounding these industries, Alice Hamilton had to discover for herself where munitions plants were located and what they produced. Her search for picric acid, while particularly dramatic, exemplified her technique. Hearing vaguely of a plant located somewhere in the marshes of New Jersey, she would follow the chemical's characteristic fumes to their source: "It was like the pillar of cloud by day that guided the children of Israel." Or she would spot orange- and yellow-stained men, known as "canaries," who would lead her to the site. By December she had visited forty-one explosive-manufacturing plants in which some 30,000 workers were exposed to more than thirty toxic substances. Using only cases of which she was certain, she identified more than 2,400 workers who had been poisoned (including fifty-three who had died), mainly by nitrous fumes and TNT.[8]

With American entry into the war, the tempo of her work increased. A second study of explosives in 1917 and 1918 revealed that, although conditions in nitrocotton and picric acid plants had improved, the manufacture of TNT and the loading of shells were still extremely dangerous processes. She made special studies as well of "dope poisoning" in the manufacture of airplanes, the effects of the airhammer on stonecutters' hands, aniline poisoning, and women in the lead industries. She also found time to serve on a committee of the War Labor Policies Board concerned with women's work in hazardous industries, on two subcommittees of the Committee on Labor of the Council of National Defense (one on women in industry, the other on industrial diseases, poisons, and explosives), and on the Illinois Health Insurance Commission. She was one of two members of the commission to submit a minority report because the majority did not endorse compulsory state health insurance.[9]

Her investigative work was done on her own, as it always had been.

Gertrude Pond Hamilton.

Montgomery Hamilton.

The Hamilton sisters. Left to right: Norah, Margaret, Alice, Edith.

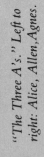

"The Three A's." Left to right: Alice, Allen, Agnes.

The Hamilton cousins. Left to right: Norah, Edith, Margaret, Jessie, Taber, Katherine, Allen Williams, Alice, Agnes, Allen Hamilton, Creighton Williams.

Medical clinic, University of Michigan. Alice Hamilton, third from left, front row of benches.

U.S. delegates to the First International Congress of Women, The Hague, 1915. Jane Addams, front row, second from left; Alice Hamilton, second row, second from left, standing.

Alice Hamilton at fifty (1919).

Alice Hamilton around the time of her retirement from Harvard (1935).

Margaret and Alice Hamilton in Hadlyme, with portraits of female ancestors, around 1959.

Alice Hamilton at ninety (1959).

Then in April 1918 the Division of Medicine and Related Sciences of the National Research Council (NRC), a section of the National Academy of Sciences established to aid in coordinating scientific and technical resources for the government's military program, approached her about its interest in conducting research on war industries. She proposed a thoughtful program that combined laboratory and clinical investigation with follow-up studies of workers who had been severely poisoned. She also recommended that qualified persons be placed in several new shell-loading plants where they could administer a urine test that measured TNT absorption and evaluate the success of various methods of preventing dust and skin contamination. The NRC established a Committee on Intoxications Among Munition Workers, which oversaw this work in cooperation with the Bureau of Labor Statistics and the Ordnance Department. In addition to serving on the committee, Hamilton supervised the work of medical students stationed in six shell-loading plants under NRC fellowships and continued to suggest new problems for research. The project continued until shortly after the armistice, when munitions plants shut down with great rapidity and the Public Health Service took over from the Bureau of Labor Statistics responsibility for field investigations in the industry. Although Hamilton feared that at the war's end the medical profession might cease to take an interest in industrial diseases, she later concluded that the war had at last made industrial medicine respectable.[10]

In view of her disapproval of the war, this was an ironic conclusion. Alice Hamilton claimed that because she wanted to continue her work for the Department of Labor she did not protest the war as conspicuously as she might otherwise have done. But the head of her bureau knew her views and refused to dismiss her despite protests against allowing a pacifist to enter munitions plants. Her opposition to the war increased steadily. She found it grotesque that men should die while producing goods designed to kill others and was also repelled by the men and women she sometimes encountered in Washington who found "joyous release"—and often handsome profits as well—in the war effort. Worst of all, the war unleashed an orgy of hatred, intolerance, and censorship incompatible with liberal values and her own deepest convictions. Hull House, which despite Jane Addams' opposition to the war served as an army recruiting center, provided a rare instance of tolerance.[11]

Elsewhere, mobs attacked pacifists and socialists who opposed the war and in many localities tried to root out all traces of German culture,

201

even to the point of renaming sauerkraut "liberty cabbage." Within three months of the U.S. entry into the war, the federal government, aided by new espionage and sedition laws, banned fifteen publications from the mails, some simply for criticizing the administration. It also imprisoned individuals like socialist leader Eugene Debs who cast doubt on the merits of the war. In its most sweeping action, the Justice Department in September 1917 rounded up nearly two hundred leaders of the Industrial Workers of the World (IWW), a radical syndicalist organization that spearheaded strikes in several industries considered vital to the war effort. The Chicago trial of about a hundred of these individuals began the following April. Contrary to Alice Hamilton's predictions, they were found guilty and sent to prison.[12]

50 To Edith Hamilton

Hull-House
Saturday afternoon
[May 25, 1918]

Dear Edith

I am at the Cordon Club where I have just had a two-hour conference with Dr. Wells over this medical research work and presently I shall go on over to the City Club, to an immigration dinner.[a] It would be the easiest thing in the world for me to keep quite decently busy with no writing at all and I can see why so many people do work on for years with loads of valuable material and never stop to write it up. It is fatally easy to find excuses for not doing that hardest kind of work.

I took the one o'clock from Baltimore and spent the morning at the Hopkins interviewing girls. It was a curious and surprising experience. I met eleven of them and put before them the job in the munition plant with all the eloquence I could command: as a patriotic duty, as a piece of real war work and yet not the destructive side of war but the saving of life: as a piece of original research and as a new, difficult, independent job. I know that in their place I would have jumped at it. Not one did. Two are considering it, but I have little hope. Apparently the traditional duties of women have still a stronger hold than any new call. Six of the eleven refused on family grounds, mother wanted them at home or near home this summer. Three refused because they believed they would get better experience else-

202

where and the two who are doubtful are considering it from that same point of view. The thing that made me meditate seriously on their mental attitude was that none of them showed any response to the appeal for definite war work. Don't you think it is strange?

I have got into an atmosphere of revolt here and as usual my involuntary response to it is very great. I cannot help loving the rebel. There are so few of them, there are so many of the ordinary. One of them is Ashleigh, a young English poet and one of the I.W.W. out on bail.[b] He is simply a burning idealist, who started as a Fabian and has gone on till now he is a syndicalist of the most radical kind. He is a slim nervous creature who was going to pieces in jail till Roger Baldwin got bail for him.[c] Roger Baldwin is here now too. I don't know if I ever told you about him but I have known and liked him for years. He is of a good Boston family, a Harvard graduate, now in the early thirties, a perfect charmer and always joyously espousing the most unpopular causes, just now, of course, free speech. Through him we are hearing of the unspeakable conditions in the County Jail where the I.W.W. have been for seven months. It makes us feel guilty and cowardly, not to have done anything about it. You see they could not get a trial till after Russia signed the Brest-Litovsk treaty with Germany. It seems a far cry to Brest-Litovsk from the Chicago jail but Roger Baldwin had it from Colonel House that as long as there was any hope of Russia standing by the Allies, they did not dare antagonize her by trying these men.[d] So though they will probably be acquitted they will have spent almost a year in jail before the end comes. Meantime every ship that sails from our Atlantic ports is being loaded by the followers of these jailed men, for these are all leaders, the rank and file are left at liberty. Surely if their doctrines are so dangerous, their disciples should not be entrusted with the safety of our soldiers and all our cargoes.

I am meditating a good deal nowadays on the psychology of pacifism, for always when I waver I find myself returning to that path with a feeling of having found my real place and left muddled thinking behind. So I ought to be able to get at the inwardness of it somewhat. I haven't thought it all out yet by any means, but I know that with me part of my inability to feel about the war as you do is due to my lack of faith in the Allies. When you spoke of Miss Thomas having her heart in the battle line in the west I saw that she must be thinking of it as a battle between the forces of right and the forces of evil and that victory will seem to her a wonderful, glorious thing.[e] Now I feel

203

ever so dubious about victory, about what it will bring, while I am certain that war is now bringing evils that we may never see disappear in our life time. I am not so absurd as to feel that it makes no difference which side wins but I do believe that, even if we win a great victory, good will come only through something else, revolution probably. Certainly not Taft's League to Enforce Peace, but something terrible and destructive and "gründlich."[f]

Then I believe I am provincially minded, what the English call a Little Englander. I am much more stirred by comparatively small things over here than by big ones over there, because we are the sinners in the former case and suffering from the sins of the Germans in the other. I do care more about the disgrace of our shutting men up in this jail for political crimes and treating them with injustice and hatred than I do about the prison camps in Germany, just as I felt the Eastland more in a way than I did the Lusitania.[g] And the longer the war keeps on the more I am going to feel that way, for hatred and intolerance are growing so fast that it makes one sick. The Governor of Iowa has just forbidden the Germans to worship God in the churches in their own tongue, just as if he had not the example of Poland and Bohemia and Slovak Hungary before his eyes. In one of our suburbs people must not sing German songs in their own homes. You will say that this is not a terrible hardship and really that is not the side of it I am thinking about, it is the effect on our own selves. Poor people do not act that way, thank Heaven. Mrs. Cesario came to see me yesterday morning, her son has just sailed, and she said in a whisper "I pray for all of them, the Germans and the Austrians just like the others. All mothers are the same. And the boys too. What could they do. God permitted it and so it had to be, who knows why." The lovely thing about poor people is that they are never afraid to say simple and obvious things.

Another interesting rebel I have been meeting is a young Russian-Frenchwoman, now married to an American soldier. She was in "the movement" in Russia from the time she was ten years old and at sixteen she was exiled for life and met Mme Breshkowsky in prison.[h] She escaped through China. She is a Menshevik, but knows most of the Bolsheviki, the leaders at least. She says that Ellis Island and the East Side of New York are the cradle of Bolshevism and that if she had stayed in New York she would be Bolshevik too. She went west, to Topeka, and there she saw what America really is and what we mean when we talk about democracy, but those that stayed in New

York, "in the toil and the great tenements and the ugliness and isolation," they said to themselves: "If this is democracy it is not good
enough for our Russia. When the revolution comes we must find
something more beautiful for her." It seems quite natural, doesn't it.
She thinks Russia may go as far back on the road to reaction as to
accept a constitutional monarch, but that never in the end will she
submit to Germany or to such a government. She thinks that just as
Napoleon set up kings here, there and everywhere, so Germany will,
but that they too will topple when the people rise after the war, for
she too is looking for a European revolution.

This is an absurdly long letter and it is now Sunday morning and
I must go over to Bowen Hall where we are to have a sort of church
to end up the settlement conference. Do ask August Detzer in for
Sunday.[i] Dorothy was here and says he is to be put on a patrol soon
and go over in convoy service.[j] Mary Wilmarth is on the ocean.[k]

Yours ever
Alice

George Mead was here the other evening, at a meeting of the conference.[l] He did not come to dinner and he introduced Miss Addams'
speech as briefly and drily as possible and after the meeting he escaped
so quickly that I could not speak to him. I am going to see him.

a. H. Gideon Wells, professor of pathology at the University of Chicago and director
of the Sprague Memorial Institute, had charge of the laboratory studies for the NRC
investigation of TNT poisoning then just getting under way.

b. Charles Ashleigh was one of the IWW leaders rounded up by the Justice Department.

c. Roger N. Baldwin, a settlement worker who had been active in juvenile court
and probation work, was instrumental in organizing the Civil Liberties Bureau, fore-
runner of the American Civil Liberties Union, with which he was closely associated
for the rest of his life.

d. The treaty of Brest-Litovsk, which ended the war between Soviet Russia and the
Central Powers, was signed in March 1918. Baldwin had met with Colonel E. M.
House, President Wilson's closest adviser, earlier in the year.

e. M. Carey Thomas (1857–1935), president of Bryn Mawr College, was also a
founder of the Bryn Mawr School in Baltimore, in which she played an active role.
An advocate of American entry into the war as early as 1914, Thomas had criticized
Jane Addams and other delegates to The Hague Congress.

f. The League to Enforce Peace, founded in 1915, had endorsed both "peace and
preparedness," maintaining that Germany was the greatest threat to peace. The League,
of which ex-president William Howard Taft was a prominent member, also campaigned
for a postwar international association.

g. The *Eastland* was an excursion steamer that overturned in the Chicago River in July 1915 as it was about to depart with some two thousand passengers for a day's outing on Lake Michigan. The death toll reached 812, victims of overloading and insufficient water ballast.

h. Catherine Breshkowsky (or Breshkovskaya), a Russian socialist revolutionary who was imprisoned and exiled for four decades, visited Hull House on a trip to the United States.

i. August Detzer was the youngest child of a Fort Wayne family close to the Hamiltons. A student at Annapolis, he went on convoy duty during the summer months.

j. Dorothy Detzer (later Denny), 1893–1981, August's sister, lived at Hull House, to which she had first been taken by AH as a teenager, from 1918 to 1920. She supported the war but later became a pacifist and served as executive secretary of the Women's International League for Peace and Freedom for many years.

k. Mary Wilmarth, an early patron of Hull House and a member of its Board of Trustees, was active in the suffrage movement and in many civic activities.

l. AH was a close friend of George Herbert Mead, an influential professor of philosophy at the University of Chicago.

"**M**Y PACIFISM is certainly no help to me," Alice wrote Edith less than two weeks later. "It is only a hindrance, but I can't change now." She elaborated on her views: "It is as if we had stood at the edge of an awful swamp and seen our friends struggling in it, and the militarists had said we must jump in and help them out, and the pacifists that we must try to build a bridge over, though no bridge had ever been built before." Acknowledging that it was too late to talk of a bridge, she regretted that one had not been tried, for "the swamp is deeper and more deadly than we knew."[13]

The bridge analogy appealed to her and she used it often. Disinclined to believe that conflicts were irreconcilable, she instinctively looked for ways to bring opposing parties together. This was as true in personal relations as it was in her attitude toward international or labor strife. But despite a tendency to see both sides of any situation and a distaste for the rhetoric of violence that often attends radical movements, when a choice had to be made she stood with labor rather than management, with the underdog rather than the establishment. Her pacifism, like her commitment to civil liberties, reflected deeply felt emotional preferences: a belief in the sanctity of human life and in the supremacy of individual conscience over any form of external authority. Her views, though idealistic in part, also rested on a disinclination to grant any individual or group coercive power over others. In an article written in 1915, she unequivocally opposed euthanasia, arguing that it was the physician's duty to preserve life in all cases: "Human nature is not wise enough nor good enough to have such power over others given into its hands." And years later, when Felix Frankfurter facetiously called her Judge Hamilton, she replied: "I wouldn't be a judge for anything."[14]

51 To Margaret Hamilton

Hull–House

June 22nd 1918

Dear Margaret,

Clara's telegram to Miss [Enella] Benedict came after she had left, but she is to spend a week with her brother at this address: 75 Bethlehem Pike, Chestnut Hills, Philadelphia, and then some time with her sister in New York, so she would not be going to Hadlyme for ten days anyway. Perhaps by then you will feel more settled, though I should not think you would need to be settled for her. She won't be able to come later, for she cannot afford the journey extra and she stays at Westport later than you stay.

Mary [Rozet Smith] is in her fourth day and all goes well, though of course not happily. She still does not want to be talked to she only looks at me and half smiles and then shuts her eyes again, and she says she feels as if she would not want to see anyone for months. Only Miss Addams and I go in and we just sit and knit. It is wonderful weather, very cool and bracing. I suppose I must be starting for Mackinaw by the end of the week though I hate to go and leave her.

Your telegram came Friday evening and yesterday morning I found a notary and got two affidavits, one of which I sent the Red Cross, Foreign Service Department, the other to Quint in Julia [Lathrop]'s care. Fancy his not giving you an address. I am so thankful he has it and of course you are too. If he and Mary come to Hadlyme to stay till he sails, I think we had better have Uncle Fred and Aunt Louise [Jennings] at the same time, for it would be so much easier to keep Uncle Fred busy.[a] Unless of course the house were not ready. But I shall be there in time to plan all that.

I spent yesterday evening at the hospital for Mary was having a horrid time with enemas and then when I went for dinner to Salamo's, at about nine, I noticed a crowd outside Bowen Hall and got uneasy and went over. I found the house packed with Russians, a fine lot of men and women, mostly young, and at the moment in great excitement, on their feet demanding that a man who had been arrested be released. At the end under the gallery was a group of plain clothes men, big heavy-jawed, stupid-looking brutes, and of course my heart sank at the sight. They are the ones that make all the rioting really. Rachel Yarros was the only resident there and I must say she did wonderfully. While I was trying to find out what had happened—it was all in Russian—she mounted the platform and told them that nobody had been arrested and quieted them down. Then she came back and told me she had made the police release a man they had dragged out of the gallery for no reason than that they knew him to be an I.W.W. How she did it I don't know, for the police were in an ugly mood. One of them told me that we had no business letting that bunch in, they were all pro-German and I.W.W. and another told me that the whole lot ought to be lined up against a wall and shot. There they stood, big, stupid brutes and I looked at them and looked at the audience of eager-faced, intellectual idealists and thought that it was absolutely in the power of those creatures to pick out any man, drag him off to the station, kicking him all the way, throw him into a cell and refuse him access to his friends or to a lawyer, bring him before a judge as stupid as they and send him for fifteen years to the penitentiary. That is not imaginary, it might have happened any minute, it was what they wanted to have happen. They were agents provocateurs and so was the group of reporters who hovered round like vultures waiting for something to happen that they could make headlines of. Of course the audience was excited and turbulent, for about three quarters of them, including the speaker, an old Revolutionist, were Mensheviki, and the rest Bolsheviki who were more noisy than all the others. But there was no harm and I would not have been really uneasy if it had not been for those miserable police. They sent for reinforcements from Maxwell street and I heard them say there was a patrol wagon on the corner and there were six men outside and about six inside in one group and then others scattered around the room, and they made themselves so prominent and wanted to hustle people and interfere. I never had a more nervous time than those three hours, for it lasted till midnight when the lights went out and at any moment we thought a riot might start. You remember the time we came back from Waukegan and found the meeting at Bowen Hall that howled Miss Addams down. Well these people were superior to those but I was afraid all the same. I think I sympathize so much with the

I. W. W. because sometimes I feel just like them. Last night I certainly did.

I have had four fellows given me to do work in the munition plants and though I am very glad to have them, it does seem to make a lot of extra things to do and will mean trips to supervise their work. But it is well worth while. I am troubled about conditions in plants and the Ordnance Department's evident fear that I am going to interfere and hamper production. It is a blessing that Felix Frankfurter is on the Labor Policies Board, for he can probably help me as my Bureau has never been able to do.[b] I really hope women will soon go into the plants, then something will be done to better things.

I have to go to Mrs. Bowen's to lunch, which does make me cross. She is alone and can't get Miss Addams and I am a stop gap and there are so many things I want to do. Do, do write to me, somebody. I have had no letters.

Yours ever
Alice.

a. Quint Hamilton worked with the Red Cross in France during the war. In 1915 he had married Mary Neal, a Wellesley College graduate of 1905; she was an organist and pianist and the author of *Music in Eighteenth-Century Spain* (1937).

b. Frankfurter chaired the War Labor Policies Board (WLPB) which had recently been constituted to standardize labor conditions in wartime.

ONE OF THE MEMBERS of the NRC Committee on Intoxications Among Munition Workers was David L. Edsall, Jackson Professor of Clinical Medicine at Harvard. Edsall had a long-standing interest in industrial medicine—he called it his "hobby"—and had established an industrial disease clinic at the Massachusetts General Hospital before the war. Soon after he became dean of the Harvard Medical School in 1918, the school launched a degree program in industrial hygiene, the first of its kind in the United States. An admirer of Hamilton's work, Edsall had heard that the concentration of the government's industrial hygiene activities in the Public Health Service might soon put her out of a job. In December he asked her whether she would be interested in conducting a five-year study of health conditions of employees in department stores in connection with the new program. Noting that Harvard would be willing to offer her an appointment as assistant professor of industrial medicine, he added: "I think it is the first time that the proposition has ever come up to have a woman appointed to any position professorial or other in the University." Edsall hoped she would accept not only because he wanted her for the program, but because he thought it "would be a large step forward in the proper attitude toward women in this University and in some other Universities." He had already cleared the matter with Harvard's president, A. Lawrence Lowell, whom he had informed that Hamilton's studies "stand out as being unquestionably both more extensive and of finer quality than those of anyone else who has done work of this kind in this country." To underscore the point, he added, "I would emphasize the fact that she is greatly superior to any man that we can learn of for such a position."[15]

Harvard's desire to harbor the best evidently overcame its prejudice against hiring a woman. Lowell had no objection "if she is really the best person for it in the country." (Later, however, he attempted to play down the appointment and informed a magazine that Hamilton held "a research position and is not a member of any faculty.") But Hamilton did not jump at the offer. She informed Edsall that she did not want to give up her connection with the Department of Labor and was not, in any event, interested in the department store study. Although Jane Addams thought she should accept because "it is important for women in general," Alice Hamilton could not "feel that strongly enough to take it for that reason alone."[16]

There matters rested when she departed with Clara Landsberg in early January for the copper-mining region of Arizona. Following publication of a government study documenting the injurious effects of the air hammer on the hands of stonecutters (in which she had participated), Alice Hamilton had been asked to determine whether the vibrations from the pneumatic jackhammer used by copper miners posed a similar health hazard. (She concluded that they did not, and thought the real danger was dust.) The trip was "the most wonderful" she ever made.[17]

The copper-mining region still bore the scars of one of the most dramatic incidents of the war. An IWW-led strike had been broken in July 1917 when the sheriff of the Bisbee district, acting at the request of officials of several mining companies, deported nearly twelve hundred striking copper miners to the sweltering desert. There the miners remained for two days, without adequate supplies of food or water, until federal troops removed them. The President's Mediation Commission, which Felix Frankfurter served as counsel, reported (without condoning the strike) that the deportations were entirely unjustified

210

and that the fear of violence that had allegedly precipitated them was unwarranted. Arriving more than a year later, Alice Hamilton found labor and management in two armed camps.

52 To Edith Hamilton

Hotel Reardon
Clifton, Arizona
January 21*st* [1919]

Dear Edith,

I am sitting in the corridor of the hotel just outside our door, the room being too small for a table. This is our second copper center and our last, from here we shall go to El Paso and then home. My dear child it is so wonderful, so beautiful and great and still when one is off on the desert, so intensely interesting when one is clambering about in the bowels of the earth, or talking unrest with managers and miners, that I have almost forgotten about Harvard and quite forgotten about anilin dyes and lectures at Chicago University. You say you will never be content till I have gone to China, well, I shall never be content till you have seen this land, stayed with Allen and Marian, had a day in their desert, motored over the Apache Trail to Globe and then from Globe to Solomonville and here. It is a thing to make one young again, indeed I feel young as I have not in years. It is partly the proud consciousness that I can do breakneck stunts in the mines without hesitating or showing how frightened I am, but partly the air and the bigness and the wonder of it all. Why not leave civilization behind next Christmas and come out here for your vacation. You could do quantities in two weeks.

We went down another mine yesterday, up on Copper Hill three miles from Globe and after it we politely declined three invitations to a ride and walked home down the mountain, past Apache tepees with fires burning in the center and the heat—there seems no smoke from an Indian fire—coming out of a hole in the roof. We called on an enormous, good-natured looking squaw who sat comfortably on the ground beside her fire kneading a sort of wide noodles which she dropped into boiling water. It seemed so much pleasanter a way to cook than to stand over a hot stove. The open fire and the door and the hole in the roof seem to make a perfect system of ventilation and the tepee had no trace of human odor. This morning we left at half

211

past eight and travelled four hours eastward along the Gila River, which is always called Hila. The San Carlos Valley through which we passed is one of the loveliest spots on earth. Always there are these grays, golden gray on the cottonwood trees, silver gray on their trunks and in the sage brush, green gray in the mesquite and greaseweed,[a] and then purple and red mountains and a sky as blue as Italy's. I should like to live in the San Carlos Valley a while but almost better would be a little adobe house in Solomonville, beside the irrigation canal, with a border of cottonwoods along it. Solomonville is already in this new country a left-over village. It is almost all adobe and in the center is a long rambling whitewashed adobe hotel, which looks very Southern. This used to be filled, for it was the county seat and now Clifton has stolen the honor away and Solomonville is dying away. We took the stage there, a crazy Ford driven by a mighty skillful man, and we had one of the loveliest of our desert drives. About midway a wheel came off and we had no extra one, but after half an hour of vigorous work on the part of our driver it was put on again and we proceeded to rush to make up time, a little recklessly I could not help thinking, for there seemed no reason why it should not come off again and some of the places we passed would not have been a bit adapted to such an accident. However this seems to be a country in which one takes risks quite naturally and we came through all right. As we neared the railway track we saw a sign "Gila" and our driver told us that was where we were to take the train. There was absolutely nothing else there and at that very moment the train appeared around the bend. He rushed to the track and stood right in the way of the train, waving our suitcases frantically. The train slowed up and we were hurled on and our bags after us and everyone took it as quite in order though we were a little breathless. Clifton, which we reached late in the afternoon, is charming. Instead of a big, shabby straggling town, a poor imitation of eastern towns, it is a compact little village, built in a canyon of the Gila, the houses clinging to the sides, some of them adobe, some frame painted bright blue or green, Mexican I suppose, others really pretty low bungalows with English ivy in abundance growing on the north sides and palms in the yards. They call it a "camp" which I think means that it is company owned. The Phelps Dodge mines are in this neighborhood and the Arizona Copper Company's and it was one of the centers of the strike of 1917. I saw one of the managers this afternoon, an underbred, faded little Englishman, and I am to go out to one of the mines tomorrow morning. So far

212

we have never taken more than one mine in a day and I think I will keep to that rule, for it is very tiring work. However I do feel that we ought to be making for home at the end of the week. I have missed six lectures already and ought to be back for next Monday's.

They are such a fine lot of men that I meet out here. The managers of course are just like the employers one meets everywhere, but the mine superintendents who guide us around are a splendid lot of young men, resourceful, controlled, alert, quick to decide and to act, giving you the impression that no emergency could ever find them at a loss. One of them told me that the engineers who made up Carey's brigade were mining men and I could well believe it. If I had a son I should like him to go in for this sort of a job, only I would insist on his spending two years at Hull House before he started managing men. That is their blind side. They are thinking all the time of achievement, of results, of making good and when some fool workman comes along and wants to throw a monkey wrench into the machinery they have no patience with him. To their minds he is already earning more than he is worth and ought to be thankful that the new mechanical inventions have not yet thrown him out, as they inevitably will do some day. You never get the idea from talking to them that they are working for some distant capitalist off in New York who must be kept satisfied with large profits, you feel that they are absorbed, like Kipling's Anglo-Indians, in doing a big job well. When the miners talk to me about them I feel that they are quite as far from understanding them as the bosses are from understanding the miners. That is the real tragedy out here, the big gulf fixed between two lots of men who are engaged in getting copper out of the mines and who are really two hostile armies working under a temporary armistice which may be ended any moment at the will of either. I am meditating an article for the New Republic on the unrest out here and on the fact that only the Federal Government can intervene and explain the two sides to each other. It is wonderful to see how both armies trust it. The State is always partial, the Governor is a labor man or a copper company man, but the Federal War Labor Board is accepted as on neither side. If a strike and a bad one is to be prevented, Washington will have to do it.

I am so sorry Mrs. Reid is ill.[b] You do not say whether she has been carefully examined. It does sound like tuberculosis, though of course that sort, pleurisy and at her age, would not be alarming. But it would mean a kind of care that she probably will not get. I wonder how you found Doris this time.[c] I cannot believe she will really lose her music, not for always. But it may be that emotional excitement of some kind is necessary to bring out the best of her music. If she really has it in her, it can't suffer more than a temporary eclipse, it will come back. Norah's last letter sounds as if she were tired and I wonder if she will not find Mary Neal a bit too much. I have written Mary to come to visit me when I get back, but I have no idea where I shall put her. Perhaps luck will provide an empty spot somewhere. Have you had any word at all from Quint. I write regularly on my date but it gets to be a little uncanny not hearing anything in return.

Yours ever
A.H.

a. AH meant greasewood.

b. Edith Gittings Reid, a writer, was a close friend of Edith Hamilton's. Her husband, Harry Reid, taught geology at Johns Hopkins.

c. Doris Reid (1895–1973), daughter of Edith and Harry Reid and a former student at the Bryn Mawr School, was at this time an aspiring pianist. She later became a stockbroker.

53 To Agnes Hamilton

Hotel Paso del Norte
El Paso
January 26th [1919]

Dear Agnes,

This is the end of my trip and I don't want to go north again and let it all sink into unreality without first writing you about it. It has been so wonderful, in every way. Seeing Allen and finding him so much more prosperous and unworried than when he was in Taos, having long, long drives with him when we could talk about everything under the sun, except the war. Marian had warned him about my pacifism and he had taken an absolute resolve not to touch on the subject, for he feels so strongly about it that he will not even read the New Republic. Then it was lovely to see Marian so well. She does about two thirds of the work Allen does, except for one day in the week when she rests practically all day. Each morning at breakfast they plan the day and divide the work and at night they talk it over. It is a quite perfect marriage I think and though Allen is not as frankly in love as Marian is I expect it is just as true of him as it is of her that nobody else counts very vitally except of course Russell.[a] He is a dear. They have brought him up so wisely. I am sure he is not naturally courageous and self reliant but they have made him so. They always assume that he can do anything that any boy can and that there is nothing to be afraid of in as friendly a place as an Arizona desert full of Gila monsters and rattlesnakes and horned toads. He and Marian have camped out quite alone, with only the Airedale dog and Marian's revolver, right out in the desert, and Russell has no idea that it is an unusual thing to do. Once when they were out Marian found a rattlesnake poised to strike, just at her feet, and shot it. Russell came running up, startled and ready to be scared and Marian told him she was so sorry she had not had time to call him so that he could do the shooting, but he should have the next one and they would hunt all around the camp to find one. Which they did, and Marian was reassured as to the safety of the place because they could not find another, while Russell quite forgot to be afraid, especially as Marian let him help her dissect it and then fried some steaks of rattlesnake meat for dinner. Russell showed me your Christmas present, which he loves.

Next to seeing them the wonderful thing has been the desert and the mountains. The Grand Canyon is a marvel and makes you gasp, but it does not take hold of you and sink into you as does the desert with the barren mountains. They are, the older ones whose outline has been worn down, like Monte Subasio at Assisi. And the desert changes all the time, not only in the different lights. First it will be all silver green with sage brush, then olive green with grease weed, then prickly pear will come, with tall yucca palms, and then all of a sudden the giant cactus, which I can never get used to, it is so strange and beautiful. I shall from now on keep on hounding every one in the family to go out and see it.

Of course even if I had been only touristing I should have loved it, but it has been much more worth while having a piece of work to do. Getting in the lovely things by the way makes one care more for them than if they were not sandwiched in with the work. And the work has been more interesting than any survey I ever made. Just the mines themselves are not as interesting as smelting, except for the risky things one has to do, the way one has to climb steep ladders down into black holes, or scramble up through low caves on one's hands and knees, or pick one's way over rails laid across a deep dump or be hauled up a rock that has no foothold. But the industry itself is simple and nothing to learn, not complicated like lead smelting. It has

215

been the human side that has been different from any situation I have ever met before. You see the bitterness of the strike of 1917 is still here and everyone, miners and managers and doctors and all, feel that they are living over a sleeping volcano and that it may burst out any minute. I came out here thinking of it as a fairly simple thing, on one side the managers wanting to crush every bit of independence on the part of the men, violating state laws, cutting wages and making big profits, and the men striking for the most elementary rights and being ruthlessly suppressed and deported. Well some of that is true but a lot more is true too. On both sides there has been ruthlessness. I talked to the wife of one of the mine superintendents who had to escape at midnight in a motor across the desert to a distant town, and to officials who had been besieged in another plant for days and had to man the pumps themselves to save the mines from flooding, and to others who had hidden on ore cars and so got out of a lonely canyon camp where the passes were in the hands of Mexican strikers. Then I talked to miners who are black-listed and cannot get work anywhere in the southwest because of the spy system that follows them up. I had an amusing instance of this myself. We had been in the Globe-Miami camps and came by a long roundabout way to the Clifton-Morenci region. I went at once to call on the general manager, Norman Carmichael, an elderly Scotchman, very courteous and shrewd. He greeted me with "How do you do Miss Addams—I beg your pardon, I meant to say Miss Hamilton." Naturally I asked why he had called me that and he said he knew perfectly well I was not Miss Addams but he had her in mind at the moment and made the mistake. So then I asked why he had her in mind. He said he knew I lived with her and had for many years. I said "How did you know anything at all about me, how did you even know I was in this country?" He smiled and said he knew all about me, had heard when I came into the state and knew what my errand was. It made me feel a bit like Belgium.

Their situation is one that I have never met before. It is two armed camps under a temporary armistice trying to get copper out of the earth without first exploding into war. There is no other class, no innocent bystander, no interested public, everybody is either company or labor. And they are off in remote spots where if anything happens it may be long before outside help can come. When labor is largely Mexican the gulf between the two camps is even greater, it makes one think of some of Kipling's tales of Englishmen in India. Yet with it all I did feel a very genuine desire on the part of the managers to keep from trouble with the men and to do it by understanding and conciliation, not by this eternal "stamping out" that the newspapers are always talking about. They are more moderate and more enlightened in their attitude toward labor than are the majority of employers I know in the east. But it is all ominous. You see there is no strong union with which they can deal. They smashed the union. Of course it was a pretty bad one, it had a record of violence and unreasonableness, but then so had the companies. And the result is that instead of dealing with a union which after all does want to settle matters and get back to work, they must deal with the I.W.W. which says frankly that it does not mean ever to settle anything till industry is in the hand of the worker and till then it means to make industry impossible. So now what the managers want, and most of the miners too, is Federal control of the mines, with continual mediation and with full publicity. And that is a pretty enlightened program you see.

It is half past eleven and our train leaves at noon. I can't tell you what I would not give to have you come out here.

Yours ever
Alice—

a. Russell Williams (b. 1908), the son of Allen and Marian Williams, later received a B.A. from Harvard and an M.D. from Johns Hopkins. He practiced medicine in California.

ON HER RETURN to Chicago Alice Hamilton found a letter from Edsall offering her a half-time appointment to teach industrial medicine at Harvard and freeing her of any responsibility for the department store study. She sent Edsall's letter to Edith, with a short note of her own typed at the bottom.[18]

54 To Edith Hamilton

[Chicago]
[late January 1919]

Dear Edith,

Isn't this wonderful. Just as I had made up my mind that I had lost my chance by being too demanding this comes, doing away with every single objection and making it as easy as possible to accept.

Send it back to me when you have read it. Of course I have written him that I accept with joy. Only what am I to do for six months of each year in Boston. It appals me to think of it.

We are just back and full of things to do.

Your ever
Alice—

IN APRIL ALICE HAMILTON left for Europe to attend the second International Congress for Women. The congress had been planned to coincide with the peace conference ending the war. But when the conference was scheduled for Paris, capital of the most vindictive winning nation, the women, who came from vanquished as well as victorious countries, chose the neutral meeting ground of Zurich. Stopping off in Paris, Alice Hamilton and Jane Addams talked with many Americans and Europeans assembled there to draft a peace treaty. From them they heard disturbing rumors about the harshness of the impending peace settlement and tales of starvation in central Europe and Soviet Russia, then under blockade by the victorious powers. With Addams, Lillian Wald, and former congresswoman Jeannette Rankin, Hamilton also toured the devastated region of northeastern France. She was moved by the desolate battlefields, with their mass graveyards, and by the ruins of the small stone houses, destroyed by German heavy artillery: "It is like killing kittens with machine guns, they are so small and helpless." After the nervous excitement and self-indulgence of Paris and the devastation of the French countryside, she welcomed the quiet calm of Switzerland; in Zurich there were food shortages, but it was a comfort that "rich and poor fare alike."[19]

Branding the famine, pestilence, and unemployment in Europe "a disgrace to civilisation," the women's congress called upon the nations meeting in Paris to end the blockade and permit the distribution of food to the starving. The congress was probably the first public body to condemn the proposed Versailles Treaty, which it presciently declared would "create all over Europe discords and animosities, which can only lead to future wars." The women explicitly denounced the secret treaties, the violations of national self-determination, the one-sided disarmament clause, and the harsh economic burden imposed on the losers. Divided on the question of the League of Nations—many, including Alice Hamilton, opposed it as a "League of the Victors"—the congress adopted a compromise resolution that endorsed

218

in principle a league based on the common interests of humanity while criticizing the proposed Covenant of the League for its failure to promote Wilson's Fourteen Points. The congress also called for the inclusion in the peace treaty of a Women's Charter that would recognize that the "natural relation between men and women is that of interdependence and cooperation" and would grant women full rights, including suffrage. Before disbanding, the delegates organized themselves into a permanent body, the Women's International League for Peace and Freedom (WILPF), and elected Jane Addams as president and Emily Greene Balch—who had lost her job at Wellesley because of her opposition to the war—as secretary-treasurer.[20]

55 To Mary Rozet Smith

Holland-America Line
S.S. Noordam
Sunday evening
[April 13, 1919]

Dearest Mary,

Emily Balch is reading "Eminent Victorians" aloud to J.A. so I am going to make the rash experiment of writing a letter in the writing room though I am pretty sure it will not work. The steamer is acting quite well but no steamer is ever really well-behaved except when it is safely in dock. J.A. had a rather bad bilious time Friday and Saturday and took calomel and was quite wretched but today she is all right again. It is all perfectly comfortable and we have not had a single rough day, so that for the first time in my life I have not even had my meals on deck but have gone down to every one and Miss Addams has too. The steamer is full. The Red Cross girls number eighty-eight and there are about a dozen Y.M.C.A.s. Then there are a lot of Dutch people and nine Boers from South Africa. These last are heavy, serious men, on their way to the Peace Conference to demand self-government, complete independence, for the Transvaal and the Orange Free State. One of them talked to me yesterday. He said they had no complaint to make of English rule, only they wanted to run themselves. I am afraid they want to run the natives too and I have heard that their ideas on that subject are quite like those of our Southerners before the war. The Red Cross girls are nice, quiet pleasant young things but the Y.M.C.A.s, male and female, are awful. They are "entertainers," and apparently of the cheapest vaudeville class, hard and vulgar. I have not been to their shows in the evening but the people who have are quite sick to think that Y.M.C.A. money is being spent to send stuff like that way across the ocean to our soldiers. And they have the expensive cabins while the Red Cross girls are three in a room on our deck.

Our party is most amicable, with occasional sudden alarums and excursions on Mrs. Kelley's part which are very refreshing. She fills Mrs. [Alice Thacher] Post with apprehension. That poor lady is convinced that we are being watched and listened to by everyone and that if anything imprudent is said we shall all be taken off at Plymouth, and never get to Berne. So she is always trying to surround Mrs. Kelley with so many of us that it will serve as a safe padding and nothing will escape to the listening world. Of course it does not work. There are four members of an industrial commission of the Y.W.C.A. on board, very nice women, exactly the sort one would like to have represent us in Europe. I find them more companionable than our crowd except Mrs. Kelley and Miss Balch. Mrs. White is, I grieve to report, quite as silly as ever, and rather pretentious as well, but that queer, blowsy, distraught-looking Miss Nichols is rather nice.[a] Mrs. Mead has decided that Esperanto is the hope of the future and one has to listen to her a great deal.[b] The colored lady, Mrs. Terrell, is affectionate and very autobiographical.[c] She really looks so little like a Negro that I doubt if people mostly know she is one.

It is eleven o'clock and Miss Balch has come up to say J.A. has gone to bed.

Thursday evening

There were two roughish days but not bad and both of us have got on wonderfully, indeed it is much the easiest voyage I have ever had. The passengers are very chummy and it is difficult to read much on deck for they drop down beside one and talk. Besides Mrs. Kelley has come on board with a most appalling lot of manuscript which she insists on my reading, so I have only managed so far to read the two Walpoles and Eminent Victorians. Every night there is some sort of an entertainment but we go only when there are speeches and we think we must. Caroline Dudley, Dr. Dudley's youngest, is one of the Red Cross girls and I enjoy her. She is a radical young thing and quite refreshing for Eliot Wadsworth and Mrs. Draper are very unquestioningly fervent and pretty dull.[d] The Y.M.C.A.s behave fairly so-so by day, but there are many tales of goings-on at night. J.A. sleeps a lot, sometimes twelve or fourteen hours. She will really be in first class condition when she lands. We are to go to Paris for about a week, then to Berne. I am going to cable Quint and Lewis Gannett to get us rooms when we reach Plymouth tomorrow.[e]

This, and one letter to my family is all I have written on shipboard, but that is doing pretty well for me. Of course I will write you from Paris and needless to say I shall long for a letter to tell me about Eleanor [Smith] and yourself.

Love to you both
Alice—

a. Grace White, a suffragist, was a leader of the Washington branch of the Woman's Peace Party. Rose Standish Nichols was a Boston landscape gardener and writer.

b. Lucia Ames Mead, a pacifist and suffragist, had long been active on behalf of various plans for international arbitration and organization.

c. Mary Church Terrell, the first Afro-American woman to be appointed to the Washington, D.C., Board of Education, was active in the woman suffrage and civil rights movements.

d. Eliot Wadsworth was vice-chairman of the central committee of the American National Red Cross. Helen Fidelia Hoffman Draper, a prominent volunteer worker for the American Red Cross, went to Europe to assist in demobilizing the women who had worked with the Red Cross there.

e. Lewis S. Gannett, a journalist who had served with the American Friends Service Committee in France during the war, was on the staff of *The Nation*.

56 To Norah Hamilton

Zürich
May 14th [1919]

Dear Norah,

Three letters came today, from you, from Margaret and from Jessie, such very welcome ones. I am much luckier this time than when I was last over here for then I went for weeks without one. This one from you told of your unwell time which I suppose I ought to have expected, only I didn't. I wonder if you are to have any more at all. I do hope not but perhaps it won't stop suddenly with you. Clara did not. Don't, please, think another thought about my health. I was very unusually tired just before I sailed but the nine days on shipboard rested me entirely and since then I have had hardly one tiring day. Here we are staying in a "Christliches Hospiz," a simple, spotlessly clean hotel, right next door to the Congress meetings, and life is not at all strenuous for me, though it is for Miss Addams.[a] To me it is a very deep and moving experience—not so much the actual meetings of the Congress, though they are interesting enough, but meeting the women and hearing from them what four years of war has meant to them, almost five years really, because over here nobody talks of the war being over yet, with fighting in a dozen places. These women, some of them, have been through repeated revolutions, those from Munich especially, and as pacifists opposed to all violence they have been in danger from revolutionists as well as from militarists before the armistice. One day I hear from a Hungarian woman of the triumph when the Soviet government was founded in Hungary and their dreams of the future, and then how Roumanians and Slovaks with French help overran Hungary until now she is almost prostrate. And the next day I hear from a Roumanian woman of the century-long oppression of Roumanians by Hungarians and of the Russian Bolshevist invasion of Roumania and how they fear Bolshevism. Then it is an Alsatian woman, who says they want autonomy in Alsace and they have protested to Wilson against the wholesale expulsion of German Alsatians. But most dreadful of all are the tales of starvation. One can't bear to hear them. You will all think I have an obsession on the subject and indeed I believe I am beginning to have. This morning a man undertook to explain to me what a difficult matter it all was, how it was not only hatred that is holding up everything but all sorts of financial complications. America can't send food to Germany free, Congress will not let us, so it must be paid for. English supplies too must be paid for. But the French say that money must be paid as indemnity to her ruined peasantry and that America's and England's humanitarianism is only a cloak for enriching Manchester merchants and Chicago pork packers. Somehow the gold reserve comes in and a lot of lifeless things that men are always putting before life itself. Meantime we are punishing tiny mites of girls and boys for the sins of statesmen. They say the weakening will affect more than this generation. One Austrian woman keeps saying that it is not the body only which has starved, it is the soul and mind of people. Kindliness and unselfishness and idealism are gone, instead there is a bitter fight for the barest necessities and a sordid absorption in the primitive needs of the body to the exclusion of everything else. She is a pitiable looking creature, emaciated, with a blotchy skin and eyes so full of tragedy one can't bear to meet them. She made a speech at the public meeting

222

last night which reduced many of us to tears, not so much about their sufferings as about the joy of meeting at last with friendly faces and hearing of brotherhood and reconciliation. She said "For the first time in four years I heard the English tongue and it was not in denunciation of our crimes, it was in denunciation of the hunger blockade." And then as I left the hall I spoke to the wife of a Swiss professor and said how much moved I had been and she said: "Yes, doubtless to you Americans it seems very terrible. But we Swiss have seen so much. For months in my Red Cross work I travelled across Switzerland with the repatriated French who were driven from invaded France to Germany, kept in camps and then returned because Germany could not feed them. I assure you since then there are no new horrors for me." And I felt as if this world were too much to be borne. Isn't it strange how much worse it is than men themselves are, how much more wicked a nation is than the people that make it up?

The lovely thing is the spirit at the Congress. Especially the British women impress me, with their beautiful high courtesy to their enemies, the most truly chivalrous attitude for the conquering side. Of course we Americans are just as much so, but then we have not had five years of fear, suspense and privation. I think the British are splendid losers and splendid winners, the French are wonderful losers, but not noble as winners. However the French women who could not get their passports have sent a greeting to German women which I am going to copy when I get a chance.

I am thinking of joining the Quakers. They seem the only really Christian body left in the world.

Yours ever
Alice—

I do wonder if you cannot go to Hadlyme after you get better, and stay. Margaret's letter is full of the beauty of it and the comfort and how easy it is to get on, without a servant. But perhaps it would be too lonely. I mean to make the experiment some time when I am tired in Boston.

I liked what you said about the mountain without a history and its poetry yet unwritten. History is mostly ugly and I am glad we have so little of it at home. Yesterday a Roumanian lady from Transylvania said to me "For three hundred years the Hungarians have held us and have made us use their language and refused us our freedom" and I said "Three hundred years seems a long time for a grudge to last," but she said, "That is not long for Europe." I suppose it is this hideous war that has made all the submerged old hatreds boil up to the surface. Someone was saying how strange and disappointing Paris seemed and this same Roumanian remarked that she wondered how people who had read history could talk of a moral regeneration through war, it is always a reaction of irresponsibility and recklessness.

I don't know what my next plans are. If I do what I want and what will be most useful I must do it quietly[?] and through the Ministry at Berne. I have a note to our minister from Mr. Morgenthau.[b] Letters can go to either Paris or London, for they may have to wait.

Goodby dear child. I hope you are stronger by now.

Yours ever
Alice.

Get better.

a. A Christliches Hospiz is a temperance hostel or hotel.
b. Henry Morgenthau, former United States ambassador to Turkey and a friend of Felix Frankfurter's, was serving as technical consultant on Turkish problems at the Paris Peace Conference.

57 To Jessie Hamilton

Glockenhof
Zürich
May 15*th* [1919]

Dear Jessie,

Never was there a nicer letter. I feel like a proud mother who hears her first-born at last adequately appreciated. You see Hadlyme is really my child. Margaret is more like a father that is awfully fond of the baby and proud of it but I am the mother who actually conceived it and brought it forth. And I do think you see all the charms and delights of it as thoroughly as a most exacting mother could wish. I was sorry it wasn't appleblossomtime, but perhaps it was just as well. Norah likes the bare time best. She says we are too fat and green, too English, when the leaves are all out. I have followed you and Agnes in my mind's eye as you went into each room and wished so that I could have luxuriated in your liking of it as Peggy did. I must be there when you first see it in the Fall, for then it has such a very different loveliness.

It is unfair that I am the one to have this wonderful time over here, when already I have had ten times as much as all my family put together. This experience is greater than any I have ever had, it is much deeper than 1915. Then the women came together impulsively, to make a protest against a war that had so overwhelmed them they could hardly find words to speak of it, but now the same women, with others who then had not begun to be articulate, have had four years of experience such as we can only guess at, and what they have to say is very tremendous. The Hungarian and Bavarian women have been through revolutions with Red Terrors and White Terrors, the Berlin women have seen even more, while the Viennese and the Würtembergers have seen quiet bloodless overthrow of government and an orderly, if not very efficient republic set up. They have not been passive onlookers, they have gone with one or the other of the revolutionary parties but always they have opposed violence, they have been true to their pacifist principles. You have no idea how intensely real it makes all the things that at home seem only "foreign news." Yesterday a Munich woman spoke on non-resistance and begged the women never to let themselves be carried away by their brother revolutionists who would thrust bayonets into their hands and tell them that revolution was worth bloodshed. Another Munich woman, who looks like a Melozzo da Forli angel, told of the course of the revolution in Bavaria, of the birth of the Soviet without bloodshed and then the murder of Eisner and the four weeks of terror that followed.[a] Yet she is what is called a Red, for she feels the violence is no essential part of it, only set up by the attempt of the aristocrats to stifle the revolution by killing their leader. Almost the most interesting are the Hungarians, one a follower of Karolyi, the other who looks like the Adams memorial, a Red.[b] She does not know what will happen to her when she returns to Budapest, for the Roumanians and French are advancing and when she gets there the White Guard may be in control. She says it takes a very short time to put the Reds out of the way, in Finland thousands were killed in a week. They do not always shoot the big leaders, but they always do the little leaders and she is a little leader.

I feel so small and narrow and comfortable and safe beside these women, who are, you know, not prominent people in any way, just school teachers or wives of middle class men or doctors. If I were weighed in the balance as they are I know I should fail ignominiously. The British women too have gone through a lot, though it has been only social ostracism and petty police nagging, still it is far more than

225

any of us Americans has had to endure. They have such courage too, they are organizing all the time mass meetings of protest against the hunger blockade and they go out in the streets late at night to placard them with protest posters while the police are not looking. They say feeling in England against the starvation of Central Europe is increasing very much among the working people. I thought food was going into Austria, our papers say it is, but the Vienna people say only the rich can buy enough American flour to make a normal ration, the rest are still getting only one fourth of a normal ration, for you see Hoover is not allowed to give any food to the Central Powers, it must be sold, and our price of $2.40 a bushel, guaranteed to the farmers, puts it way beyond them. Congress passed that measure, they say Hoover was terribly disappointed. At least the Quakers are going in soon and that is a comfort, for though they can only reach a few it is something. I have such a respect and admiration for the Quakers. Everywhere they are received as the only people in the world whom everyone trusts, whom everyone knows to be disinterested, and all through the war they have worked on helping whomever they could reach and never admitting that any man was their enemy. And it has been such good, thorough work, houses people can live in, fields plowed and planted, chickens and rabbits raised to restock farms. I think some day I'll be a Quaker.

Zürich is very delightful. It is radiantly clean and fresh, the sky is blue and so are the tramcars, and the houses are almost as white as the mountains, and the blossoming fruit trees and millions of forget-me-nots make more blue and white. The people are all comfortable middle-class, nobody looks poor or rich. As for the young girls, they are too delightful. Their cheeks are crimson and I think there isn't as much face powder used in the whole city to last an American typist one day. They look like water colors, while our girls at home look like rather dusty pastels. Really I have not seen one vulgar or conspicuous girl on the streets.

We are staying in a "Christliches Hospiz," which is rightly named, with a tiny, spotless, bare little two-bedded room. Our food is of the plainest, heavy black bread, no butter, coffee with skim milk and sometimes saccharin, three meatless days a week and always a meatless kind of a porridge soup for dinner. Once it was an oatmeal gruel made with water. We get almost no fat but I don't miss it, though some of the British and Americans go and buy cheese to make up for it. We have to take our food cards everywhere, for they are quite

226

strict. But the things they do for their poor neighbors is really wonderful. Of course you know about their care of the wretched repatriated French children. That is over now and they are taking Austrian children instead, to feed them for four weeks at a time. Six hundred are coming to Zürich on Monday.

One feels so sorry for Switzerland too, for she is in a bad way, tiny between such big fierce neighbors. The reason we have no cream and butter is that during the war she could get no coal from France and had to get it from Germany and Germany would take nothing in return but cattle. Poor Switzerland had no choice but to stop industry or sell her cattle and she sold a lot and now France demands that an equal number of cattle be sent to devastated France from Switzerland as a penalty for trading with Germany. As it is, there is still so little coal that the locomotive which drew us was burning wood and we can have a hot tub bath only Saturdays and Sundays. Then Switzerland is dreading the flood of emigration when peace is signed. Here she is, a normal, normally taxed little country surrounded by big countries terribly overburdened with taxes. I asked one of them if they would not pass laws restricting immigration and she answered with some bitterness "We will do what our powerful neighbors allow us to do. We are not free." She said too that it was not that they disliked having Germans, French and Austrians come in, only they are so afraid of being swamped. They are only three million and no matter how good the immigrants might be, they could not be really Swiss "they could not think Swiss."

I am utterly up in the air as to what our next move is to be. The Congress closes tomorrow. We shall stay on three days more and then perhaps go back to Paris on our way to England or we may be able to carry out a plan which would go a little way to help the misery one feels over the starvation blockade. That I shall have to write about later.

The stories of the starvation of children are bad enough but, perhaps it is because I have never had children and did have Mother, that I feel even more the starvation of the old. A German Red Cross woman said yesterday that in Austria they have reached the stage of primitive savage tribes who leave their aged to die because they are no longer of use. The Austrians say there is not food enough for children and for the old, the children must be saved, the old must die. And they are dying. There is a huge cemetery for the poor and every second or third night a thousand bodies, many of them from the asylums for

227

the aged, are wrapped in heavy paper, loaded on trucks, drawn to the cemetery and buried in one huge trench. They wait till a thousand are dead and dig a trench for all at once.

I do so want to tell you again how dear your letter was. Hadlyme seen from Europe seems so much lovelier than ever and your letter brought it up vividly before me. I like so to think of it over there. Even if people do say hateful things over home, they are not actually now doing hateful things as people here are. I can't ever believe that if they actually knew what it means to real little boys and girls and real old men and women, they would forbid Mr. Hoover to give food to Germany. It is the cruelty that comes from lack of imagination, like a child's.

Please give love to Aunt Phoebe and Katherine.

Yours ever
Alice—

a. Melozzo da Forli was a fifteenth-century Italian painter. Kurt Eisner, organizer of the revolution that overthrew the monarchy in Bavaria and first premier of the new republic, had been assassinated in February.

b. Count Michael Karolyi became premier of Hungary in 1918 following the dissolution of the Austro-Hungarian monarchy. His efforts to maintain a balance between the extreme right and left in the new republic failed, and in March the Hungarian Communists, led by Béla Kun, took power.

THE CONTRAST between what she had observed in Europe and the "golden angels of victory" that lined Chicago's Michigan Avenue in celebration of a victory loan moved Alice Hamilton to write a short article that appeared in *The New Republic* in June. She used the angels of victory as an ironic symbol of the distance that separated Chicago from Zurich, "from a fairy tale to a very ugly reality."[21]

58 To Mary Rozet Smith

Glockenhof
Zürich
May 19th [1919]

Dearest Mary,

Your joint letter came last night and I am so grateful for my part of it. It did seem strange having nothing from you when two letters had come from Mrs. Bowen and I could not help fearing that Eleanor [Smith] was ill and you too distracted to write. It was an awfully nice letter, full of the things I wanted to hear about and I am so thankful to hear that Eleanor is better. I am going to send her a song the Swiss girls sang for us at the banquet last night, for though the words are not much the music is an adaptation of an old Swiss folksong. It does seem strange to hear about Victory Loans. Mrs. Bowen's letter was enthusiastic about the beauty of Michigan Avenue with its Angels of Victory. Somehow the last thing one would think of over here is an angel of victory, whether one is thinking of starving Germany or desolate eastern France or the bloody wars in Hungary and Russia. The only suggestion of angelic things is here in this kindly, peaceful country that hasn't had any victory over anybody.

The Congress is over and was a success in every way. J.A. got in bed last night at some time after midnight and couldn't sleep for over an hour, partly from the excitement, partly from tiredness. She has had a wonderfully happy time, she has been just lapped around with enthusiastic devotion, she has carried through a very difficult piece of work—handling a meeting of women from eighteen different countries and at least three languages—with unvarying success, and she has had an intensely interesting experience not only in the meetings themselves but in between, talking to the women from all over Europe. Then the public success has been great. The Zürich people were dubious about that, for this city is in fear of radicalism and of anything that can be construed as unneutral, and also woman suffrage is a burning issue just now. But our four public meetings in the evening were crowded. They gave us first the big half of the University, then when that was too small they gave us a big church and Miss Addams presided in a pulpit eighteen feet above her hearers. And last night there was a banquet with the Mayor and the Commissioner of Education making speeches and presenting J.A. with a book of prints of Zürich. She is tired now, of course, but she will rest for a few days before going back to Paris.

I suppose you are wondering what it has all amounted to. I think it has been tremendously worth while. None of us from the Allied Countries can help now doing all we can to get the food blockade raised and have the troops withdrawn from Russia and Hungary. And it has done us good to be able to show the other women that we didn't feel toward them as enemies, to really speak out our detestation of the hatred and intolerance the war has brought. Of course I don't

229

know that anything practical will come of it directly but then what comes from these great medical, or educational, or feminist congresses, yet they are very worth while. We did send two telegrams to Paris, one to Wilson signed by J.A., the other to the Big Four. The first protested against the food blockade and that one Wilson answered to J.A., sympathetically though not very hopefully, saying that the practical difficulties were great. The other, protesting against the peace terms as a source of future wars, hasn't been answered. Of course Wilson's reply was a nice little addition to J.A.'s prestige here.

It is really amazing how little nonsense and even how little undigested radicalism was talked, when one considers all the newly enfranchised and revolutionary women there were here. Really the most foolish ones were the Australians who talked a good deal of half-baked nonsense, but luckily they arrived very late. There were some intensely interesting times, one when the women from Bavaria, Austria, Würtemberg, Prussia, Hungary, described the revolutions that they had themselves lived through. Another was when one of the German women told of the protests they had sent to the Government against the invasion of Belgium, the annexation of Belgium, the deportations, the Brest Litovsk treaty and the offensive of 1918. Naturally they were silenced, no paper could publish the protest, their mail was held up, telephone service denied them and they could hold no meetings even in private, and had domiciliary visits of the police over and over again. But we were all so thankful that they did protest. Of course their tales of the hunger blockade have been heart-rending.

Nevertheless, to be quite honest, I must admit that they are a bit difficult, these German women. They may be excellent but the best of them are dense. All the first days we Americans and the British were almost over-doing it in our eagerness to make them feel we were against the treatment being meted out to them since the armistice. We sympathized and we pitied and we passionately declared that our governments were cruel (this the British said) or culpably yielding (this we said). And then little by little the atmosphere changed. We grew a bit tired of having all the repentance on our side. The Germans lapped it up eagerly and begged for more, but never a word came from them of any "mea culpa" on their side. I don't think that in the nicest of them it was more than denseness but that is just it, the nicest are dense. The last day of the Congress a Frenchwoman arrived, a lovely, sad-faced woman from the devastated regions. We gave her a great welcome, of course, and one of the Munich women stepped forward and gave her her hand and said "A German woman gives her hand to a French woman and hopes that together they may heal the wounds the men have made." The French woman received her with much gentle dignity and went on to make a really beautiful speech and we were all greatly moved, but if only the German woman could have put in a little of the other thing, if she could have said "help you heal to [sic] wounds our men have given you," people would have welcomed it. Then at dinner today a very sweet German woman from Wiesbaden said that they were now under French occupation and could realize at last how Belgium felt. Well, really, you know Wiesbaden is not Louvain. And one would have thought nice German women could realize it without a personal experience. Of course this isn't enough to give the least sense of friction, it is only that the Germans just are different and one wishes they weren't.

Herr von Borosini is here and we have seen him twice and he comes again tomorrow.[a] He looks shockingly, not ill, apparently he has not been really ill, that was all cleverly managed in a way well known to soldiers, but something has happened to change him from a happy-go-lucky boy to a really tragic looking man. We have always taken him rather as a joke and I should never have given him credit for deep feeling, but he has gone through very deep waters, nobody can look at his face and not know that. He did not suffer in his English camps, it isn't that, he was always liked and kindly treated. But it was something, perhaps just the loss of his liberty, perhaps his country's defeat. I hope his wife will go to him as soon as she can. He is working for the consulate now and has five hundred francs a month and he says it only costs him three hundred and fifty to live, the rest he spends on German prisoners. He is amazingly reticent, he really tells one almost nothing.

I think before you get this you may have had a cable about J.A.'s change of plans. This is the way it has come. She dined with Mr. Hoover in Paris and he told her if ever she wanted to go into Germany or Austria to help about food he could send her in and would be glad to. Then the Quakers saw him about feeding Russia and he told them they could not go to Russia now, but he would like to have them go into Germany and either he suggested J.A. going with them or they suggested it and he took it up. Anyway the Quakers are to go in and facilitate the selling of food at a low price and the distribution of clothing. Hoover cannot give away food there, Congress won't let him, but the Quakers have money and Hoover will sell them the food and deliver it. The thing to do is to get it into the hands of Germans who will distribute it rightly. There are to be two Quaker women, Mrs. Lewis and Miss Carolena Wood, and us, if we go.[b] I want very much to do it. It will be a great comfort to feel one is helping even if ever so little to bring food to starving people. Of course, if I can do it as I go, I will find out all I can about aniline poisoning, but after all that isn't as vitally important. When I was over here before, the going around Europe was only an exciting adventure, but this would really mean something, and I know I should always be glad I had done it. Then too, one can speak about it in quite a different way if one has really seen it and I shall want to speak of it when I get back home. We go to Paris Wednesday probably, for J.A. and quaint old Mrs. Despard and the efficient Miss [Chrystal] Macmillan and an Italian and a Swiss lady, have to present our Congress' resolutions to the Big Four.[c] There we shall see Mr. Hoover and talk it all over and I'll write you positively. I don't know how long it will take, but I shouldn't be back by July 1st as I hoped. However my family does not really need me, though I do want terribly to be with them for a good long summer.

J.A. has just come to call me to a meeting. Goodby and much love.

A H

a. Victor von Borosini, an Austrian, was a former Hull House resident.

b. Lucy Biddle Lewis did not join the Quaker mission to Germany, but Carolena Wood, an American Quaker whose "spiritual enthusiasm" was something of a trial to AH, did.

c. Charlotte Despard, a British suffrage leader who had been imprisoned three times, was active in the Labour Party and in the international peace movement.

A FTER THE CONGRESS, Jane Addams and Alice Hamilton spent nearly two months in Europe waiting for permission to enter Germany as part of a Quaker mission charged with distributing $30,000 worth of food plus supplies and with investigating the effects of the famine. Herbert Hoover, head of the American Relief Administration, fearing the possibility of revolution, insisted that they wait until Germany had ratified the peace treaty. As the two women rather fretfully bided their time, Addams found Hamilton a "perfect trump" of a companion. They met with statesmen, writers, and pacifists in France and England. Hamilton also investigated dye works in both countries and saw her brother, Quint, in France.[22]

On July 7, little more than a week after the signing of the treaty, the Quaker mission finally arrived in Berlin, where it divided into two parties. Alice Hamilton and Jane Addams, along with their traveling companions Carolena Wood and Aletta Jacobs, concentrated on the industrial cities of Saxony, Leipzig, Halle, and Chemnitz, where conditions were worst; Hamilton and Addams also visited the villages of South Saxony as well as Frankfurt. It was one of the most searing experiences of Alice Hamilton's life. In the crèches, child welfare clinics, playgrounds, and outdoor sanitaria she saw, instead of the healthy red-cheeked Germans she had known, listless children with pallid gray faces, swollen bellies, matchstick legs, and shoulder blades like wings. She also found a high incidence of famine-related diseases, including rickets, scurvy, and especially tuberculosis. Death rates among the old were staggering, for when a family faced a choice of which members to feed, it was usually the old who were allowed—or who chose—to die. Hamilton was impressed by the lack of bitterness among the German people; many insisted on adhering to their allotted ration of food—about one-third of the normal caloric intake—and refused to buy black-market goods. Anna Edinger, widow of Hamilton's former professor, attributed her husband's death to such scruples.[23]

In contrast to her detachment four years earlier, Alice Hamilton had come to Europe hoping to "get on the inside of things here and find out how much suffering there is." Her trips through wartorn France and famine-stricken Germany, and her conversations with women from many nations, gave her this knowledge. Convinced that Americans shared responsibility for the famine, she resolved to do everything in her power to bring relief to the German people. Only this determination, she later maintained, enabled her to witness the horrors of starvation. She began her mission on the homeward voyage by writing an article on the famine with Jane Addams. With characteristic scrupulosity and fairness, she refused to blame any individual or nation for the blockade, although evidently invited to do so by Paul Kellogg, editor of *The Survey,* which published the article in September.[24]

59 To Paul U. Kellogg

Hadlyme, Conn.
August *27th* 1919

Dear Mr. Kellogg:

I am so glad you like our German report. It was written on the steamer coming across, on a borrowed Corona, which accounts for the looks, and it was revised after we got home and the American atmosphere startled us into a new caution. It pleases me to see that one American anyway thinks us over-cautious.

It seems to me—I cannot consult with Miss Addams as she has gone home—not safe for us to try to explain the situation in Germany, for we should be doing it at second hand and might easily fall into decided inaccuracies. For instance Alonzo Taylor's assertion that you quote, as to the fault resting on the German Government.[a] That may be true, we could not assert it or deny it, we could only repeat stray and sometimes conflicting statements we heard about it. We were told that the present government is inexpert, that after devoting a milliard of marks to lower the price of food they succeeded in their object for three days only, after that all the cheap food had been bought up by profiteers and was being resold at the old prices, because no rule as to quantity distribution had been made. Others would call that deliberate dishonesty, playing into the hands of the profiteers, but we should not feel justified in repeating either version.

Your questions about the embargo interest me very much, but if I tried to answer them I should be retailing the gossip we heard from our newspaper men in Paris and from some of the American Commission experts. I think you ought to get authoritative statements about the situation for I find everyone here puzzled and at sea over what really happened and why. My answers cannot be authoritative in any way and I give them only for private perusal, but you should be able to get the real facts from the English society, the "Fight the Famine Fund" of which Lord Parmoor is chairman.[b] I do not know his address, perhaps no more specific one is needed, or perhaps you might write to Lady Courtney of Penwith, 27 Cheyne Walk.[c] She is in it too and could refer your letter to the right person.

So far as we could find out, the blockade was maintained out of fear that, unless Germany was starving, she would not sign the sort of peace treaty the Allies purposed to force on her, and the facts justified this belief. Starvation brought the country at large to the point where the Weimar government dared not refuse to sign, since a refusal meant clamping down the blockade at once. The threat of a march on Berlin was not nearly so effective as the threat of cutting off the food. This explains why the Allies did not keep their promises at the time of the armistice—for that matter they kept none of the promises made then. The shutting off of the fisheries in the North Sea followed upon the armistice because only then, as has often been pointed out, could the blockade be made complete, since Germany's fleet was then surrendered. The removal of herds is part of the restitution programme, Germany having driven off as many cattle from Belgium and France. The Red Cross has, so far as I know, done nothing in Germany and I have not heard that they plan to. As for food, you remember that an act of Congress forbade Mr. Hoover to *give* any food to the Central Powers. All must be sold and with industry almost at a standstill people cannot buy, on no wages or scanty wages, our fabulously expensive flour, meat and fats. Germany needs, above all things, fatty and albuminous foods, next she needs warm clothing, then rubber medical goods, and absorbent cotton, and vaseline, and soap. I do not know how America is to help. I have no idea how many people would be willing to do so, but certainly so far the Quakers are the only ones who are doing it openly and with the connections which we built up in German cities the help sent through them might be very extensive and still be handled with the utmost efficiency.

You did not enclose Mr. Lasker's note so I cannot answer that, but as to your question about finding someone to write on the new constitution in relation to social matters, I believe I could give you the name of a German who would do it very well. I hope you could offer him a little money for it. Remember that there are sixteen marks to a dollar now, so a few dollars would seem a great deal to him. He is

> Dr. Polligkeit. Zentralstelle für Privatfürsorge
> Stiftstrasse
> Frankfurt am Main.

Another would be

> Frau Geheimrath Edinger
> Leerbachstrasse. Frankfurt am Main.

She would need no payment—

Miss Addams will probably send you a list of the social workers,

administrators and others whom we met. If she does not, I will let
you have the names in my notebook.

Yours sincerely
Alice Hamilton

a. Alonzo E. Taylor, a member of Herbert Hoover's staff, had played an important
role in the U.S. Food Administration during the war. In 1919 he was investigating the
food situation in central Europe and the Balkans.

b. Lord Parmoor (Charles Alfred Cripps) had opposed British participation in the
war and had championed the rights of conscientious objectors.

c. Lady Kate Courtney, Lord Parmoor's sister-in-law, and her husband had also
opposed the war. She had helped arrange Jane Addams' interviews with British statesmen in 1915.

TOWARD THE END of her life, Alice Hamilton looked back on 1919
as her best year. It had begun with the exciting investigation of the
Arizona copper mines. The deeply moving experience in Europe had
completed her transformation from a sometimes bemused bystander
into an effective political person. There was also the invitation to join
the Harvard faculty in the fall.

VII

THE HARVARD YEARS
1919–1927

"**G**OING TO HARVARD is very grand," Alice Hamilton wrote in March 1919. "If one could wear it as a decoration, like the Order of the Garter, I would love it." Even before her appointment as assistant professor of industrial medicine, she had been asked to deliver the Cutter lectures in Preventive Medicine and Hygiene, an invitation, she noted in her autobiography, that had astonished her: "Harvard was then—and still is—the stronghold of masculinity against the in-roads of women, who elsewhere were encroaching so alarmingly." Her appearance in Boston in April to deliver the lectures attracted considerable attention, eliciting such headlines as "Very Feminine is Dr. Alice Hamilton, the First Woman to Break Down Sex Barrier and Join Harvard Faculty" and "The Last Citadel Has Fallen." Asked what she thought of her appointment, she attributed it to the newness of the field and to luck, an explanation often advanced by successful women. But she added, "I am not the first woman who ought to have been called to Harvard."[1]

Though she never complained of it publicly, Hamilton found her position as Harvard's only woman often frustrating and disappointing. By arranging to teach only half the year, she undermined any chance she might have had to play a leading role in the division of industrial hygiene. But her anomalous situation and her temperament made it unlikely that she would have done so in any case. She received a succession of three-year appointments, which in the early years she could not be certain would be renewed. Never a seeker after power, she had an instinctive reluctance to put herself forward, particularly if by so doing she risked alienating others. She wondered too whether

237

she was living up to what was expected of her. At a particularly discouraged moment she wrote: "Really I have never had such a deep sense of failure as I have this year. It is the end of my sixth year here and I cannot think of one thing that has gone well and I know that the fault is nobody's but mine, that I came brashly into a milieu to which I was not adequate, and tried to fill a place which needed more brains than I have." In her autobiography she could write humorously of the stricture about football tickets, but at the time she often smarted at the slights, whether it was the open disdain of a young instructor, "an ex-football champion, a 'he-man,' " she thought had "a complex against treating a woman as an equal," or the more customary experience of being excluded.[2]

Despite her tenuous situation, she made important contributions to the work of the department. She helped secure many articles for the *Journal of Industrial Hygiene,* which was edited at Harvard. One editor later claimed that Hamilton, an associate editor, had "literally kept [the journal] going." Most important, through her initiative the presidents of several lead companies funded a three-year study of lead poisoning, with no strings attached. Headed by Joseph C. Aub, a young physiologist at the Medical School who had a distinguished career as a researcher, teacher, and clinician, the lead project was Harvard's most important investigation of industrial diseases during the early years. Aub and his associates discovered how lead was absorbed, stored, and eliminated from the body and also developed an effective treatment for lead colic. This work conclusively demonstrated that lead was more toxic when inhaled than when swallowed, a finding with important implications for prophylaxis that ran counter to the conventional wisdom among manufacturers.[3]

It was laboratory and clinical work of this sort, rather than field studies, that brought academic prestige in the 1920s. Aub, in whom Hamilton sometimes confided about Harvard matters, later acknowledged that she had never received the recognition she deserved from her colleagues. He attributed the tendency to "minimize her efforts" partly to the fact of her sex and partly to prevailing academic disdain for practical achievement. Aub himself believed that no one in the world, with the possible exception of Sir Thomas Oliver (whose 1902 textbook had initially inspired Hamilton's interest in the subject) knew more about industrial diseases.[4]

Hamilton's move to Harvard coincided with a new phase of industrial medicine in the United States. After the war, the danger of industrial diseases increased markedly, keeping pace with the nation's rapid economic expansion. The burgeoning chemical and automobile industries were especially potent sources of novel poisons, the effects of which were little understood. But if the task seemed larger, Alice Hamilton believed that it was in some respects easier. Workmen's compensation laws and insurance companies provided incentives to employers to reduce their accident and disease rates, a goal of the business-affiliated National Safety Council as well. Physicians, who had formerly ignored industrial diseases, developed an interest in research; investigations like the Harvard lead study provided essential information that made prophylaxis or early treatment more feasible. The Public Health Service too undertook studies of health hazards in industry and developed new techniques for assaying exposure to dust. So great was the change, Hamilton reported, that participants in the 1925 International Congress on Occupational Accidents and Diseases credited the United States with leading other nations in research. Even so, Hamilton did not believe that sufficient funds were being allocated for research, especially to the Public Health Service. As a result, workers still served as "experimental laboratory animals" for the testing of new chemicals.[5]

Certainly in the application of existing knowledge the United States lagged behind other industrial nations. Despite the best efforts of progressive reformers to enlarge public responsibility for health and welfare, the nation clung to a system of voluntarism that left to private interests the responsibility for finding solutions for the human wastage of the industrial system. Federal agencies had authority to make surveys but none to enforce their findings, and during the 1920s the conservative political climate doomed efforts to extend federal authority in matters of welfare. The situation in the states was little better. Workmen's compensation laws, a major legacy of progressivism, were directed principally to the problem of accidents; by 1935, only fifteen states provided compensation for occupational diseases, and some did not include all of them. Even these few programs were often administered by officials chosen for their political connections rather than for their scientific or administrative capabilities. The system permitted the sovereign states, in Hamilton's words, to "maintain their right to neglect the safety and health of their wage workers if they wish to." Reviewing her nineteen years in the field, Hamilton concluded in 1929: "The system remains, in spite of recent improvements in its workings, essentially an industrial feudalism, wise and

239

kindly for the most part, but surely an anomaly in a modern democratic country."[6]

As the field expanded and became more complex, Hamilton's role changed from that of pioneer investigator to that of codifier, troubleshooter, and consultant. She had chosen to teach part-time in order to continue her field studies. Immediately after the war she investigated aniline dye and carbon monoxide (technically an asphyxiant rather than a poison) for the United States Department of Labor, and, with Harvard colleagues, she subsequently studied mercury poisoning in the felt hat industry and in quicksilver mines. Altogether she found the era of constructive change, as she designated the postwar years, less compelling than the early days. But although she never made much of her later achievements, her career continued to reveal the commitment and innovative spirit that had characterized her earlier work. Her ultimate goal remained constant: to secure greater knowledge about and control over industrial poisons.[7]

In *Industrial Poisons in the United States* (1925), the first American text on the subject, Hamilton codified knowledge in her field and consolidated her position of leadership as well. Written mainly for industrial physicians, the volume demonstrated her unparalleled firsthand acquaintance with the poisonous compounds used in American industry and her encyclopedic knowledge of the American and European scientific literature. Reviewers commended its lucid style, its objectivity in appraising all sides of controversial questions, and its extensive and up-to-date coverage. George Kober, whose entry into the field had preceded Hamilton's, considered it "by far the best publication on the subject in any language," while Sir Thomas Oliver himself, in an appreciative letter, not only praised the book but noted his longstanding admiration of her work.[8]

In the absence of an effective program of control, Alice Hamilton maintained a constant vigilance over developments in her field. Serving as both clearing house and watchdog, she received numerous requests for information and assistance: from hospital physicians describing unusual cases, from employees of state labor departments seeking advice on factory protection or workmen's compensation, from businessmen asking how they might prevent industrial diseases. Following up all such requests, sometimes in person, she provided information or recommended changes in plant procedures.

She kept after those in authority, mainly businessmen, managers, and chemists, tactfully appealing to them to do the right thing. Following a visit to a Pittsburgh sanitary ware manufacturing plant in 1927, she suggested to the president ways in which he might protect his sandblast workers from silica, a problem, she noted, in which she knew he had been interested since 1911, when they had first met. A few days after she sent her letter, a notice went out directing plant managers to have the lungs of all sandblasters examined at least once every six months and to remove the men whose lungs had been affected by sandblast dust to other work until the trouble had cleared. The directive said nothing about x-ray examinations or protective devices—respirators and helmets—and she wrote twice more in detail about these. Sometimes she met with more conspicuous success. She was elated when she persuaded a chemist who worked for a shoe factory near Boston to forgo the use of benzene: "I think it is the first time I have known such a thing to happen, to give up a dangerous thing before it had done any harm in the man's own plant. It does greatly relieve my mind."[9]

In addition to her efforts with manufacturers, Hamilton prodded others to take up the broader issues of investigating and controlling industrial diseases. During the early 1920s she attempted, unsuccessfully for the most part, to raise funds from at least five foundations for several research and field studies and for an impartial consulting service. She attributed her failure to the likelihood that the results would be too hostile to capitalism. But even a fund dedicated to radical causes had little interest in this work, proving her contention that the prevention of industrial disease was one of the least popular of causes. With greater success, she recommended investigations of specific industrial poisons to one or another federal bureau. Most notably, she was among those who urged the Surgeon General to call two national conferences, one on tetra-ethyl lead in 1925, the other on radium in 1928; she considered these events (and the ensuing reports by the Public Health Service) the high marks of voluntary cooperation among the federal government, industry, and medicine (labor was less well represented). She also worked with a union-based health organization and with several branches of the Consumers' League on investigations of dangerous trades.

Hamilton was one of the few individuals willing and able to work closely with all concerned parties, with business and labor as well as with government, the medical profession, and public-interest groups. Not only did she insist that a spirit of impartiality was essential for success in the field, but she also believed that when it came to industrial

241

diseases none of the interested parties had a monopoly on virtue. If businessmen often minimized the importance of diseases of the workplace, labor unions (where they existed) also tended to ignore them. She reserved some of her harshest words for industrial physicians who put the interests of their companies above their responsibilities as doctors. And always she was dismayed by the legal system, adversarial in nature, that made straightforward presentation of the scientific facts so difficult. To a representative of the Consumers' League who was about to testify, she wrote: "The only advice I can give you is to be perfectly cool and never let the lawyers of the other side bully you or get you confused, which is what they all try to do . . . Never let yourself forget for a moment how contemptible a part you think they are playing; that will give you the right manner. Your strong point is your entire disinterestedness; the jury cannot help being impressed by that."[10]

Both at the time and later Hamilton expressed optimism about the progress that occurred after the war. In view of the conservatism of the times, and the unlikelihood that any of the interested parties would press for major change, she placed ever greater reliance on an aroused public opinion as the best hope for real change in her field. "Publicity is a wonderful thing," she wrote before the conference on tetra-ethyl lead. "It may be the pebble with which David will kill Goliath." This was an optimistic faith, one she shared with an entire generation of reformers, and it made her somewhat complacent about the long-range impact of passing episodes such as the Surgeon General's conferences which, while briefly focusing public attention on a new industrial poison, did nothing substantial to regulate these substances.[11]

If she expressed satisfaction with half measures, it was also because she was realistic about the difficulty of gaining absolute control over industrial diseases. Returning often to the successful elimination of phossy jaw, she believed that the ban on the importation of white phosphorus had occurred only because a safe substitute had been found and because the terrible appearance of the victims had elicited unusual sympathy. Few cases were so dramatic. Hamilton favored the elimination of unnecessary substances (such as lead in enamel) and those employed in the luxury trades, among which she included potteries and granite tombstones. But for the most part she believed that an agent would be eliminated only when a substitute was discovered or when it was so dangerous that even the best plants could not offer adequate protection to their workers. Under such circumstances, she readily accepted piecemeal change and small victories.[12]

Freed from the constraints she had felt as a government employee, and more certain of herself, Hamilton took a more active role in the women's reform and peace network than she had in the past. Working through such organizations as the League of Women Voters, the Women's Trade Union League, the National Consumers' League (of which Hamilton was a vice president) and the Women's International League for Peace and Freedom (of which she was a national board member), women remained one of the most active forces for social change in an era of increasing political conservatism. Indeed, their continued advocacy of social welfare measures provided a vital link between the goals of the social-justice wing of progressivism and the social welfare measures of the New Deal. Alice Hamilton enjoyed her association with women's groups, noting after her first meeting of the League of Women Voters: "I have had to attend so many men's conferences but this was the first women's I had ever been to, in this country, and really women do it better. We talk more in private but not nearly so much in public. Then too it was nice to be one of a crowd and not 'the only woman.' "[13]

Hamilton was an important member of the women's network, not as an organizer or political infighter, but as a persuasive advocate. Talking only on subjects about which she knew first-hand, often speaking extemporaneously, she was by all accounts an extremely effective speaker. The prominent Boston reformer Elizabeth Glendower Evans emphasized the moving power of Hamilton's voice, which had "a depth and a measured rhythm and a quality of breeding which makes her every word impressive." Dorothy Detzer, executive secretary of the WILPF, later recalled: "I couldn't now tell you whether it was the substance and structure of her speech, or the charm with which she made it—that was the most appealing, and absorbing. She always *held* her audience. You knew at once that she spoke from a full mind and that whatever the subject, she had never exhausted it when she sat down." Her professional eminence, unusual for a woman, gave special weight to her testimony. This she frequently offered on such subjects as child labor, the eight-hour day, the minimum wage, workmen's compensation, protective legislation, and, with the advent of the depression, health insurance and old-age pensions.[14]

In the heated postwar climate, two issues increasingly concerned

243

her: international affairs and domestic infringements on civil liberties. Her interest in foreign affairs had been awakened during the war, and she kept informed through wide reading and assiduous attendance at Foreign Policy Association meetings. She also gained first-hand knowledge as a member of the League of Nations Health Committee from 1924 to 1930 and as a visitor to the Soviet Union in 1924 and to Nazi Germany in 1933. Distressed by the nationalist and militarist drift of the times, she deplored the "transparent hypocrisy" that prompted Americans to endorse the Kellogg-Briand Pact outlawing war while actually increasing military expenditures and refusing to join other nations in the World Court or even in efforts to outlaw the use of poisonous gas and germ warfare.[15]

The self-righteousness and nationalism she objected to in U.S. foreign policy she also condemned in the nation's treatment of dissenters and aliens. Long after hostility to Germany had abated, the intolerance unleashed by the war, compounded by reaction to the Bolshevist revolution, still made itself felt. The DAR, the American Legion, the Chemical Warfare Service, among others, attacked Jane Addams and her associates as dangerous radicals bent on destroying traditional moral codes. Hamilton's name appeared on several of the lists of subversives compiled by self-appointed patriots; some of these, in the form of "spider-web" charts, graphically conveyed the connections among many of the reform organizations to which she belonged. Though she sometimes responded to attacks on these organizations, she was much more distressed by the plight of the poor and foreign-born who stood to lose their freedom and even their lives from such intolerance. At Hull House shortly after the war, she assisted aliens who were about to be deported for their political beliefs (a practice that remained a lifelong concern), and in Boston she was drawn into the Sacco-Vanzetti case, which she viewed as the ultimate miscarriage of justice. Hamilton's empathy for aliens was also apparent in her opposition to Prohibition, which, unlike many of her Hull House associates and other friends, she came to view as a piece of class legislation that permitted the rich to indulge while depriving the poor of a comfort; it offended her belief in personal liberty as well. (Also unlike many of her colleagues, she supported Al Smith in 1928.)[16]

From 1919 to 1935 Alice Hamilton divided her time among Boston, Chicago, and Hadlyme. She spent the fall semester of each year teaching at Harvard, much of the spring at Hull House where she relieved Jane Addams of some responsibilities, and summers in the settled domesticity of her own home. Her later claim that her life after 1919 was one of "steady routine" scarcely does justice to her wide ranging interests and commitments, though it does reflect her feeling that she had finally gained control over her life. In addition to preparing her Harvard lectures, writing numerous professional articles and two books, and keeping up a brisk business correspondence, she traveled extensively in the United States and Europe. She visited factories, addressed professional associations and reform gatherings, and lectured at several colleges, where she picked up some of the income she had sacrificed by arranging to teach part time. A good neighbor in each of her three residences, she participated in discussion groups and sat on committees, took tea with friends, and was never too busy to accompany a friend to a doctor.[17]

Family ties remained important. In 1920 Alice persuaded her cousins Agnes, Jessie, and Katherine and their mother, Phoebe Hamilton, to purchase a home in Deep River, Connecticut, a few miles from Hadlyme, thereby recreating the tradition of "the other house" that had originated in Fort Wayne and Mackinac. Alice's political leanings were distinctly to the left of those of other members of the family: her pacifism angered Allen Williams, and Agnes was not only a fervent prohibitionist, but also considered the American Civil Liberties Union, of which Alice was a staunch supporter, "a committee of parlor Bolsheviki." But even her conservative aunt Phoebe wrote of Alice: "The more you see of her the more wonderful you find she is." In the late 1920s and early 1930s, members of the younger generation took their place in the family circle, as the Hadlyme home of the Hamilton sisters became a refuge for the children of their male cousins and other young people of college age.[18]

Alice Hamilton arrived in Boston on September 21, 1919, shortly after state troops had broken up a police strike that many Bostonians and Americans had viewed with panic as a harbinger of Bolshevism. A few days later she attended a meeting of the newly organized League for Democratic Control to review the facts of the case. There she met a group of women who were to become friends and political associates. One of them was Katherine Bowditch Codman, a woman of buoyant spirit and easy grace who became one of Alice Hamilton's closest friends and a co-worker in radical causes, among them birth control and the abolition of capital punishment. During her years in Boston, Hamilton lived at 227 Beacon Street, home of Katy Codman and her

245

husband Ernest Amory Codman, a surgeon. A nonconformist, Amory Codman had aroused the opposition of Boston's medical establishment by his advocacy of the "end result system," a plan to standardize hospital care by following up each patient to determine whether treatment had been successful.

60 To Agnes Hamilton

Women's City Club of Boston
October 13*th* [1919]

Dear Agnes,

Thank you for the nine dollars. I hope the chintz cost as much as that, if it did not I am cheating you. I can't in the least remember.

I did not care much for Gilbert Beaver's tract.[a] I cannot believe in the effect of prayer except on the person who prays. But I have been reading some Quaker things which Miss Addams has been sending me lately and I am going to put one of them in with this letter. The whole Quaker theory appeals to me, the absence of fixed dogma in matters of faith and the insistence on strict dogma in matters of conduct, the emphasis on individual liberty, the reverence for the inward light of others as well as one's own, the repudiation of authority. I don't quite know if I want to join them and yet if I feel I ought to join all these associations founded to uphold political and economic reforms that I believe in, maybe I ought to openly join people with whose religious views I am in sympathy.

I am just back from one of my investigating trips, for Harvard does not give me any work to do and I have to make jobs for myself. I can't get used to this freedom and I can't yet settle down to a feeling of leisure. Ever since I started doing explosives in April 1916 I have had a sense of hurry, of racing with the calendar and it seems so queer to find myself looking at it now and not caring whether it is the 14*th* or the 24*th*. After all, I am fifty now and it is time I took life more quietly.

Boston is beautiful in this Fall weather, but abysmally lonely. I have arranged to take the third floor of a house on Beacon street and take my meals with the people who own it, a doctor and his wife. I haven't even seen the rooms nor the man, but his wife is nice, very, and anyway I know now that after twenty-two years of settlement life a bachelor existence is not for me. Here at the Club I am perfectly independent, can take my meals when I please, read a book while I eat or let some woman converse with me. It sounds nice but it is not. It makes me think so much about the details of life, meals and how to spend the evening and how I am feeling. I want things to be decided and arranged so that I can conform to them without thinking and I want demands on me so that solitude and leisure are rare and precious as they are at Hull House. Sometimes I try to talk at the table and am rewarded by one woman telling me that she only hopes the German women and children are starving to death but she is afraid it is too good to be true, and another saying that Congress ought to pass a law making trades unions criminal and give orders to the soldiers to shoot down any workmen who insisted on joining them. Then I take refuge in the pages of the Transcript where I find the same sort of thing, only I can skip it there. The doctor and his wife are radical, thank Heaven. She is a little like Marian Walker and I know I shall like her. Anyway it is better than living alone and fussing about meals.

We have a plan, Peg and I, to meet in Wilmington for Thanksgiving Day. I have to do two plants there and I could do them on Tuesday and Friday. If we find we can, you will have to come too.

Yours ever
Alice—

a. Gilbert Beaver was the first chairman of the American Fellowship of Reconciliation, an organization of religious pacifists founded in 1915. AH became a member in 1917 and remained one until her death.

In the autumn of 1919 Alice Hamilton gave frequent talks about her recent experience in Germany to raise funds for the Quaker relief centers there. At the beginning of December, Frederick C. Shattuck, professor emeritus at the Harvard Medical School and a recent member of the Board of Overseers, asked her to cease these efforts. Shattuck, the major fundraiser for the new industrial hygiene program, told her that a financial backer of the medical school had threatened to withhold all contributions as long as a "pro-German" was on the faculty. Alice Hamilton's letter of December 9 was her answer. In her autobiography she noted that the matter had been pursued with Dean Edsall, who refused to interfere. Shattuck continued to express concern about Hamilton's politics, particularly after she visited the Soviet Union, but by then she considered him "an old dear."[19]

61 To Frederick C. Shattuck

Journal of Industrial Hygiene
Boston
December 9*th* 1919

My dear Dr. Shattuck:

You will remember that I promised to think over very carefully the things you said to me last Monday and to write you. Needless to say, I have thought a great deal about it, one does not take such a thing lightly, but I am afraid I cannot write you what I know you wish to hear from me, that your statement of my relation to Harvard University and its bearing on my espousal of an unpopular cause has made me decide to drop all efforts to help the children of Germany. What I have said in public about Germany and the Germans is based on what I saw there last July. It was a pitiful, heartbreaking sight and all that made it endurable was the thought that by seeing the misery with my own eyes I should be able to tell people over here about it with more convincing force and perhaps induce some of them to help. I cannot bring myself to believe that this is something I ought to give up because of the prevailing feeling against Germany here in Boston. If the women who feel that German children should be left to their fate could actually see the little things they would feel quite as much pity as I do, they could not help it.

I think you will say that this is beside the point, that you had no desire to change my views, only to induce me to refrain from expressing them. But it would be quite impossible for me to enter into a relation with any institution if by so doing it was necessary for me to detach myself from purely human problems and to take no part in questions which are of the deepest importance to me as a human being, not as a member of a faculty or a physician or anything that represents but one side of me. Up to now this question has never arisen. My chief in the Government has always known of my attitude toward war, he knew of my object in going to Germany this summer, but nobody in the Government has ever challenged my views nor held that they rendered me less useful. For more than two years of the war the Government entrusted me with the investigation of all factories making explosives and airplanes, showing that they believed me not disloyal.

I do very much regret that your work for the College should be made by even a little more difficult because of me, but I am afraid I cannot change.

Sincerely yours
Alice Hamilton

AFTER TEACHING at Harvard during the fall term, Alice Hamilton usually headed for Chicago in February. She often stopped en route to visit factories and to lecture at Bryn Mawr College, where from 1920 to 1943 she was special lecturer on industrial poisons at the Carola Woerishoffer Graduate Department of Social Economy and Social Research.

62 To Katherine Bowditch Codman

Bryn Mawr College
Friday morning
[February 27, 1920]

Dearest Lady,

I know you expect my first letter to begin "I forgot." It would not sound natural any other way. I explored my overstuffed bag when I got on the train and found there these little envelopes which I had done up so nicely for the maids and then spoiled it all by going off with them tucked away with tickets and things. Will you give them— and say goodby for me?

I am waiting for my train to Philadelphia and then west to Fort Wayne. Certainly I have earned my lecture fees here. My three lectures turned out to be two hours apiece and they ran in two extra "talks," so that now I feel like a squeezed out orange and a very silly one too.

I hope you will remember to miss me, you know you promised to and I am counting on it. I wish I had it in me to tell you just how grateful I am to you for this winter but I cannot. I think you have the gift of kindness, it seems to come without any effort and without the least trace of fussing. All through these months I am quite sure you have never failed to do the kindest possible thing, the quite un-necessary thing often, and yet you never made me feel—in spite of my natural suspiciousness—that you were bothering about me a bit.

249

It has been a happy winter because of you. It would have been a nice one and an interesting one anyway, I should have liked Boston and found people new and pleasantly different, but it is you that have made it really happy and given me a warm feeling when I think back on it all, and made me feel I belong in a niche, instead of being just pasted on the outside. I suppose you have done this sort of thing to lots of people before—you betray a practiced hand, but indeed you never did it to one who was more grateful than I am.

Yours always
Alice H.

IN EARLY MAY Alice Hamilton attended a meeting of the California State Conference of Social Agencies in Riverside. In addition to giving talks on "Ethics and Hygiene," "Industrial Hygiene," and "International Standards of Child Welfare," she spoke to groups in Los Angeles and San Diego, several times on dangerous trades and once on military training.

63 To Katherine Bowditch Codman

The Mission Inn
Riverside, California
May 4th [1920]

Dearest Lady,

This is silly paper, but all the "Inn" affords. It is a wonderful place, but rather affected. I suppose the best thing a bran new country can do is to copy something old and beautiful and this must be a copy of an old monastery. Only when it comes to putting a most modern bathtub into a room built carefully to resemble a monk's cell, it does seem silly. All the same I was glad of the tub after three unbathed days on the train and I remember the real monks' cells in Amalfi had no tubs. It has been very longest railway journey and I rather dread taking it again so soon, especially the Nebraska-Iowa-Illinois end. I could look out at the desert for days on end and never get tired of it.

It seems to be a State Conference of Charities that I have come to, only of course it has another name, we don't speak of charity any more. I am in the hands of a very, very good woman, who has the gentle fervency of Mrs. Richard Cabot, is very like her indeed.[a] She wants me to make speeches embodying her ideas and they are good ideas, the ones I understand seem to be, but they are not mine and I can't cram up and then get them off as she wants me to. But it is difficult refusing and what with her unswerving conviction that it is my duty and my own feeling that I ought not to have let them pay my way out, I feel shaky, and am almost ready to say I will speak on Freudian theories of prostitution and other impossible subjects. They have given me my own subject fortunately, for our meeting, and starving German children for a big church meeting.

It is very beautiful here—and they expect you to say so the moment you speak to them, don't they? and act as if the Lord Almighty had had nothing to do with the making of it, only themselves. Coming over the Sierras it quite took my breath away, it was so beautiful like Switzerland. Only of course not the villages. Tidy little Ladies-Home-Journal bungalows are healthful but they are not beautiful and neither are brick grocery stores or tanks of gasoline. But the orange trees are beautiful and the flowers are as abundant as everyone told me they would be. Mary Smith said I had not enough clothes for California, which is very dressy she says. I knew I had not and almost meant to buy some but then I went to a tea at Adela Barrett's and met a lot of women of about my age whom I have been seeing for years and they had wonderful clothes—some of them—but they didn't look beautiful, and it came over me that they ought to realize that their day for beauty was over and that really it wasn't worth the bother to work so much for what was left. Of course there are exceptions, Adela Barrett herself is one, you are very much one when you get on your blue dinner dress, but for most of us tidiness is quite enough.[b] So I didn't buy anything at all and I don't care—but maybe I shall later on.

I am writing nothings because I am lonely, and it is not quite bed time. Thank you for your typewritten letter and for the cartoon. It is almost unbelievable that even the Transcript can do that. About 227, I shall cling to the hope up to the very last moment. It was dear of Miss Olivia to say she would take me, but I would rather have one tenth of a chance to come to you than a whole chance to go anywhere else, so I can't even think of her offer because I am not going to give up hope of yours till midnight September 20*th*.

Yours ever
Alice—

This is a sketch of my finger for Dr. Codman.[c]

251

a. Ella Lyman Cabot taught and wrote about ethics, religious education, and child psychology; she was also prominent in Boston civic life.

b. Adela Barrett, a wealthy Chicagoan, was a good friend of Norah Hamilton's.

c. AH drew a finger that appears to be black and blue at the base of the nail.

In the early 1920s the Hamilton sisters experienced an unaccustomed strain in their relations. The ostensible cause was Edith's increasing dissatisfaction with her situation at the Bryn Mawr School, as well as her developing friendship with Doris Reid, a former student. In fact, Edith seems to have been in the throes of an acute personal crisis that first manifested itself in illness and then led to her resignation from the school, to greater distance from her sisters, and, ultimately, to a more fulfilling life. The crisis strained family harmony principally because of its impact on Margaret, who had lived and worked with her sister for twenty years and who feared that Edith's headlong confrontation with M. Carey Thomas would cost her the work that she, unlike Edith, loved dearly.

Edith Hamilton's relationship with Thomas had been deteriorating since 1919 at least: there were differences over whether some students should be allowed to take regular college boards rather than the Bryn Mawr College examinations required of all and over the organization and composition of the school's Board of Managers, which Thomas had long dominated. There were also mutual accusations of disloyalty and lack of trust. Thomas, who scrutinized school finances closely and liked to have her own way, had always been difficult. But in her early fifties Edith Hamilton found the situation intolerable. She had never liked the work, she wrote her cousin Jessie, and had had to drive herself to do it through the years. Though fearful that it would be "very, very easy" for her to be lazy (like her father), she looked forward to a less pressured existence that would include plenty of reading and studying Greek. In October 1920 she announced her intention of resigning in two years, but the Board refused to accept the resignation or to act on it. That year her health kept her from fulfilling many of her school obligations. According to Alice, who had it from Edith's physician, her sister suffered mainly from hardening of the arteries and "occupational fatigue" and badly needed a rest. To Margaret, Edith seemed to have "so many bitternesses and upset, even morbid, feelings inside of her."[20]

252

64 To Margaret Hamilton

62 Washington Square
[New York City]
Tuesday morning
[May 17, 1921]

Dear Margaret,

I wake up in the middle of the night and wonder how things are with you and all day long I have down at the bottom of my mind a queer unquiet feeling. Do write me. I think of Edith and wonder if she is hurting you and if you are worried too about her state and I wonder if Norah seems all tired out and you worry about her. And on top of it all, the rush and work of the last weeks. Do write me that it is not going badly. Mabel Kittredge is the one person who realizes how little I want to go over and frankly pities me for having to, yet sees why it is that I feel I ought.[a] She herself is not going, she says it is too dreadfully expensive.

I wish you could be off here in a completely detached, impersonal atmosphere. That is what I feel most about your state there in Baltimore, the choking personal atmosphere. The only comfort is it cannot last now for more than two and a half weeks.

Work goes quickly and well here. It is really an advantage to be conspicuous when it comes to seeing factories, although a disadvantage when one is planning to snoop around as I shall want to next week in Danbury. Yesterday I was sent around the Brooklyn hat plants in the automobile of the Commissioner of Health and under the escort of one of the heads of departments. Today I meet him again. Also I am going to the Sage Foundation to see if I can coax the money out of them.[b]

Write me a bit of a note.

Yours ever
Alice—

a. With Jane Addams and Mary Rozet Smith, AH attended the third international meeting of the WILPF in Vienna in July. They traveled first in Italy.

b. AH was hoping to raise funds to study mercury poisoning in the felt hat industry.

ALICE HAMILTON actively opposed the Equal Rights Amendment, which was put forward in 1921 by Alice Paul and the National Woman's

Party (NWP) in order to remove the still numerous legal disabilities under which women labored, among them discriminatory divorce and property laws and restrictions on office holding. Like most social reformers of her generation with an interest in labor, Hamilton believed that the amendment would nullify the protective legislation for women that the reformers had struggled so long to attain. Hamilton favored such legislation for men as well as women, but, always a pragmatist, thought it "better not to give up the ferry-boat before the bridge is built." Given the conservative political climate of the times, she observed: "It will be a long time before American legislatures will take seriously a proposal to pass laws regulating hours and wages for men."[21]

Alice Hamilton debated prominent members of the NWP on the platform and in print. As a physician she spoke with authority on women's special health problems—the heavier toll of tuberculosis on women in industry than on their male counterparts, the more severe effects of lead poisoning on women than on men (the prevailing view at the time), the substantial risk of miscarriage, stillbirth, and sterility among women lead workers. Nowhere was she more impassioned on behalf of working women than in a letter to Edith Houghton Hooker of Baltimore, soon to become editor of the NWP journal, *Equal Rights*. Her argument was based not on physiology but on the exceptionally precarious situation of women workers under a hostile economic system. It is likely that the opinions to which she refers are those of the Lawyers' Council for the NWP, which held that the amendment would not jeopardize protective legislation, a judgment that went contrary to the views of both liberal and conservative lawyers.

65 To Edith Houghton Hooker

Industrial Hygiene
Harvard Medical School
January 16*th* 1922

My dear Mrs. Hooker:

Thank you for the copies of legal opinions on the proposed constitutional amendment which I find here on my return to Boston. The reason why I do not value them is that they are the expressions of impartial men who have no interest in overthrowing protective leg-

254

islation. The opinions I do attach importance to are those of the men before whom will be brought the cases against manufacturers, laundry owners, hotel and restaurant keepers, who will have seized the chance to repudiate the minimum wage laws and eight- or ten-hour laws. These judges will be selected because they are known to be unfriendly to labor and friendly to capital. They will belong to the class of Federal Judge Anderson of Indiana and that Brooklyn judge who recently said that the first duty of the judiciary was to protect capital. It is possible that these men may in the end lose out, but it will be only after a long and costly fight, financed easily on one side, with blood and tears on the other. And while that legislation pends and perhaps even after it has been decided, it will be impossible to pass any further laws safeguarding women in industry.

I am quite well aware of the feebleness of such an argument when one is addressing a feminist of the "Manchester School" like you. You told me very frankly that you wanted all such laws rescinded and women left free to make their own fight and conclude their own bargains. I could not discuss it with you then, for I felt it too deeply and one cannot combine a friendly cup of tea with hot rejoinders such as those which flew to my mind as I listened to you. I could not help comparing you as you sat there, sheltered, safe, beautifully guarded against even the uglinesses of life, with the women for whom you demand "freedom of contract." The Lithuanian women in the laundries whom the Illinois law—for we are very scrupulous about restricting women's freedom in Illinois—permits to work seventy hours a week on the night shift; the Portuguese women in the Rhode Island textile mills, on long night shifts—Rhode Island has practically no labor legislation,—the great army of waitresses and hotel chambermaids, unorganized, utterly ignorant of ways of making their grievances known, working long hours and living wretchedly. To tell them to get what they should have by using their right of contract is to go back to the days of the Manchester School in England, when men maintained that there must be no interference with the right of women and children to make their own bargains with their employers in the cotton mills or at the pitheads. It is only a great ignorance of the poor as they actually are, only a great ignorance as to what is possible and what is impossible under our supposed democracy and actual plutocracy, that could make you argue as you do. The vote? Why we are beginning to find that we can get almost nothing with it, as a labor group. American labor is too feeble to do what English labor ean, for reasons too many to outline. And the women who need protection most are not politically conscious, they are not even rebels. They are just weary toilers who hold desperately to their jobs, knowing that they are not valuable, that any demand, any protest, means their discharge, and discharge means starvation for the children.

The liberation of women from disabilities due to sex is desirable and I think a feminist organization such as yours should undertake it. But if you carry it through in the way you have started, it will be as if you began to drain a swamp by first taking down the frail bridge we have thrown over it, a bridge built by years of toil on the part of women who cared supremely about the cruel inequalities of life and wanted to make some effort to compensate for them, to at least secure for women less fortunate than themselves the right to health and some leisure and a living wage. To my mind, if you succeed in rescinding all the laws in the country discriminating against women and do it at the expense of present and future protective laws you will have harmed a far larger number of women than you will have benefited and the harm done to them will be more disastrous.

You see, this was not a conversation to be held over the teacups and I am grateful to you for giving me the opportunity to carry it on by letter. Remember, when you think me over-strenuous, that I have lived for twenty-two years among the poor and that for twelve years I have studied trades employing all sorts of labor, from the skilled and highly paid to the unskilled and meagrely paid. The working woman is a very real person to me.

Sincerely yours,[a]

To return your courtesy in sending me the opinions gathered by the National Woman's Party I am sending you some gathered by the Women's Trade Union League. I think you underestimate the standing of that body. I have followed its career with the greatest interest from its founding and I know that, although the help of women of the employing class is accepted, the League is by an enormous majority made up of girls employed in the trades. If they feel apprehension as to the effect of your proposed amendment, it is only fair to listen to them. The members of the Woman's Party would, I am sure, be distressed if my prophecies came true—at least many would—but not one of them would live a bit less softly nor sleep an hour the less because of it. Not so with the working girls. To them it is a question of losing what little they have gained and sinking back into low wages, long hours and night work.

a. This is an unsigned draft.

EDITH HAMILTON, who had spent the summer of 1921 in England with Doris Reid, did not return to Baltimore until two months after the beginning of the fall term. Still in poor health, she kept to a one-quarter schedule and stayed with the Reid family; Margaret moved into a flat with Clara Landsberg, who was now teaching Latin at the Bryn Mawr School. At a meeting of the school's Board of Managers on January 27, 1922, Edith Hamilton offered either to resign at the end of the current year or, if she received an immediate leave of absence, to return for one final year. The Board accepted the resignation for the current year and, in Alice's words: "Baltimore got very much excited over it, for the impression got about that Edith's hand had been forced and they all do dislike Miss Thomas so much."[22]

66 To Margaret Hamilton

Women's City Club of Boston
Thursday
February 9th [1922]

Dear Margaret,

I came in here for half an hour on my way to lunch on Mount Vernon St. with Bishop Williams of Michigan. I have just been reading over Edith's letter and reading between the lines too. It does distress me dreadfully and although I know I could do nothing I wish I might go down to Baltimore and see you both. Sorry as I am for her, I am much sorrier for you, except that you have not yourself to blame and I cannot help feeling that part of her worry is a realization that she went too far, presumed on the hold she had over the Board and sees now that she gave herself up into their hands. But for you it means a sudden sweeping away of the foundation and you did not have anything to do with the disaster but it is a much greater disaster for you than for her. Edith wants another year's salary, you want your work. Of course I will never say such a thing to anybody but you but I can't help seeing that because of this affair with Doris Edith first destroyed your home and now may destroy your work. It all comes back to the Doris affair in the end. Doris and she have talked it over and over, Doris is young and her judgment poor, she wants Edith back in England where she was better and where she was free from worries, she eagerly insists that the Board will never let Edith go, that she can make her own terms, she talks about the plans to close the house and go in April, and all the serious, business side of which a man would think first, and which you and I would consider most important, all that is lost in a sea of personal relations. I know quite as well as you do that Edith will be frantic if you do lose out, will be for doing all sorts of desperate things when it is too late. And I know you won't be resentful but we may as well face things as they are. Don't you issue any ultimatum. Miss Thomas is quite capable of calling your bluff and making a clean sweep and a new start. My advice is to keep perfectly quiet and not in any way intimate that your action is bound up in Edith's, affected by it at all. You cannot possibly tell whether or not relations with Edith's successor would be impossible till you had had a year with her, so that means putting off the whole decision. Nor would I ask for an increase of salary this year, wait till everything has quieted down. Oh I do so wish Edith's temperament were a little stiller, a little less tempestuous and extreme. My heart sinks when I look ahead for her. No matter what happens to you, you are much, much better off.

Clara wrote me a splendid long letter which filled me with gratitude and which I will answer. You must tell me about Bryn Mawr too. It sounded so nice. I wonder where Lucy [Donnelly] put up two of you. It is interesting that Clara noticed the change in Lucy and thought she would make a good president. Katherine Drinker, who had never liked her at all in college, came back from Bryn Mawr ever so much impressed by her grasp of the college problem and her standing with the Faculty.[a] Marion Park came in to tea the other afternoon but I am quite sure she had not yet made up her mind.[b] She said she had asked for time.

I went yesterday to Connecticut College in New London to speak. You know the grey buildings that look down over the river as you pass through on the train. Polly Packard was there, to my surprise, for I thought she had stayed home and come out.[c] She was nice and friendly and took me over to a mite of a room where she and three quite charming young girls gave me tea. They were such well-bred

258

girls that I was glad to see Polly with them. The college is attractive and I had a big audience of townspeople as well as students and faculty and they paid me $25.00 and my expenses, but it was a chore. They make you do so many social things in addition to your speech that you almost perish.

I have been planning for next year to get all the outside lectures I can and make up my income in that way. What I can confidently count on is this: (à la Quint)

Harvard	$2400.00	at least if I were not
Bryn Mawr	100.	to be reappointed they
Radcliffe	200	would certainly have told
Wharton school	50	me before this.
	2750.00	
What I want to add is:		
Bryn Mawr	400	
Chicago Univ.	200	
	600	
	2750	
	$3350	

Katherine Drinker gave fourteen lectures on hygiene at Bryn Mawr for 500. She is having a baby and probably will not want to do it next year. I would not be willing to make seven trips as she did but I will offer to go five times and get in ten on hygiene and four on industrial hygiene. Then I have written Dr. Jordan of Chicago to ask if he would like a short course in March and April.[d] I won't give more than ten lectures for that sum—I used to give twenty-six. If I got all those I need not worry about another job and there are plenty of industrial explorations I could make around Chicago.

This is a full week, for I go to Hartford tomorrow, to speak for the Consumers' League. I felt I could not refuse and they will pay my expenses. I am to spend the night with Mary Buckley which will be nice.[e] New London and Hartford make me homesick for Hadlyme. I am going to send seeds to Maria Selden soon. And Simpson must be told to transplant your grapevines and cut back the blackberries. That last is essential. If they get two years tangled growth we shall have to get Wilcox again.

259

Do write me. I think so much about you and wonder and "wearry" as small Taber says.[f]

Love to Clara
Alice

a. Katherine Rotan Drinker, a physician, was a research assistant in the department of physiology and, with her husband Cecil K. Drinker, served as co-managing editor of the *Journal of Industrial Hygiene*.

b. Marion Edwards Park, dean of Radcliffe College, accepted the presidency of Bryn Mawr College, succeeding M. Carey Thomas in the fall of 1922.

c. Polly Packard was the daughter of Caroline and George Packard, old friends of AH's from Mackinac and Chicago.

d. Edwin Oakes Jordan was chairman of the department of hygiene and bacteriology at the University of Chicago.

e. AH probably meant Mary Bulkley, an active member of the League of Women Voters and president of the Connecticut League from 1926 to 1931.

f. Taber Hamilton was the oldest child of Taber and Abigail (Gail) Gillan Hamilton.

THE "FUSS" about the Bryn Mawr School continued. In March the Baltimore papers gave prominent coverage to the dissension, reporting on parents' protest meetings and detailing charges and countercharges, among them that Thomas had forced Hamilton's resignation and had threatened to close the school. These Thomas denied, while also explaining her opposition to Hamilton's choice of a successor. At the height of the furor, Margaret Hamilton felt obliged to offer her resignation as head of the primary school. She soon withdrew it after parents, alumnae, and the Board asked her to stay on; she was one of five faculty members named to smooth the transition until a new headmistress could be appointed.[23]

Edith Hamilton felt "completely beaten" in her fight with Thomas. But after a period of recuperation and readjustment she moved to New York City with Doris Reid, with whom she lived the rest of her life. Impressing literary and theatrical circles there, she soon launched her career as a writer. For a time she harbored resentment against Alice and Margaret, perhaps because she felt they had not given her unqualified support during her ordeal. In February 1923, Alice wrote Margaret that Edith had taken the position "that until I can assure her that I do not and never did consider her in any way abnormal mentally, there can be no coming together again, and as she does not see how that can happen she thinks we never can." By 1925 the sisters were again on good terms, and early the following year Alice wrote Norah that Edith "was so like her old, old self." It is unlikely, however, that Edith was ever again as intimate with her sisters as she had been; in later years at least she was "company."[24]

During the early 1920s, liberals and radicals mounted an amnesty campaign to secure release of all political prisoners still serving sentences for having opposed the war. At the request of Tyrrell Williams, a cousin of Allen Williams and a professor at Washington University Law School who was working on the case, Agnes Hamilton looked up the wives of two imprisoned IWW men in March 1922. One of the prisoners was E. F. Doree, a labor organizer, who had been sentenced in 1918 by Judge Kenesaw Mountain Landis. As part of the campaign for clemency, in the spring of 1922 Kate Richards O'Hare, a socialist who had herself served time for opposing the war, and her husband Francis Patrick O'Hare organized the Children's Crusade, a march on Washington by the families of the political prisoners. Alice Hamilton had initially been skeptical of the plan, but when some of the relatives stopped off at Hull House she was deeply moved by their plight: "It stirs me up more than anything in the world and makes me angrier, this imprisoning people for their opinions. As if the majority had not always been wrong." The last prisoners were finally freed by President Coolidge in December 1923.[25]

67 To Agnes Hamilton

Hull-House
April 28*th* 1922

Dear Agnes:

Tyrrell writes me that he has asked you to write Judge Landis about the Doree case and has told Doree to write to him also, sending the letter to me to take to the Judge personally. Doree has not done it so far but if he does I ought to have a few more facts than those Tyrrell gives in his letter to Judge Landis, a copy of which he encloses to me. In it he says that you found things even worse than Murphy had written. Won't you tell me just the worst things so that if I do go to see the Judge I can make out the case as pitiful as possible. If I can only get a few newspaper men to stand around and then make out a good "sob story" which they can headline, I believe I might succeed. I have met prostitutes and murderers before this and not minded particularly, but I shall mind meeting and speaking to Judge Landis, more than to any one person I can think of this minute. Once I had to dine with an old villain of a doctor whom I knew to be cruel, neglectful, lying and utterly callous to anything but the interests of the company he was working for. That is the only time I can remember feeling a pharisaical desire to draw my skirts away from contamination, but I shall feel with Judge Landis that I do not want even to look him in the eye and I shall hate to touch him. I have known some of his victims too well.

Elisabeth Gilman is looking after the O'Hare crusaders in Baltimore and trying to rope Clara in to help her.[a] Clara and Margaret had a lovely and much-needed holiday down in Virginia and were motored all the way back by Frances Howard so they did not have the fatigue of driving themselves. Margaret wrote that she spent all the first days sleeping or lying out in a sunny part of the beach. She looked so very badly after the school fuss that I was really shocked and I shall be much relieved when June is here and school closing. Mr. Simpson writes that the freshet was way beyond the usual line and washed away the steps of Norah's studio and all Peg's roses. She will have to choose a higher place for them. But worse than the flood was a wind which blew down my beloved crab apple tree, the one up under the elm against the terrace, a lovely thing which threw itself back against the hillside and was the center of the view from my attic window. I cannot bear to think it is gone.

This is the end of a long day at the typewriter and my shoulder is aching and I must stop.

With loads of love and with rememberings to Mrs. Bradford and the Lewis'.

Yours ever
Alice

a. Elisabeth Gilman, a Baltimore reformer and socialist, was a cofounder of the Maryland Civil Liberties Committee.

In November 1922 Gerard Swope, who had recently become president of the General Electric Company, raised with Harvard's new School of Public Health (to which the industrial hygiene program had been transferred) the possibility of conducting a survey of health conditions at G.E. Cecil K. Drinker, professor of physiology, and Roger

I. Lee, professor of hygiene and at the time also acting dean of the school, took charge of the negotiations for Harvard and hoped to talk Swope into a large-scale study that might cost as much as $100,000. The Administrative Board of the School of Public Health designated Lee as the person who should direct the survey. Swope discussed his plans independently with Hamilton and in mid-December asked her to make a preliminary investigation. She so informed Lee, indicating that this survey would not take the place of the larger study contemplated by the school. Later that month, Swope asked Drinker and Lee whether the work could be done by Harvard's experts, "stating that he only knew himself of Dr. Alice Hamilton of whom he had the very highest regard." The following day Drinker wrote Hamilton of the conversation and promised to keep her better informed about negotiations than he had in the past. A team from the school that included Hamilton, Lee, Cecil Drinker, and his brother Philip Drinker, a chemical engineer who specialized in ventilation and illumination, visited the G.E. plant in Schenectady in mid-January 1923. About a week later Hamilton received a letter from Swope "insisting" on her investigation.[26]

68 To Clara Landsberg

227 Beacon Street
January 24*th* 1923

Dear Clara:

The meek do inherit the earth after all. I have my faith restored. This letter from Gerard Swope will explain, but you must send it back at once, for it is an important document. That is why I send it to you and not to Margaret, because you are so much more trustworthy. It came yesterday and I took it to Cecil Drinker. To my surprise he said at once "That's all right, go ahead and do it." I said that I had supposed it was a most important matter for the school to have the inquiry put into its hands, that I thought they were all rather upset when it appeared that I had been asked to do it. He said that was true, but that I had played absolutely fairly with the school—"and I don't mind saying that I think you were treated pretty badly. Now it's plain that Mr. Swope wants you to do it, the School has no

263

business with what you do during your free time, it is for you to decide that yourself." Which shows a change of attitude as great as that of Dr. Lee who used not to know of my existence and now defers and flatters and is most honeyed. Dr. Drinker told me that things were to be quite different from now on, that Dr. Lee had put me on the executive committee and I would always be consulted. I said "Well, great is the power and glory of the General Electric Company" and he got red and stammered and then laughed and said "Well. Maybe. After all that's the way Roger Lee is."

I have just come back from showing these letters to Dr. Lee. As soon as I came in his secretary was sent to hasten tea and toast an English muffin expressly for me, the things not being quite ready yet for a committee meeting which was to follow. Then he read the letters, laughed genially and said that that was what he had understood all along. I said that I did not at all understand it, that I could see no use of two investigations and that if Mr. Swope accepted our report of Schenectady and consented to the scheme of having us do all the other plants, it would surely be foolish for me to carry on a separate investigation. I said, however, that the only answer I had made to Mr. Swope was that I should be glad to see him when he comes and that I was showing him, Dr. Lee, the letters as a protection to myself so that there could be no misunderstanding. At that he almost grovelled in his anxiety to reassure me that such a thing was impossible—Joe Aub having told me that Roger Lee thought I had deliberately double-crossed them. So here I am, with all kinds of kudos, with the whole thing back in my hands again, and with never a suspicion of fight or unpleasantness. If I had so planned it, instead of blindly blundering along it could not have come out better. I have gone over it so often in my mind and been tempted to do something to vindicate my rights, even to write Gerard and tell him please to see that I was given a proper share in the thing, but now I am so thankful that I decided not to do one single thing which could appear underhanded to them and over-touchy to Gerard. Dr. Drinker told me this morning when he came up to beg me to try to get the bigger plan across, that if I went in on it during my free time I must of course ask a proper salary. I had made up my mind to do that but I was glad to have the suggestion come from him.

It really was nice to have a letter to show which had this intimate, personal tone. At all the meetings Dr. Lee and Dr. Drinker would go on about their interviews with Gerard, of the manner of man he seemed to be, of what one might perhaps think he would do, and I said never a word, being determined not to make a display of my intimacy. But to a man like Roger Lee this letter is like a patent of nobility.

Whichever way it comes out my worries and my resentments are over and it is a relief, for there is quite enough worry about Edith. Her letter in answer to mine was like her first one. She said a good deal about the many years she had given money to Mother without interest and evidently has quite forgotten that it was paid later. The rest is bitter and assumes that I am a hard, unsentimental person to whom feelings do not count. I have not answered it yet. I am so much interested in all you write me about Edith Reid. It is the only inside news we ever get.

I cannot find my letter from Mary Smith but it was just about the same as yours. Miss Addams had been having a lovely time in England, visiting some of the people I have met with her. She wrote from Paris and kept the letter open to tell me that Mary had arrived, very well and quite free from asthma. They must now be in the Red Sea, for they sailed from Marseilles Jan. 12th. I still long to be there but of course if I had gone I should be coming back to Harvard next year in an even more unsatisfactory position than ever, while now my status is entirely changed. It is really not only Gerard. The trip to Schenectady showed them that I know my job and am easy to work with and both Drinkers confided to me their hope that next time we three could go to the plant without having to bother with the others. So it would have been a mistake to go.

I am glad January is ending. The von Brandts go and my lectures stop and I can catch up with myself again. It has been a very full month. Along in the middle of February I have Smith and Greenwich and Bryn Mawr, all amounting to $170. which is not much for the bother. Bryn Mawr must pay more next year.

Yours ever
Alice

WHEN SWOPE met with a group of experts from the school on February 8, he requested that they visit the G.E. plant at Pittsfield, Massachusetts, but mentioned nothing about a larger survey. That evening at dinner he asked Alice Hamilton to visit as many G.E. plants as possible, saying that "he was not ready to enter on any such extended program as Dr. Lee had put up to him, till he knew more about actual conditions." Swope insisted that Hamilton receive $25 per day plus expenses. She thought it was too much, but bowed to the "temptation to earn large gobs of money." (The income helped compensate for the severing of her connection with the Department of Labor under the Harding administration.) Swope never pursued the matter with the Harvard group, and Hamilton later reported that "he does not seem to have taken them seriously at all." Working as a consultant for G.E. for ten years, she found the company's health program admirable, but considered the work tame. Going from mercury mines in California to G.E. factories, she observed, was "like going from a strikers' mass meeting to the meeting of an infant class in Sunday School."[27]

Alice Hamilton was an active member of the advisory council of the Workers' Health Bureau (WHB), a New York City–based organization that from 1921 to 1928 sought to increase union interest in and responsibility for workers' health. Headed by Grace Burnham and Harriet Silverman, the WHB surveyed health conditions among workers, distributed literature about industrial diseases, and established a workers' health clinic. Its affiliated union locals contributed a per capita membership fee, but these funds were never sufficient to support the ambitious program to which the directors aspired. Though she considered the WHB "violently prejudiced on the side of labor, so that it cannot see straight," Hamilton tried on several occasions to raise funds for the organization from the American Fund for Public Service, popularly known as the Garland Fund. Managed by a group of prominent radicals and liberals, among them Roger Baldwin, Elizabeth Gurley Flynn, and Norman Thomas, the Garland Fund had as its top priorities the support of organized labor and farmers and the protection of the rights of minorities. Although the board did not act favorably on the request for funds in Hamilton's letter to Lewis Gannett of February 26, 1923, it did assist the WHB over the years, mainly with emergency funds to help cover the organization's deficits.[28]

69 To Lewis Gannett

Harvard University
School of Public Health
February 26, 1923

Dear Mr. Gannett:

Last week I went to New York to meet Dr. Hayhurst of Ohio, an authority on industrial poisoning, and with him to go over the examinations made by the Workers' Health Bureau for the Painters' Union.[a] We met at their office and later we spoke to the painters assembled in the 62nd Street Armory and told them what we had found. I think it was the most impressive audience I have ever addressed. There were something over 3000 men there and, as Dr. Hayhurst said to me, it was a demonstration of the possibility of something that sanitarians have always held to be impossible; namely, arousing the interest of the working people in their own health problems.

I want to urge very strongly that the Garland Fund treat the Workers' Health Bureau with generosity during the next two years until it can get on its own feet. If it has not succeeded in doing this, then I suppose you might think yourselves justified in dropping it. As I see it, Mrs. Burnham and Miss Silverman have organized a service of the greatest usefulness and have done it on amazingly little money. When they showed me the installation of their clinic I would have guessed that they had spent $7000 or $8000 on it, and they tell me they got it for $4000. The meeting the other evening showed how successful they are in making contacts with the unions, but I have had too long an experience in this sort of work to believe that any group of men in any class can be so vitally interested in safeguarding their health as to sacrifice an appreciable sum of money for preventive work. This mass meeting was greatly helped out by the fact that poisons in the painting trade could be linked up with the fight for the 40 hour week.

The gist of the whole matter is this: the Garland Fund is I understand to be distributed to unpopular causes, and with that principle I fully agree. What I want to impress upon you is this, that the Workers' Health Bureau has a task at once highly useful and unpopular. If the unions in this country were established as they are in England, we might ask them to stand on their own feet in protecting themselves against disease, but as you know they are still fighting for their basic rights and they need their money for things that seem to every man more important than his health. As for getting the money from the general public, can you imagine a cause less likely to appeal to them than that of a trade union belonging to the much execrated building trades? I think that when it comes to unpopularity the organized building trades need yield to nobody. Radicals and Conservatives alike detest them.

May I beg that you will bring this letter before your governing board, most of the members of whom I know, ask them to help out the Workers' Health Bureau for two years, and then reconsider the whole matter.

Sincerely yours,
Alice Hamilton

a. Emery Roe Hayhurst, professor of hygiene and chairman of the department of public health at Ohio State University, had worked with AH on the 1910 Illinois Survey; he had investigated poisons in brass manufacturing and zinc smelting.

70 To Margaret Hamilton

Hull–House
March 30*th* 1923

Dearest Margaret:

It is Friday evening, just before dinner. Your letter was here when I came back from a busy but rather futile day and it was so nice to get it that here I am sitting down to answer it although the bell will ring in a few minutes. This has been a bothersome week. I have not dined at home since last Saturday night, which is a dreadful record. Monday I had to speak before the social workers' club, Tuesday I went to Rockford and spoke to the women's club, for which they paid me fifty dollars and my expenses. Then Mrs. [Anna Lathrop] Case had a little dinner party that night too, ever so nice, with Arthur Ruhl as the guest of honor. He is a journalist who has been in Russia under the A.R.A. [American Relief Administration] and is very sympathetic and nice. He has a profound distrust of Marguerite Harrison, says there is not the slightest doubt that she went in as a secret agent for the Government and certainly double-crossed either her own land or the soviet government. Apparently the latter does not trust her for she was arrested in Chita and taken to Moscow.[a]

Wednesday evening I dined with the Meads. That is one of the things I do always as soon as I can, for I do want to hold on to as old a friendship as ours and I am fonder all the time of Helen.[b] George grows increasingly less intimate and less interesting. He is a fanatical adherent of the League and assumes that all the woes of Europe will be over when once we have joined it. So we discuss perfectly colorless things and I find myself talking more and more with Helen. It is not so much that we disagree but that he seems to me to take so much more superficial an attitude toward things, seems never to want to talk about real things as he used. We drove all the way home together—he had to go down town—and we discussed Chicago city politics. When I think of the way a talk with him used to stretch my mind I feel as if I had lost something.

Thursday night I had to speak again, this time to the Tuberculosis Institute, but now I think there are no more speaking engagements for a good while. Today Norah and I had Caroline and Polly [Packard] to lunch at a new French place, to talk over Italy and their trip and pay back some of the nice things she has done for Norah this winter. Then Norah took them to an exhibit of new art and I came back here, and found your letter. I have not been writing to Edith. Perhaps I should, but I cannot see clearly why and Norah says she thinks it is worse to keep up a pretence than to let it go. I have had to write notes to Doris, first because she wrote me about seconding her for the Cosmopolitan Club—I don't believe in her place I would have asked that—and then about some dresses Edith wanted to use for her portrait. Doris' letters are as brief as possible but she always mentions Edith, and I answer perfectly naturally. But when I think of writing to Edith I seem paralyzed, there does not seem anything to say. I cannot keep on pretending that all is right between us, when so plainly it is not. Yet I believe that when I go to New York next time I will go and see her, not staying there of course, but to see and talk to her. I must do it from time to time, in case there should be a softening. It was dear of Aunt Louise to take all that trouble for us, but it will not do any good. Edith always did live to somebody, always felt and thought of things as they would seem to Lucy [Donnelly], and now it is to Doris and I believe that whenever she thinks of her relations with us she instinctively begins to see how it would appear to Doris and what she would say. And Doris, I feel sure, always takes the stand that we are wrong and she alone is faithful and loyal. So just where any change is to come I cannot see.

I am so sorry about Miss Kelly, but it is such a universal fault,

269

almost universal, I mean.[c] The more unsure she feels of herself the more she will want to ignore your importance and the less she will want to admit any debt to you. It is small, but the great majority of the Harvard faculty suffer from that same smallness. I have so often felt myself pushed into obscurity and passed over that I have almost ceased to fuss over it. I wish I could say, never mind, in the end you will get your just dues. Unfortunately often one does not get one's just dues, they are grabbed and one cannot grab back. What I have made up my mind to is this. I might possibly get juster treatment if I fought for it, but that is very questionable. On the other hand there is no question that if I fought I should lose my pleasant relations with the men, and I would rather cling to the pleasantness than lose it and perhaps not get anything else.

A queer thing did happen to me just lately and I felt I must protest. An article came out in the Journal of Industrial Hygiene, of which I am one of the editors, on benzole poisoning and every bit of it except the description of experiment was stolen from the article I wrote for another journal last year.[d] I sent the two articles with the corresponding paragraphs marked, to Cecil Drinker who had accepted it and he writes back in great horror and mortification. He says the article was by the professor of pharmacology of Yale Medical School, one of the most prominent men in the country, and one of his students. He has written to the professor and will send me his answer. He said he never suspected such a thing could happen. I think it probably was the student, but I am eager to see what Professor Underhill says. The words are hardly changed at all, it is a ridiculously naive steal.

My plans are all in the air because of this cold Spring. I do go to Minneapolis this Sunday night, to do the General Electric factory there, and I must go to Buffalo for the 11th because the Academy of Medicine asked me to last Fall, and are paying my expenses. I expected then to go on east, do factories in New Jersey and in Bridgeport and spend a week end in Hadlyme planting seeds. But all I can hear about New England seems to show that the Spring is very very late and I am afraid I must put off that part of it, perhaps till May. I could stop in Cleveland on my way back and get in some four factories. Norah will not go west till she has the art school at least started.[e] She is getting backing from Mr. Aldis and Mr. Shaw and Mrs. John Carpenter but she has to round it all up and I am afraid that will mean several weeks yet.

I do hope you will have fun at Virginia Beach. It is so nice you are going, if it doesn't hurt your rib.[f]

a. Marguerite Harrison, a journalist, had been arrested in the Soviet Union in 1920 and later released through the efforts of the A.R.A.

b. Helen Castle Mead, who grew up in a missionary family in Hawaii, was acclaimed by Jane Addams as an "absolute democrat" in her freedom from racial prejudice.

c. Amy Kelly was Edith Hamilton's successor as headmistress of the Bryn Mawr School; she later wrote the popular *Eleanor of Aquitaine and the Four Kings*.

d. "The Influence of Benzol upon Certain Aspects of Metabolism," by Frank P. Underhill and Benedict R. Harris of Yale, appeared in the February 1923 issue of the *Journal of Industrial Hygiene*. It closely resembled AH's article "The Growing Menace of Benzene (Benzol) Poisoning in American Industry," which had been published in the March 4, 1922 issue of the *Journal of the American Medical Association*.

e. Norah Hamilton was in charge of the new Hull House Art School, a unit that reorganized and enlarged the previously disparate children's art activities.

f. The letter is unsigned.

In February 1924 Alice Hamilton was appointed to the Health Committee of the League of Nations. She was one of two American members (the other was Surgeon General Hugh S. Cumming) and the only woman. After the announcement of her appointment, she wrote Agnes that she had been startled to receive an invitation to a mass meeting in New York to express her faith in the League because she was not yet "a strong pro-Leaguer." Her experience during her two terms on the committee brought her around. She never quite got used to the nationalistic sentiments that dominated the proceedings of the League's Assembly, but the Health Committee, which transcended both politics and nationalism, made a deep impression on her. Its work quickly became one of her favorite lecture topics.[29]

Hamilton attended her first session of the Health Committee from September 29 through October 4, 1924. She had arrived early to get her bearings, and said of her first day: "I thought it was discipline enough in humility to be on the Harvard faculty but this is much worse. I am an utter ignoramus and can only sit and listen and blush if I am asked a question." Things went better subsequently, and she spoke on three occasions.[30]

271

71 To Julia C. Lathrop

League of Nations
Health Committee
Geneva
October 3, 1924

Dear Julia,

I thought of you several times this morning when there was a lively discussion in this committee on the proper answer to be made to the resolution which I enclose.[a] The French and the Dutch delegates were for sending in a statement to the effect that all questions with regard to child welfare are predominantly questions of hygiene. While they were talking Dr. Cumming, who has just come, whispered to me that the two Americans had better consult on this matter and stand together and I agreed. But then it became plain that the thing must be decided at this meeting, the last meeting we hold, and when Sir George Buchanan spoke against the French statement saying that child welfare in England meant many things with which hygiene has little or no connection, and that the laity would resent the doctors seeming to grab it all, I signaled to Dr. Cumming that I wanted to back him up and he nodded.[b] I don't know if he really liked it, but Sir George certainly did and the French statement was changed—the Frenchman is an irritating person who talks endlessly and loses his temper if he is opposed—and after the meeting I pretended that that was what Dr. Cumming had wanted and he quite fell in with it. Now I am going to lunch with Dame Rachel Crowdy and I hope she will tell me the inwardness of it all.[c]

It has been an extraordinarily interesting experience, much more so than I expected. I think it is partly because I came a week in advance and attended all the preliminary committee meetings, when they went over the reports of field studies done by experts and by members of the committee. I spent days listening to accounts of malaria in Italy, old Serbia, Albania, Dalmatia, Greece and Russia, of sleeping sickness and tuberculosis in equatorial Africa, and of cholera in Persia. The formal meetings of the Committee would have been rather dull, I think, if it had not been for this earlier orientation. Some time I must tell you about the curious results of the study of cancer mortality among women in Great Britain, in Holland and in Italy, and among their descendants in the United States. The men are ever so able and the work has gone smoothly and quickly, except for the Latin passion for eloquent speeches. The French and Portuguese are the worst sinners, the Spaniard comes next, the Italians are much better. But the best man on the whole committee is old Professor Nocht of Hamburg, head of the institute of tropical diseases there, respected more than any other man and never speaking unless he has something weighty to say. They have been very nice to me and I have struggled to behave as they expect, but I rather think that the mouselike rôle I chose was not right, that they expected much more bowing and smiling and shaking hands all round, regardless of the disturbance I might make. What Anglo-Saxons regard as proper consideration and modesty I suspect Europeans of thinking only a want of manners. Anyway they can't think me a chatterer. I spoke only twice, and not till the tenth and eleventh days.

Of course the exciting meetings of the Assembly have been going on, but I had to get them at second hand from Mabel Kittredge. I wonder what you think about the protocol.[d] We are eager to see the comments on it from home. McDonald of the Foreign Policy Association thinks it puts off our entry into the League for a much longer time and so does Sarah Wambaugh who is a League expert and devotee.[e] I think the English are pretty dubious about it, they say the French and Czechs got all they wanted and the British the task of seeing that they can keep it. Alfred Zimmern says it means that they have at last made up their minds that we are not coming in and so they can go ahead and do openly what they—the French and the succession states have always wanted to do but dared not, for fear of scaring us off—put teeth into Article 10 and Article 16.[f] But he is in favor of it because it is at least an agreement, an acceptance of the principle of openness and of arbitration and it is certainly better than war.

We leave Geneva Monday and, if there is no slip-up anywhere we go through Berlin, seeing the Lomonossoffs, to Warsaw for three days and then on one of the biweekly trains to Moscow.[g] I have letters to the Government people in Warsaw, to let me see factories there, and I am hoping to see them in Russia, but the men on the committee which went into Russia to study malaria say that one is taken in charge by Soviet officials and personally conducted to see what they choose to show, nothing else.

We have our return passage from Havre Nov. *22nd* which brings me back in time to begin lecturing Dec. 1*st*. Isn't anything going to bring you to Boston this winter?

Yours always
Alice H.

a. This was a recent resolution by the Assembly of the League of Nations about reorganizing the League's work on child welfare.

b. Sir George Buchanan was senior medical officer of the Ministry of Health in London.

c. Dame Rachel Crowdy, a nurse, headed the Social Section of the League of Nations; its most important work concerned the suppression of the opium trade and the traffic in women and children.

d. The Protocol of Geneva was a document designed to clarify and formalize League policy on such matters as arbitration, security, and disarmament. It had recently been approved by the Assembly, but was later rejected by Great Britain.

e. James G. McDonald was chairman of the board of the Foreign Policy Association. Sarah Wambaugh, a specialist in international affairs, later became a technical expert on plebiscites.

f. Alfred Zimmern was founder and director of the Geneva School of International Studies.

g. George V. Lomonossoff, an engineer and transportation expert, and Raissa Rosen Lomonossoff, both Russian by birth, were former Hull House residents.

ALICE HAMILTON spent a month in the Soviet Union in 1924 as a guest of the Department of Health, which had asked her to survey the nation's industrial hygiene program. Both her contemporary and retrospective accounts reveal a deep ambivalence about a regime toward which few Americans were neutral. Two months after she returned to the United States she wrote: "I get puzzled over Russia whenever I try to talk about it and always afterwards I feel that I have shown one side only, sometimes one and sometimes the other."[31]

On the positive side, she found the Soviet Union, with its admirably equipped tuberculosis clinics and its impressive Institute of Occupational Diseases, more advanced in industrial hygiene than any nation she had visited. She considered it fitting that a regime that paid homage to its workers should protect them from work-related diseases, though she sometimes thought "there was more industrial hygiene in Russia than industry." She also found much to admire in the Bolshevist ideal of equality, which brought to mind the Biblical injunction: "He hath put down the mighty from their seat and hath exalted those of low degree." She liked the apparent equality of female and male physicians, the absence of materialism, and the uniform appearance of the people, "plain and shabby, never . . . a flapper or a woman with made-up face, . . . no rich people and few absolutely poverty-stricken." But on the whole she found the application of the theory deplorable: "It is the way they do it, the evil that good may come and good never does come." She noted the rise of a new aristocracy of party workers that belied the equalitarian ideal, but she was most distressed by the fanaticism of party members, the concentration of power, and the suppression of free speech and press which she thought must stifle all independent intellectual life in the Soviet Union for years to come.[32]

Despite her ambivalence, she favored recognition of the Soviet Union and in public usually stressed the positive side of her experience. At a meeting of the Foreign Policy Association in March 1925, she declared her belief "that a country in which money is the measure of success needs to get something from a country that says we shall have neither rich nor poor." A newspaper account reported that she was interrupted by "pronounced hissing." She also received considerable applause, however, and later declared of the experience that "it was really not very bad."[33]

Some of Hamilton's initial responses to the trip are recorded in the long diary letter of November 10, 1924, to her family, of which only the concluding sections are printed here. The main body of the letter was typed in Moscow; the postscript, dated November 17, 1924, was handwritten in Warsaw.

72 To Alice Hamilton's family

Moscow
November 10th 1924

Dear family,
. . .

Russia is such a strange mixture. I can't generalize about it, because one thing contradicts another. On the one hand there is the cruelty, even now, to the counter-revolutionaries, but then it is not fair to dwell on that because both sides were cruel, the Bolsheviki more so only because they came out on top. Everyone here assumes that if the Whites had won they would have exterminated the Reds so far as they could catch them. And I have been told by Whites that in the matter of brutality, of killing prisoners, and hostages, of torturing and the rest, there was nothing to choose between the two, only that one

275

woman who was a Red Cross nurse under both sides, said that she blamed the Whites more, because they were the highest in the land and one expected more of them than of the lowest. But it is only turned upside down. The spying and arresting, and exiling and imprisoning are now done by the under dog to the upper instead of the other way round. It is all force and tyranny. There is less free speech and free press than under the Tsar, for absolutely nothing is allowed that is not communistic. Even so insignificant an organisation as the Fellowship of Reconciliation is not per[mitted?] and Tchertkoff's World Tomorrow is confiscated.[a]

I don't know how these people ever succeeded in building up their power, unless, as some say, it was just sheer greatness, genius for leadership, on Lenin's part, but I do feel that it is impossible that they should be downed. There is absolutely nobody to take their place, it would be chaos if they fell, but then there is nobody strong enough to make them fall. On the one hand it seems impossible that they should keep on, against all the laws of finance and economics—if that is true—and on the other it seems impossible that they should be upset. I suppose they will gradually change, as they have begun to, and as they come in contact with outsiders. The silliest thing of all the silly things the rest of us did was to blockade Russia and drive her into carrying things to such an extreme. If we had only kept up relations with her she could never have gone so far.

This letter will have to go out with me. It is not the sort of thing one sends through the mails.

Warsaw
November 17*th*

We are out of Russia and I feel—Mabel does too—as I did when we left Belgium under the German occupation, as if at last I could draw a long breath, at last I could speak without whispering and looking over my shoulder and could feel free from a hundred vague fears and dreads. I did not venture even to write it down while I was there, that Russia is a terrible, terrible country. No matter what the theory of government is it is carried out by means of privilege to a small group in which all power is concentrated and by terrorism over the rest of the population, by spies and agents provocateurs—arrests at midnight, farcical trials for crimes that cannot be proved or disproved, like "unconsci[ous?] espionage" "moral and intellectual disloyalty," imprisonment with no appeal, and exile worse than under the Tsar, for wives may not now go with their husbands. We did not feel this at first, we were meeting officials and people like Anna Louise Strong, Gertrude Haessler, Marie Yarros, who are enthusiastically loyal to the Soviet, but little by little the other world began to come up to the surface, through interpreters, people I met in the laboratories, teachers Mabel met, and the women who stole in to see us after seeing our names in the papers.[b] The stories are heartrending, there is nothing worse in Kennan's book, and it is not only isolated instances, it is a whole class.[c] One feels it is more than fear that prompts the persecution, it is the joy of the underdog, now on top, to repay the kicks he got when he was down. A government that can hold its own only by denying the right of free speech, even of secret balloting, and that makes opinion a crime must be wrong no matter what its theoretical aims are. Soviet Russia is like Spain under the Inquisition and I don't suppose it thinks itself any wiser and more infallible than Spain did. But if it succeeds I believe it will do so at the expense of its intellectual life. It is simply an aristocracy which holds in its own hands every bit of the power. The trades-unions, the cooperatives, the peasants' soviets, all have Communists in actual control, they are never self-governing. Under the Tsar, the press was censored, under the U.S.S.R. there is no printing at all except by the government. No clubs or societies are permitted, except the Communist ones, and if more than fifteen people come together for any purpose, the police may break up the meeting. And it is not the policy of the Party to get people to join it, on the contrary it is as exclusive as any aristocracy. They want to keep the number small and so a candidate must first show that he is well-born—of proletariat grandparents as well as parents—and then must pass an examination in Marxism and atheism and then submit to a long probationary period during which he is spied on continually. When he joins, it is like becoming a Jesuit, he renounces all individual ethics and takes the good of the party as the only test of right and wrong. He may never criticize the party and he must be ready to sacrifice his nearest and dearest at its behest. He also gives up luxury and frivolity and he may never grow rich, but he attains a position of power which means far more than wealth to most men. Of course, as with the Jesuits, there are sincere fanatics among them, people say perhaps a third, and there are the men who want the power and privilege and care nothing for ideals. It is a hideously wrong theory. Marie Yarros keeps on by losing herself in work, very necessary work, and not thinking of anything else, so do the Quakers. But to people

277

only looking on, as we were, it is a dreadful, a terrifying land and we are thankful beyond words to be back in shabby, kindly, safe, unenlightened Poland. I will mail this in Paris.

Yours always
Alice—

a. *World Tomorrow* was published by the Fellowship of Reconciliation. Earlier in this letter AH described a jail visit with a conscientious objector, which she had made at the request of Vladimir Tchertkoff, the "son of Tolstoy's oldest friend."

b. Anna Louise Strong, an American radical, settled in Moscow in the early 1920s and became an apologist for the Soviet revolution. Gertrude Haessler, a graduate of the University of Wisconsin, was earlier described by AH as "a charming American girl who is said to belong to the extreme left group of the Communist International but does not look it." Marie Yarros, a Russian revolutionary who had been a "hard labor prisoner" under the tsarist regime, was a former Hull House resident.

c. *Siberia and the Exile System* (1891) was a sensational account of the tsarist prison system by George Kennan (1845–1923), which had an enormous impact in the United States.

IN JANUARY 1925 nine women's organizations sponsored a Conference on the Cause and Cure of War, which met in Washington, D.C. Carrie Chapman Catt, former head of the National American Woman Suffrage Association, initiated the idea for the conference and presided over it. Alice Hamilton attended for a day and delivered a talk entitled "Medicine as a Factor in Internationalism." Her behind-the-scenes report to Jane Addams reveals some of the tensions among women's organizations in the 1920s. Catt not only had excluded peace groups from the conference, but had prevented the WILPF from distributing its literature. The organization's reputation for radicalism was undoubtedly the issue. At a time when the women's movement was under attack as unpatriotic, the WILPF seemed especially suspect to the more conservative. During its fourth international convention, held in Washington the previous May, the WILPF had been condemned by the Daughters of 1812 and other patriotic groups, and subsequently the DAR called for the group's resignation from the National Council of Women, a prominent umbrella organization. After several months' delay, on January 20, 1925, while the Conference on the Cause and Cure of War was in session, Jane Addams announced the withdrawal, declaring that the WILPF's program "is perhaps too far advanced for some of the organizations of the Council."[34]

73 To Jane Addams

227 Beacon Street
January 24, 1925

Dearest lady:

Well, I went to the Washington meeting and behaved like a perfect lady. Mrs. Hull and Mrs. Lynn Haines caught me outside the meeting room when I arrived for the morning session and implored me to say something in defense of the W.I.L. which, it seems Mrs. Catt had attacked by name a day or so before, though she made it also the occasion for a warm personal tribute to you which they said received prolonged applause.[a] Then they wanted me to talk about Frederick Libby and the decision of the Washington School Board that he is not fit to talk to school children, and then Jeannette Rankin came and said I must talk about perverted patriotism and all that, and then Esther Lovejoy's friend told me that Mrs. Catt had refused to let the Women's Hospitals literature—the Near-East and Russian relief work, you know— be shown, and somebody must talk like that.[b] I had a nice paper, neatly typewritten, on the permanent health committee of the League of Nations and I clung to it till I went in to the meeting and heard Raymond Fosdick steal all my thunder.[c] His League speech was practically all on the health committee. So I had to write a new speech, but J. Lathrop and G. Abbott told me not to say a word about the W.I.L. and that had been my own feeling. You can see from this that there were a good many people who were stirred up by the different things that had happened or failed to happen, and yet on the whole I was very favorably impressed. There was a large audience and it was much more pacifist than I am sure the leaders expected or perhaps realized. At least it was the radically pacifistic sentiments which got the most applause and the questions asked were inspired largely by the same spirit. Where the 1812 daughters were I do not know, but they never peeped. If only Mrs. Catt and the others had had the courage to trust to the peace feeling of these women they could have avoided all the mistakes they made, the biggest one of course being your absence. I was glad to hear the [sic] Mrs. Catt had at least asked you. But she isn't a big woman, none of the absolute suffragists are, they never had a chance to get broad. We all were relieved to have the W.I.L. do as you advised and resign from the Council before we were asked to do so. I suppose you heard that the suggestion was made that we stay in and undertake to get Borah's backing for a request to Congress for a ten thousand dollar appropriation.[d] Mrs. Hull told me that and said it was strange that the little W.I.L. could raise thirty thousand and this great body could not do as well.

I had a nice time with Grace Abbott and Julia and Mabel Kittredge and then I got in two hours in Baltimore and found Clara looking better and really serene in her mind and resting without being bored or impatient. She has overworked for years, her doctor says, and this is just a result of nerve tiredness and needs a rest cure. And I feel sure she is glad to take it and not to have to drive herself to teach or to do anything. She is luxuriating in books and does not seem to find the days long at all. We mean to keep up the cure all summer too.

Emily Balch has had her operation and is convalescing. I hope to see her next week.

Yours always
Alice—

a. Hannah Clothier Hull, a Quaker, succeeded Jane Addams as national chairman of the United States section of the WILPF. Dora B. Haines was a writer and magazine editor.

b. Frederick J. Libby was executive secretary of the National Council for the Prevention of War. Jeannette Rankin was WILPF field secretary at this time, but resigned later in the year. Esther Pohl Lovejoy, a physician, was director of the American Women's Hospitals Service, an organization that provided medical and relief services, principally but not exclusively in Europe.

c. Raymond Blaine Fosdick, a lawyer who had represented the War Department in several capacities, had been under-secretary-general of the League of Nations in 1919–1920.

d. William E. Borah was chairman of the Senate Foreign Relations Committee.

BY THE MID-1920s progressive reformers were under heavy attack by such groups as the American Legion and the DAR for their alleged radicalism and lack of patriotism. At the same time they received several severe political setbacks. One of the most disturbing was the defeat of their efforts to outlaw child labor at the federal level. Following a 1922 Supreme Court declaration that a federal child labor law was unconstitutional, Florence Kelley led reformers in calling for a constitutional amendment. Congress passed the amendment in June 1924, but it subsequently died dismally in the states. The struggle was particularly bitter in Massachusetts, where manufacturers and the Roman Catholic Church, headed by William Cardinal O'Connell, led the successful fight against ratification.

Alice Hamilton became involved—in a secondary but important way—in one of the most dramatic cases in the annals of industrial toxicology: the discovery of radium poisoning in factories making luminous watch and clock dials. In the early 1920s a number of women who had worked for the U.S. Radium Corporation factory in Orange, New Jersey, began to suffer from painful and disfiguring jaw conditions that resisted diagnosis and treatment; between 1922 and 1924 several of them died. At the end of January 1925 Hamilton, who chaired the committee on industrial poisons of the National Consumers' League, wrote the secretary of the New Jersey League that she hoped to investigate radium poisoning under the auspices of the Bureau of Labor Statistics. But she soon concluded that it would be "unnecessary" for her to do so because she believed that a 1924 study of the U.S. Radium Corporation by her Harvard colleagues Cecil and Katherine Drinker and William B. Castle would soon be published. On the basis of careful clinical and laboratory studies, the Harvard group had concluded that "excessive exposure to radiation" was the cause of the necrosis of the jaw and had ruled out other potential causes. Although the president of the corporation refused to accept the findings or to institute the physicians' recommendations, Cecil Drinker initially declined to publish the study on the grounds that the company had commissioned and paid for it. When he learned that another investigator was about to publish the results of his own study, Drinker reconsidered. He finally changed his mind after Hamilton informed Katherine Drinker in early April that the U.S. Radium Corporation had issued a reassuring statement to the New Jersey Department of Labor that falsified the Drinkers' findings. The Drinker report appeared in the August 1925 issue of the *Journal of Industrial Hygiene*.[35]

74 To Clara Landsberg

227 Beacon Street
Sunday morning
[February 8, 1925]

Dear Clara:

It is after breakfast Sunday morning, after a long night and a late rising, so that it is already after ten o'clock and Amory has gone off for the car. It is a Spring-like day, though snow and ice lie deep everywhere. It will be lovely off in the country. We have not been out for three weeks, one day because Amory had a case and could not leave and anyway it was zero weather with a gale of wind, and last Sunday because Roland Hayes, the Negro tenor, was singing in the afternoon and the Frankfurters invited me to have lunch with them and hear him.[a] It was worth sacrificing one of my Ponkapoag Sundays, and few things are.[b] He is wonderful. His voice has the quality which one tries to describe by calling it creamy or velvety, but it is never thick or dull so those words do not fit. It is only that it never for a moment becomes thin or harsh, it is always caressing or deeply mournful and always sweet. Somehow his nature seems to come through. You feel that he is never a virtuoso doing something showy, or even enjoying his own achievements, he is always reverent and simple and loving. He is coal black and his white teeth are uncanny. When he sings his Negro spirituals everyone is moved to tears. Symphony Hall is always crowded for him, not one seat to be had for weeks before.

This week has been a chore. First I made up my mind to revamp my index, making it a guide for people looking up things in connection with a trade or with a symptom or with a poison. It will be much better but it takes lots of work. Edith sent me on a little more, but she could not do many pages because it is really too hard for anyone but the one who wrote it. I feel conscience-stricken when I see how much work she had to put in on what she did complete. Then there were so many hearings at the State House which I had to attend. Of course it is far easier than in Chicago when a hearing means Springfield and two nights and a day at one's own expense, but it also makes it easier for people to insist on one's coming. First there was a bill the Consumers' League and the School of Public Health had helped get up, together with the association of dry cleaners, for clean work and prevention of fumes. That meant a committee meeting and a hearing. Then there was one on the physical examination of children who apply for their working papers and finally a two-day hearing in the biggest room in the State house, on the Child-Labor amendment. Katy stayed through both days of that but I could not so I got most of it from her. Our side had the morning of the first day and Helen Rotch made the opening speech, admirably, with all her facts in orderly array and with clearness and logic.[c] Then I came, taking up only the question of regulating the labor of the young in the dangerous trades, as they do in all industrial countries and giving statistics of excessive sickness rates among the boys and girls under eighteen in munitions work and in benzol. Then came a good employer and then I had to go. Nobody applauded and we were as decorous and dignified as possible. But the hall was packed with Catholics sent by the Cardinal and with ladies of the Civic Federation. In the afternoon they let loose and every outburst against Bolshevism, Mrs. Kelley, Moscow, Grace Abbott, nationalization of children and all the other Tommy-rot was applauded. I dropped in for fifteen minutes at the end of the afternoon to hear an elderly lawyer tell the committee in a shaking voice how much he loved his children and how hideous it would be for the State to rob him of them. The arguments were so preposterous that it was hard to believe anyone could take them seriously. Of course Hull-House came in. A Harvard professor's wife, who is surely mad, said that Hull-House was a branch of the Soviet government and its paid propagandists were Mrs. Kelley, G. Abbott, J. Lathrop, Owen Lovejoy, while Miss Addams was the most dangerous woman in America.[d] Of course the amendment is lost. Indiana rejected it last week, making, I think, sixteen states. But I do believe that in five years these arguments will seem as ridiculous as the war propaganda seems now.

I have thought a lot about that factory in New Jersey and the radium poisoning. If ever I was a lazy idiot it was last Summer when that wonderful new problem was handed to me and I refused it. It is probably coming out all right as far as the protection of the workers is concerned, better than if I had done it but I can't help wishing that I had had the excitement and glory of discovering a perfectly new occupational poison. Dr. Drinker's attitude has changed since he finds that somebody else will publish if he does not. I had a letter yesterday from Mr. [Ethelbert] Stewart, of the Bureau of Labor Statistics asking me to work it up and also some cases of phossy jaw in making fireworks and I wrote him that it would not be necessary to do anything about the former, but that I would help out on the phossy jaw cases and on radium cases in other cities, if he would have a non-medical investigator go over the field first and have cases ready for me to see. So I will have a few jobs to do when I go west and probably a few for the G.E. also, at least I hope so.

Have you any idea when Edith and Doris plan to go north? I wrote Mr. Ellsworth as Doris asked me to and he has written back telling me of an agency here in Boston where I can get all sorts of advice as to how Edith ought to start out on such a thing as a lecture tour of schools and colleges.[e] But now I don't know what to do about it. I don't want to interview them if the idea is only Doris' and I am afraid

283

that if I push it I may simply set Edith against it. I would send the letter to Doris but that dreadful habit they have of showing each other everything makes that impossible. And I can't help fearing that whatever I do Doris will use against me. What do you think?

I am sending you on a piece of Chinese embroidery which I saw in Yamanaka's window yesterday and bought on the spur of the moment, because I liked it so and it was only $8.50, marked way down, for they are having a sale. I thought we would try it on the mantelpiece of the blue room and if it does not go there, perhaps it might on the mantelpiece in the small sitting room of the Deep River house, or maybe I will give it to Mary Straus for a Hausgeschenk.[f] Anyway you can enjoy its color for a day or so and then put it down at the bottom of the chest in the dark room where I put the other Hadlyme things.

Amory has come back and is joyously summoning us to start, so goodbye. I am waiting to hear about your doctors' consultation.

Yours always
Alice—

a. Marion Denman Frankfurter, Felix Frankfurter's wife, was a Smith College graduate. AH variously spelled her name Marion and Marian, and the inconsistencies have been retained.

b. AH used to join the Codmans at their country retreat in Ponkapoag.

c. Helen Ludington Rotch, Katharine Ludington's sister, was a founder and leading member of the Massachusetts League of Women Voters.

d. Owen Lovejoy was general secretary of the National Child Labor Committee.

e. William W. Ellsworth, a friend of Mabel Kittredge's, was a writer and former president of The Century Co., a publishing concern.

f. Mary Howe Straus, a Wellesley College graduate, had done volunteer work at Hull House.

DURING THE WINTER of 1924–1925 Alice Hamilton was in demand as a speaker and dinner guest because of her trip to the Soviet Union. The "Russian agricultural scheme" referred to in her letter of February 26 was the Russian Reconstruction Farms, an organization that contracted with the Soviet government to teach peasants how to use modern farm machinery and to demonstrate the value of large-scale, collective farming. Donald Stevens, who had been a conscientious objector during the war, and Harold Ware, a member of the Communist Party who had already led an expedition of American farm youths on a similar demonstration project, were attempting to raise funds for the project.[36]

75 To Margaret Hamilton

227 Beacon Street
February 26 [1925]

Dear Margaret,

I am going to use up this old paper on you. It is a bitterly cold night, a gale that is cruel, and I have a fire which is purring pleasantly, and I am very quiet and comfortable. Amory is at his brother's and Katy has gone to her mother's for a few days, to let her sister get away. So I had a solitary dinner and propped the Nation up against my glass. Ruth Porter had just gone.[a] She has come back from a cruise around the West Indies and she says Bermuda is lovely.

I wish you would keep the Chinese silk thing if you really like it. I have some Russian towels for Mary Straus and probably I shall want to give her a really useful piece of furniture as a Hausgeschenk. I only suggested her in case you did not think it would fit 847 or the blue room in Hadlyme. The linen for Mary must wait till next Christmas. I cannot afford another present this year.

I am so shocked to hear of Grace Meigs' death. She had at least three little children and this is poor Dr. Crowder's second bereavement.[b] He lost his first wife in childbirth. I wonder what the cause was. I am planning to go first to Chicago because I am tired and want to lie low for a couple of weeks. Then I will make one trip the last of March, and another planned so as to let me reach Hadlyme for planting. The storage battery plants are in Indianapolis, Cleveland, Buffalo and Philadelphia, the General Electric are in Fort Wayne, Youngstown, Erie, Schenectady and Pittsfield. I will fit them in together. You see I shall have till the last of June in Hull-House and that gives me lots of time with Mary and J.A.

Kitty Ludington was here and I dined with her at Helen [Rotch]'s and at Felix' and she came to tea here. She is to be at Lyme all summer, she says.

I have been up in the air about this Russian agricultural scheme. First I held off from it, largely because Edith Hilles and Louise wrote me to, till I could hear full details from them and from G. R. Taylor.[c]

He has not written but they have and I enclose their letters. Meantime Donald Stevens came and I liked him very much, felt him devoted and sincere. I gave him all the advice I could and so did Katy and I asked Dr. Cabot to see him, but that was all. He simply went ahead collecting perfectly grand names for his committee, including Dr. Cabot and Dean Pound and perfectly unimpeachable social lights, for everyone likes and trusts him.[d] Finally Elizabeth Balch, Emily's sister, decided to join the colony and is going to fit herself as a cheesemaker at an agricultural school. Meantime these letters came and another one today, urging me to have nothing to do with it. Mabel Kittredge, who has gone on the New York committee, sent me the contract, and it is just as Edith says, "heads, I win, tails you lose." All the same I think it is not a hopelessly reckless gamble. The Soviet government has every motive to help the scheme instead of hindering or spoiling it and I believe all this elaborate safeguarding is partly their silly, doctrinaire way of doing things, partly their fear that the Americans may make money. Anyway I find that Donald Stevens has, quite involuntarily, put me in a place where I cannot be neutral, I must either be for the plan or against it. If I refused to go on the committee all these Boston people who know me would take it as a condemnation. So I have given them my name and when people ask me, I tell them that I know Donald Stevens is honest and so is Harold Ware but that they are taking great risks with the Soviet Government.

I had a queer evening at Mrs. Fiske Warren's.[e] She invited me to dinner, I did not know why, but now I think it was to convert me to anti-Bolshevism. It was *very* grand, fourteen of us, and there were candlesticks on the table eighteen inches high and in them candles two feet high, and all the center of the table was a mirror. Only, for that one should have young guests, not red-faced old gentlemen and an over-dressed, diamonded old woman, who were the only reflections I could see. I do not know who the others were, but there was a titled Englishwoman, and some elderly ladies Katy said were grand, and my neighbors were a rather precise Englishman, studying bio-chemistry at Harvard, and a doctor interested in psychical research, pretty boresome both of them. But when the ladies went upstairs the Bolshevist part began. It was led by Mrs. Osgood, Mrs. Warren's mother and also the mother of Mrs. Erskine Childers, and I am sure the latter is like her, for she is certainly a powerful and ruthless old thing.[f] She is very old but her voice rang out like a man's as she hurled denunciations against Russia. All the others backed her up, though they seemed feeble beside her. I sat perfectly silent, of course, and nobody asked me a question except just as the men were coming up when Mrs. Warren said hastily "How often have you been in Russia?" but I did not have to answer for the room filled at once and I was seized upon by a young Russian emigré, a charming youth with whom I had an interesting time up to the last minute. Mrs. Osgood kept a baleful eye on me all the time. I wore my grey Liberty and looked nice but old.

I am enclosing Edith Hilles' letter and one from Louise and one from Mabel. I am so glad J.A. has gone on the committee.

It is good to know that you will be leaving so soon after Edith arrives.[g]

a. Ruth Furness Porter, a graduate of Bryn Mawr College, had done volunteer work at Hull House.

b. Grace Lynde Meigs was a physician who had worked for the Children's Bureau. Her husband, Thomas Reid Crowder, was a specialist in railway sanitation.

c. Edith Hilles and Louise Lewis had accompanied AH to the Soviet Union. Edith Hilles (later Dewees), a Vassar graduate who had studied agriculture, worked in both a professional and voluntary capacity for a variety of social, feminist, and pacifist causes. Louise Lewis, a niece of Esther Kelly Bradford, lived at the Lighthouse, where she specialized in work for the unemployed. Graham Romeyn Taylor, a social worker who had investigated the famine in the Soviet Union in the early 1920s, feared that the government might confiscate the farms if they proved successful.

d. Richard C. Cabot, a prominent Boston physician and founder of medical social service, taught in Harvard's Department of Social Ethics. Roscoe Pound was dean of the Harvard Law School.

e. Gretchen Osgood Warren, a celebrated Boston hostess, wrote poetry and was interested in the arts.

f. Margaret Cushing Permain Osgood was Mrs. Warren's mother. Her daughter Mary Osgood married Erskine Childers, an Englishman who had supported Irish Home Rule and had fought with the Irish Republican Army. Childers was court-martialed and executed as a traitor by the Irish Free State in 1922.

g. The letter is unsigned.

Hotel Ohio
Youngstown, Ohio
Friday evening
[April 17, 1925]

Dear Aunt Phoebe,

It is the end of a long day and in half an hour I shall take my train for Chicago. The lamp works' inspection took only the morning and then the doctor drove me out to the Country club for lunch. It was a good lunch and he is a nice young man who had an interesting experience in France during the War. I came back and wrote my reports and letters and then it was half past four and I thought I would take a walk, but Youngstown is hopelessly ugly and there are no traffic regulations and I hated to be killed in such a squalid spot so as a last resort I went into a movie theater and saw a gorgeous Wild West play, "The thundering herd." It is thrilling beyond description and the scenery is real, not made up in Hollywood. The best thing about it is the way the Indian is treated. Of course in the fight your sympathies are with the whites but it is made perfectly clear that the Indian fought because the whites slaughtered the buffaloes and so took away his only meat. You are shown the prairies covered with the bones and the Indian children crying for food and the last picture is of a lonely Indian gazing out over his lost hunting grounds. Movies are certainly a boon to the travelling man. I will go to one whenever I am stranded.

The Capitol Limited on the B & O is a delightful train, with half-partitions between the sections, and a dressing room twice as big as usual. It was full of D.A.R.s and one of them, a sweet old lady, talked to me and made me feel quite ashamed of my prejudice against them. I have always thought of it as an organization of ancestral snobbishness and militarism but my old lady talked only about the help it gives to schools for the Southern whites. That is the thing that keeps her in it, she says. Someone across the aisle said it was a patriotic society and she accepted the term as meaning warlike and apologized for it. "We don't really think so much about war" she said "but about the sort of country our ancestors tried to found." So I shall henceforth feel more charitably toward the D.A.R.

This is really a bread-and-butter and thank-you letter, for a lovely two days and nights' visit. I do so love to come to Fort Wayne and I only wish I could do it oftener.

With much love—
Alice—

ALICE HAMILTON spent six weeks in Europe in the fall of 1925. She attended a meeting in Amsterdam of the Congress on Occupational Accidents and Diseases, the first since the war, following which she visited storage-battery factories in England and Germany and attended a session of the League of Nations Health Committee in Geneva. It was one of the few long trips she ever made alone.

77 To Jessie Hamilton

Zealand
Tuesday
[September 1, 1925]

Dear Jessie:

I have just finished eating an apple out of your basket while the rest of the passengers had bouillon and crackers, and it seems an appropriate time to write and thank you for that same basket and for your niceness in writing me a steamer letter. I have lived on your fruit and Edith Hilles', indeed on Sunday, which was a fiendish day I had nothing but fruit and a little soup. I can never be seasick enough not to want apples and grapes. These will last more than the voyage and I will put the solid remnant, oranges and apples, in my holdall.

It is the stupidest and longest trip I ever took. They say we shall not reach Antwerp till Sunday noon. The boat is comfortable, very, and the food really interesting, though I partake of it with caution. But I am frightfully lonely and not tired enough just to enjoy the rest. The passengers are largely Belgian, and out of some fifty of us no less than twenty-six are "religious," twenty-four priests and novices and two sisters. The priests sit all at one long table in the dining room and they pace the deck reading their breviaries. They are Dominicans, bound chiefly for South Africa, the Belgian Congo. They look simple and obstinate, they remind me of the Amish men as I used to see them in Fort Wayne. I sit between a big elderly nun and a middle-aged

American woman. The nun is Breton, with a rough-hewn face that makes one think of an Irish potato, and coarse red hands. She has been twenty-four years in the Province of Quebec and speaks no English at all. I tried to talk French to her but she snubbed me and she talks only to her companion sister. It is funny sitting next a nun, like sitting next a horse with blinkers, you cannot catch their eye unless they turn way round. The American woman is kind but so colorless that I have hard work recognizing her on the deck. She lives in Washington and is going all alone to study the art of Italy which she says transcends anything that books can say about it and she has read many books on it. I am afraid she has not an interesting mind and I wish I could slip somebody else in her place and let them have two glorious Autumn months in Italy. Think of the ripe grapes and the yellowing leaves. All the same, what I really think of is my own grape arbor up on the hill and the yellowing of the marsh grass, and I am so homesick that I ache. Last night suddenly when I woke up I decided I was an idiot to come to this Amsterdam congress, when I might have had three more weeks there.

There are four nice-looking women on board and two men I should like to talk to, but they have their own people with them and I don't like to begin. The woman who sits opposite me at the table is a Belgian, with beautiful French and excellent English. We walk the deck sometimes together but I think she must feel that it is ill bred to ever talk personal matters with a stranger so we stick to abstractions always. Talk at the table languishes and for the first time in my life I am glad to have a band. It is a little one but good and I watch the second violinist, a slim, fair boy with the most sensitive hands. I don't know why he is so different from an American boy but I cannot imagine our ever producing anything like him. He is at once pathetic and impudent, babyish and old, appealing and cold. I should feel quite helpless with him.

It is a sunny, warm day, still in the Gulf stream, there are no white caps and I am as comfortable as it is given me to be on shipboard, but oh I do hate it.

Love to Aunt Phoebe and all of you and thanks, thanks for the fruit and the letter.

Yours always
Alice

In NOVEMBER 1925 Alice Hamilton wrote Robert Morss Lovett, professor of English literature at the University of Chicago and a resident of Hull House, in his capacity as a director of the Garland Fund. The letter presents a succinct statement of her ideas for developing an impartial consulting bureau that could provide expert advice to both business and labor and help to offset the peculiar difficulties surrounding the control of industrial diseases in the United States. It also reflects her interest, during the early 1920s, in raising funds for a thorough investigation of benzene (then often called benzol), a new and highly volatile solvent used extensively in the rubber and canning industries and in quick-drying paints. Hamilton was one of the first Americans to call attention to benzene, which she had first encountered during the war and which she considered one of the most dangerous of all substances—it took "only a small quantity to poison or even to kill."[37]

78 To Robert Morss Lovett

Harvard University
School of Public Health
November 5, 1925

My dear Mr. Lovett:

The following is an outline of the plan for expenditure of funds in the interest of workers' health which I discussed with you in New York on October 30th.

There are two needs at present in American industry which are not being fully met. The first is scientific research in the field of the newer industrial poisons. No organization has undertaken this, although from time to time various centers of scientific research, such as Harvard, Yale, Syracuse University, and Columbia, have carried on studies of this kind, but the field is not by any means covered. One poisonous substance which came into prominence during the war and whose use has increased rapidly ever since is coal-tar benzol. An association of employers, the National Safety Council, has undertaken an inquiry into the actual conditions under which benzol is used in industry, but certain features of benzol poisoning which can only be determined by animal experimentation are still unsolved and it will be impossible to fully control benzol poisoning in industry until these dark spots have been cleared up. I should therefore suggest as one line of work the provision of $25,000 for three years to be devoted to the thorough study of benzol and I would suggest that Harvard Medical School is best equipped to carry this out.

The second need is to secure expert information concerning the dangerous dusts and vapors encountered in industry. In such industrial countries as England and Germany for instance, the central factory inspection department is responsible for this field. If any accident arises from a new compound used in a trade, the experts from this department visit the plant, examine the workmen, and if necessary assign to the pharmacological department of some university the experimental work necessary to determine the nature of the poison. If it is a question of a well-known substance which is causing unusual trouble, the experts decide what means of protection are necessary to prevent such trouble.

In the United States we have no such experts. The Federal Public Health Service has no authority to make investigation unless invited to do so by the employers. They have not even a body of experts who can pass on a theoretical question. No state department of labor or department of health is equipped to deal with such problems. They cannot even undertake to analyze a mixture used in industry in order to determine its nature. It is therefore highly important that some consultative body with access to appropriate laboratory facilities be constituted in this country. It should have its seat in New York City and it should be widely advertised throughout the country, not only to trade unionists but also to employers of labor, for it is of the utmost importance that this body of experts be known as absolutely impartial, capable of dealing with these questions in the spirit of pure scientific inquiry.

This will not be easy to arrange, and I cannot at this moment describe in detail a practical plan, but I do urge you to accept the plan in principle.

The Workers' Health Bureau of New York City to which the American fund has already contributed has already established far-reaching relations with organized labor. I have followed its work from the beginning and can assure you that it has a record of achievement which is very admirable. Their technique is as follows: They approach a trade union and establish relations with it, then they select an individual in a given plant, instruct him in the sanitary standards which apply to his industry and give him a questionnaire which he is to fill out concerning conditions in his plant. This is then checked up by one of the two directors of the Workers' Health Bureau; they then compare it with the sanitary code covering the locality and if there are violations report them to the proper authority. If, however, the sanitary code, as is often true, is inadequate for the protection of the workers, they work out in consultation with employers and employed a new detailed sanitary code and endeavor to have it accepted as a working agreement. They have succeeded in introducing a large number of these, especially in connection with the painters' trade.

They also undertake to study the health hazards of an industry; for instance, they have begun work with the stone cutters. We know already the injurious character of different kinds of natural stone so that in connection with granite, marble, sandstone and limestone cutters no examination of the stone is needed. For artificial stone cutters it is necessary to have analyses made of the stone dust. This the Workers' Health Bureau is doing. After this they must select typical groups of cutters numbering at least one hundred, and have thorough physical examination made including an X-ray of the chest and an expert interpretation of the X-ray. This is the only way to determine the actual injury done by dust.

The unions affiliated with the Workers' Health Bureau pay for its services but it is impossible to ask them to pay for this kind of research. An examination such as the above costs between $5 and $10 apiece and the stone cutters could not possibly pay for it. A similar piece of work ought to be done for the garage workers who are exposed to poisonous exhaust gases, but these men earn from $28 to $35 a week and they could not pay for such a study.

The Workers' Health Bureau needs a clinician on full time, an extension of its laboratory facilities, and a fund to pay for X-rays. Perhaps $15,000 a year would cover what it needs to extend the work as it should be extended, for it has already gone far out into the field and is capable of much further development.

A very close relation should be established between the Workers' Health Bureau and the consulting committee of experts, but the details of this will have to be worked out very carefully.

The first and the third enterprises would require $40,000 a year. The expense of the second I cannot possibly compute until the plan is worked out in detail. I hope to be able to take that up with you in the near future.

Sincerely yours,
Alice Hamilton

SHORTLY AFTER writing Lovett, Alice Hamilton read the report of a committee that had been established in 1923 by the National Safety Council (an organization of business and insurance underwriters) to investigate the danger of benzene. Headed by C.-E. A. Winslow of Yale, the committee conducted an innovative study that confirmed the highly toxic nature of benzene and recommended substitution of less harmful substances wherever possible. Hamilton wrote Lovett again on December 11, suggesting that a survey of occupational diseases in Maryland be substituted for the proposed benzene study. Despite a personal appeal at the Garland Fund's January meeting, the board turned down her request.[38]

Early in 1926 Alice Hamilton told Cecil Drinker, assistant dean of the School of Public Health, that she wanted a sabbatical the following year. Reporting the episode to Clara Landsberg, she observed: "He paid me the first and only compliment I have had from my faculty colleagues since I came here. He said 'Well of course you're entitled to it. But whom shall we get to give your lectures? There's nobody in this country who can. We shall have to get [Edgar L.] Collis from England.' I was quite breathless with surprise, but he said it as if it were a matter-of-course."[39]

Hamilton planned to leave for Europe in November, but delayed her departure because Jane Addams had not yet fully recovered from an attack of angina—her first—suffered the previous July. Hamilton often rushed to Addams' side during her frequent and serious illnesses, even venturing to Japan in 1923, when Addams underwent surgery there for breast cancer. On one occasion Hamilton had written: "What I really want is a lot of consultants on the job all the time and every possible thing tried, as if she were King George, for she is really much more important than he." As a physician, Hamilton played a special role in interceding with Addams' doctors and in persuading Addams, who tended to neglect her health, to follow their orders. Hamilton finally left early in January; she and Norah joined Margaret and Clara Landsberg, with whom they traveled through Italy and Egypt. They returned to the United States in May.[40]

79 To Margaret Hamilton

12 West Walton Place
Chicago
October 5 [1926]

Dearest Margaret:

I don't need to explain to you how hard it is to make a decision now, you understand how things are, with me and with people here. On the one side there is the perfectly natural eagerness of everybody to give J.A. whatever she wants, whatever could contribute at all to her ease and happiness, there is her own pleasure in having me about and her conviction that I never cared to go abroad anyway, there is Mary's relief in being able to share a heavy responsibility, and Norah's quite genuine feeling that nothing could compare with having J.A. really want me and being able to do things for her. On the other hand there is my own experience which tells me that I am not at all essential, that a heavy sacrifice made for J.A. is usually a mistake because it proves not to mean very much to her, there is Dr. Herrick's statement to me yesterday that there is no real reason for me to change my plans, and my own strong conviction that when Mrs. Lovett and Mary are back again, my being here will seem only a luxury, not a necessity. What I should like to do is this. Stay here till the 19th telling everyone that my plans are all vague and up in the air, go to Richmond and Bryn Mawr from the 20th to 23rd, then come back here for the last week in October. If in that time there has been no fresh attack of angina, I can go without seeming brutal and indifferent. Mrs. Lovett will have slipped into place, they will have seen how easily things went without me and the situation will have cleared itself up. Now this means definitely deciding now to sail no earlier than Nov. 16th but I believe you two would rather risk losing that two weeks for the sake of having me sail with you. Of course there is a risk, I might feel I almost must stay till January but just now I don't see that happening. It is not much of a risk. After all, I am not necessary, Mary alone is that, I am only—after she and Mrs. Lovett are back— an added pleasure. And it is not as if I were giving up J.A. and Hull-House. I shall always spend four months here, as I always have. Dr. Herrick told me that no man could possibly say when another angina might come, if ever, nor can we stave it off. All we can do is to save her heart, look out for fatigue and breathlessness. And she is being very reasonable and is much fuller of spirit and interest than I have often seen her of late years. Dr. Herrick wants her to go away in February, to some place not too high.

Could Clara plan her visit to Rochester about Nov. 1*st* and you go to New York for a couple of days then, let me pick you up, do our last shopping and go up to Hadlyme for packing and closing the house. Or, we could go straight to Hadlyme and then take a few days in New York at the end. I know I am looking at it reasonably and that in a little while it will seem perfectly natural to everyone.

Of course for this period I am simply a visitor and an assistant and I do not see any chance of doing writing of my own at present. J.A. needs somebody and I am the only one, till Mrs. Lovett comes back, about the 14*th*. Mary will be gone till the 16*th* probably. There is a bitter fight on over Lyman.[b] Charlotte is contesting the court order and is bringing in as witnesses on her side a lot of that fast English set in Sheridan. A mother usually has the advantage, in such a case. Lyman is not to appear in court nor is he to be asked to decide, but Mary says the whole thing is very bad for him . . .[c]

Yours always
Alice.

a. James B. Herrick was a prominent Chicago heart specialist. Ida Mott-Smith Lovett, a Radcliffe graduate, was a Hull House resident.

b. Lyman Silsbee Smith, son of Francis Drexel Smith, was Mary Rozet Smith's nephew.

c. Two short concluding paragraphs, which are partly obscured by an ink blot, have been omitted.

In September 1926, while preparing an article comparing storage-battery manufacturing in Great Britain, Germany, and the United States, Hamilton had asked the New York State Department of Labor to assist in improving conditions in the U.S. Light and Heat Corporation of Niagara Falls, New York, a company she considered the sole exception to her generalization that progress had been made in the industry since she first studied it in 1914. "Usually it is possible to bring about reforms through a personal appeal to the owners or managers," she wrote, "but here I despair of such an appeal, for they have shown too plainly through all these years that they have no standards at all. Only the authority of the State can do it and I hope it will be exerted, perhaps has already been exerted." The director of the state's Bureau of Industrial Hygiene replied that the company was "now within the requirements of the State Code, although undoubtedly not in accord with a number of recommendations which the Bureau of Industrial Hygiene as well as the Bureau of Inspection are prepared to make." In December Hamilton sent a copy of her report to the U.S. Light and Heat Corporation. D. H. Kelly, vice-president of the company, defended its record, claiming that, except for "the one bad spot" in the oxide mixing room, he believed that ideal conditions prevailed. He also requested "a frank letter" on Hamilton's findings.[41]

80 To D. H. Kelly

Hull-House
Chicago
June 2, 1927

Dear Mr. Kelly:

On my return from Europe I found that the Editor of the Journal of Industrial Hygiene had postponed the publication of my report on The Storage Battery Industry until I could take up with you the objections made by you to certain parts of the paper. This was quite right, for I should have been very unwilling to have the paper go to publication without first satisfying so far as possible any criticisms from the men to whom the report was submitted.

The assurance of the Editor that no one would be able to recognize your plant from the descriptions seems to me quite true. I was careful to omit anything which seemed to me likely to cause an outsider to understand which plant I was speaking of, and I am surprised that you think that the identity would be recognized. That difficulty, however, I think can be easily dealt with.

You speak of my visit as having been brief and my guide unfamiliar with the working of the plant but you see I was already quite familiar with it and the visit was in the nature of a confirmation of my former impressions. I studied it first thoroughly in 1914; then in 1918 a joint committee appointed by the Ordnance Department, the U.S. Public Health Service, the Women's Bureau and the Bureau of Labor Statistics, which last I represented, spent many days in Niagara Falls making a study of those dangerous trades in which it was desired to substitute women for men, the U.S. Light and Heat especially. My next visit was in 1921 or 1922 when I was anxious to see if the introduction of a pasting machine had done away with the evils in the pasting room. You will see, therefore, that I did not need an expert guide on this last visit.

Nor was it necessary for me to spend more than ten minutes with the plant physician, for a short conference with him gave me all that I needed to know about the scope and methods of the medical department.

Of course, my object in making this study was to bring about improvement in the hygiene of the storage battery industry, especially in the more backward plants. You ask me to tell you frankly what is wrong with yours and I shall be glad to do so. If in this way, by a personal appeal to you, I can persuade you to make the necessary reforms in the Niagara Falls plant, I shall be accomplishing all I could possibly hope to do by the publication of the article, and I should be willing to omit all the passages you object to.

In the first place I ask for nothing elaborate or expensive. In the course of my visits to the plant I have often been shown blueprints of extensive engineering projects which I was told would be installed as soon as the company could afford it, but I have always said, and still say, that what is needed in the plant is cleanliness, and without that no apparatus will be of much value. I notice that you have put hoods over the kettles in the Planté department,[a] which were probably expensive and not absolutely necessary, but you allow dangerous dust and scrap to lie all about this room, although it would not be costly to clean it and keep it clean. The casting room has had money expended in it but here too simple cleanliness would have gone farther. The dangers in these rooms are in any case very slight compared to the rooms in which oxides are handled.

These are the essentials which should be insisted on and which will do away with the worst dangers in the plant. They will not bring it up to the standard of the four best plants.

Clean up, especially, the mixing-blending room and the pasting room. The floors can be made clean and can be kept so. No oxides need accumulate on the floor, in corners and behind apparatus. The floors should be kept damp all the time, preferably by a thick layer of wet sawdust. All surfaces must be washed twice in twenty-four hours. No old, dusty oxide barrels should stand about. The pasting machine should be kept clean, no dry oxide being allowed to accumulate. It is not necessary to have an elaborate apparatus for dumping, blending, and mixing the oxides. I should like to refer you to pages

13, 24 and 25 of my report for descriptions of simple and effective methods of doing this work.

The notes made on my last visit show that in sawing, and lug trimming and cleaning, dust lies thick on benches and floor; dusty litharge barrels stand about. One lug filer has no protection; others are covered and there may be an exhaust inside but the cover protects only the little file or brush and whenever a finished plate is dropped on the pile of plates on the bench a puff of dust rises. These benches should all be slotted, provided with down suction and kept clean. Cleanliness and dampness of floors and surfaces are necessary here as in the departments just described.

My notes also speak of much unnecessary hand work in connection with dry plates which are placed on open trucks, then on shelves, and then back on trucks for pickling or forming, and wherever this is done there is an accumulation of dust. Of course it would be far better to substitute mechanical conveyors or at least to have boxes which can be handled so as to do away with the necessity of lifting plates out and in. In any case the receptacles and trucks should be kept clean. The drying room and the storage room should also be clean and damp.

My notes show several unnecessary sources of dust. For instance, oxide barrels are tipped over in a hallway and pounded to get rid of the oxide, the powder falls over the floor and some of it is gathered up and put into open cans; the rest lies on the floor. Then these dusty barrels are used all over the plant as receptacles for scrap. In the department where plates are repaired for cheap batteries, brown oxide dust lies over everything and many packing boxes stand around, open and full of dry, dusty plates. Even in the forming room, where only wet plates are handled, thick dust has accumulated and the same is true of the benches of the lead burners. These last should be slotted and provided with a down draft.

With regard to the personal care of the men, about a thousand of whom are employed by you, three things seem to me essential: Clean working clothes (oxide-caked overalls were the rule when I was there), abundant provision for bodily cleanliness and thorough medical supervision, including instruction of the men and enforcement of a certain amount of shop discipline. I was told that you have an enormous labor turnover. This must be expensive—perhaps more expensive than it would be to institute a proper system of care for the employees. At present your rate of lead poisoning is very high. The Department of

Labor tells me that 109 cases were reported in 1925 and these are all cases of actual lead poisoning entitled to compensation. You have no way of knowing how many cases of mild poisoning or of lead absorption occurred in the plant. In the model plants the records cover these two last classes of cases and men who show signs of lead absorption without any symptoms of illness are shifted to non-lead work until their condition becomes normal.

The form of medical service which is provided in the best plants and which should be provided in yours is described under the heading, "The Medical Department," in my report. My notes of the interview held with your physician read as follows:

"About one thousand are employed. The turnover is enormous and chiefly floating, unskilled labor. Dr. Hensel agrees that the fact that men stay only a short time is all that prevents severe lead poisoning. He does not examine the men for employment, nor make regular examinations of the employed. It would disorganize the work and anyway what should he look for? If he followed the principle that a man shows absorption of lead and should be shifted to non-lead work when he has a lead line, pallor, loss of weight and stippled red cells, then he would have to shift all the lead men and the plant would close down. A man who comes to the dispensary complaining that he is sick gets treatment. About 80 compensation cases have developed so far this year. Sometimes men are picked up by the police suffering from what they say is lead colic but nobody knows if it is. That happened in two cases recently. There is no instruction of the men in the dangers of their work for fear of putting ideas into their heads."

I have written you in great detail and it may be that you are uninterested in all this and consider that I am going out of my field in doing so, but I hope that you will not take this letter in that spirit but will see in it only a very great desire on my part to have a situation remedied which has troubled me for thirteen years—ever since 1914—which I have tried in vain through interviews with the management and with the State Labor Department to have remedied, and which I know from my acquaintance with other plants is unnecessary and can be remedied, provided the men in charge really wish to do so.

I will postpone any action with regard to my article until I hear from you.

Yours very truly,[b]

300

a. The Planté plate was a special form of storage battery plate of lead and lead oxide.

b. This is an unsigned carbon copy.

RESPONDING TO Hamilton's letter, Kelly wrote on June 13 that "your suggestions will be followed to the letter." He went on to claim that New York state's liberal compensation laws and the company's policy of strict reporting "account to a great extent for the number of lead cases we have." Countering her charges in a four-page letter, he stated that the company doctor gave a different version of their conversation, claimed that the company properly instructed the men about the dangers of the work, cited the chief physician's opinion that more than 50 percent of the cases receiving compensation were not in fact due to lead poisoning, and insisted that Hamilton would have had a more accurate picture if she had been accompanied by an "expert guide" who could have explained the changes in operating procedures made by the company prior to her most recent visit. He also maintained that conditions in the Prest-O-Lite plant in Indianapolis, which had recently been purchased by U.S. Light and Heat, were twice as hazardous as those in his company.[42]

Considering Kelly's invitation to visit the plant and his promise to make the changes she recommended as signs of good faith, Hamilton wrote the *Journal of Industrial Hygiene* (which had delayed publication of her article until she could respond to the company's objections) that she had "agreed to omit any reference to that plant from my article," which she thought "may now be considered quite harmless." She also deleted a sentence that had offended the New York State Factory Inspection Department: "Now that they have seen it and know my opinion there is no need to publish it." The article was published in August 1927.[43]

81 To D. H. Kelly

Harvard School of Public Health
[Boston]
June 16, 1927

My dear Mr. Kelly:

Thank you for your letter of June 13th which I have read with much interest but I must admit with a growing bewilderment. It was hard to believe as I read it that I had actually visited the Niagara Falls plant and the Prest-O-Lite plant in view of the impression I have of them and the totally contrary impression held by you.

May I take up your answer somewhat in detail?

In the first place, the varying compensation laws have nothing to do with the records of lead poisoning in a well-managed plant. In the Prest-O-Lite and in the Westinghouse, the plants to which I referred as admirable in their medical service, the physicians examine all employees and report cases of lead poisoning from the medical point of view, and as their standards are very strict their records contain cases which would not be compensable even in New York State; yet in spite of this their rate is very low compared to yours.

I would rather not comment on Dr. Hensel's report of our interview. It is, after all, not a matter of importance, all that is important is that there should be periodical examinations of the men in lead work, careful instruction and prompt shifting to non-lead work when symptoms of lead absorption are discovered.

I am sorry that I did not have an expert guide but I still feel that an expert in manufacturing processes was not necessary. For the past seventeen years I have been going through manufacturing establishments of all descriptions, here and abroad, looking for lead fumes and lead dust, and even if I had known nothing about the processes in storage battery manufacture I should still have detected lead oxide dust, the one thing I was looking for.

The two things that bewilder me most in your letter are, first, your statement that at any time in the past seven years the Company could have afforded to install any improvements necessary for the protection of the employees. Certainly the impressions gained during my own visits and from correspondence with the State Labor Department have been that the Company was in a condition which made any suggestion for unusual expenditures unreasonable and impossible of compliance. I have always been urged to be patient and give them time in view of their urgent difficulties.

The second is, the comparison you make between the Prest-O-Lite and the Niagara Falls plants. In the face of it I am compelled to reiterate my statement that the two stand at the opposite ends of the scale, if one is considering factory hygiene. Indeed, I do speak with a knowledge based on years of study of the hygiene of the lead trades, and of this trade in particular. In 1918 the Prest-O-Lite was in the same class with the U.S. Light and Heat but in 1926 I found it completely revolutionized. Since you say you have recently purchased it I can only beg you to leave it in its present working order. Indeed, if a suggestion is not impertinent, I would say that you could do nothing better for the Niagara Falls Plant than to have it visited by Mr. R. I. Hoffman, Mr. J. N. McLaughlin or Mr. C. A. Olds together with Dr. William Boyle, and to follow their advice with regard to changes in the plant and in the medical service.

I shall be very glad to accept your invitation to go over the plant with an expert guide perhaps toward the end of September, when I expect to be in the vicinity of Niagara Falls. I will, of course, write you in advance to see what time will be convenient for you.

After consultation with the editor of the Journal of Industrial Hygiene I have decided to omit all mention of the U.S. Light and Heat plant from my article, not even putting the name in the list, since my only object was to bring about reforms there and that can be better attained now in other ways. It is only fair to myself, however, to point out to you that if my description of conditions in your plant were wholly unjust and inaccurate, you would not have recognized it as applying to that plant nor would you have feared that others would recognize it.

Very sincerely yours,[a]

a. This is an unsigned carbon copy.

FOR ALICE HAMILTON as for many Americans, the Sacco-Vanzetti case was a symbol of the intolerant postwar mentality and of the defects in "the administration of what we still continue to call justice." She did not personally know Nicola Sacco and Bartolomeo Vanzetti, political anarchists who had been convicted in 1921 of the murders of a paymaster and his guard in South Braintree, Massachusetts. But several of her friends did, including Katy Codman, who visited the men in prison and befriended their wives, and Elizabeth Glendower Evans, who did much to publicize the case and served as a rallying point for the cause. It was Evans who interested Felix Frankfurter in the trial. His article in the March 1927 *Atlantic Monthly* fueled the liberal cause, forcing Governor Alvan T. Fuller to appoint a committee to investigate the fairness of the trial. In early August the committee, on which Harvard president A. Lawrence Lowell played a leading role, reported that in its view the trial had been substantially fair. Fuller subsequently refused clemency, and there ensued a series of futile efforts to halt the execution. Alice Hamilton, who had written Frankfurter that she would go anywhere or do anything that might help, was one of many prominent individuals who signed a petition asking the governor to commute the sentence or stay the execution. On August 22 she joined five prominent men in a last-ditch effort to induce the governor to commute the sentence.[44]

82 To Edith Hamilton

Hadlyme
August 25, 1927

Dear Edith,

Margaret and I reached home Tuesday noon to find three letters waiting for us, yours, M. Shearman's and M. Kittredge's, all alike, all full of nothing but the S.V. case. I should have liked nothing better than to sit down then and write you all about our day and night in Boston, but so much came in to make it impossible till today. Mrs. Wilson was still here and as we had deserted her for two nights, I felt she must be tended up to a little and then it was time to go over to the Seldens' for a big neighborhood tea which they were giving just for us. And when we got home and had had supper a telephone [call] came from Mabel Kittredge that she was over at the Beebes' and the ferry not running. It was pouring rain, as it is most of the time this summer, and Simpson's boat had six inches of water and there was no time to bail it because the steamer was due, but we did get across and brought her back though we had to wait midstream for the steamer to make the landing and go on. Mabel had been with the Parsons at Kennebunk and had got so outdone by their "law-and-order" attitude that she felt she must stop over here and let herself go. She said Margaret Deland was her only comfort up there.[a]

Well, on Sunday Paul Kellogg telephoned me from Boston asking me to meet our small committee in Boston Monday morning and make a final attempt to reach the Governor. So we went that night, stayed at my club, and the next morning went to the Bellevue. There we found Paul Kellogg and a small group, Waldo Cook, editor of the Springfield Republican, John F. Moors, a very prominent Bostonian, one of the Harvard Corporation and a close friend of Lowell's, Dr. [Edward Staples] Drown, a lovely old Episcopal clergyman, John

Elliott of the Ethical Society in New York and a few others. There was a good deal of doubt as to the Governor seeing us but by and by word came that he would and the five men and I went over, passing through a group of forty-seven policemen to get in. As we waited in the ante-chamber a large delegation of weedy, pale, rather stupid-looking youths went in. I thought they were garment-makers with a plea for S.V. but they proved to be American Legionnaires going to pledge support to the Governor in this crisis. Then we were summoned, to meet a vigorous man, in the forties I should say, blond, thick-set, with fiery blue eyes. He was in a state of obvious excitement, holding himself in with an effort. And as he talked to us his florid face would crimson and his neck swell and he had every appearance of hot anger. Then two or three times he would pull himself together, become again the suave automobile salesman, say he was treating us as intimate friends, laying all his cards on the table, and so on. But it was a terrible interview. We had expected him to give us five minutes and had agreed that Mr. Cook and Mr. Moors should speak briefly and no one else, but he kept us forty minutes. We tried to keep him to the discussion of commutation of sentence, which was all we had asked for, but it was impossible. He would talk of the sinister propaganda which was behind all this agitation and which had as its inspirer Felix Frankfurter and a hundred and fifty thousand dollars. We, of course, were well meaning dupes who did not know who was pulling the strings. If we could know the things he had heard in that room we would be as convinced as he. When Mr. Cook said that if his decision was made on secret evidence he could not expect the world to accept it, he said that there was no other way of getting at the truth. He has read practically nothing of what has been written in protest. When we spoke of the wide-spread doubt as to the men's guilt, he retorted "Haven't I seen that dirty rag the Defense Committee gets out, with abusive headlines all across the page." We explained that we meant protests from eminent lawyers, professors, clergymen—at which he smiled as if that was palpably absurd. His secretary—a most baleful looking creature—sat there and we all knew that he shows Fuller what he chooses and Fuller reads nothing else. Mr. Moors told me that this secretary was the worst element in the whole case. Fuller has a fanatical belief in capital punishment. We simply could not make him discuss commutation. With him it was pardon or execution. Twice he turned, with a real fury, on the two Massachusetts men and demanded to know if they were among those who are coming before the Legislature next Fall to seek for the abolition of capital punishment. We beat against that wall in vain. We could make not the slightest impression. Nor could anyone. He was in a state where it was psychologically impossible for him to let a new idea into his head. As we left him I found that all the others felt as I did, that any appeal, from anyone, would fall on deaf ears. He listened to them all day long. [William G.] Thompson, the lawyer for the defense, came to us at the City Club after midnight and told us he had reviewed the case from half past ten to midnight and the Governor had listened. I think it has been that throughout, he has spent three months seeing people and he has never once changed the mind with which he began, but he was determined to make the world see how conscientiously thorough he had been. Mrs. Glendower Evans saw him early in May and told Katy Codman then that the men were as good as dead.

When we left the State House a thin, blonde man, shaking with excitement clutched Mr. Cook and asked what had happened. He was John King of the Boston Transcript who has not been able to write his side in the paper, but they have given him all the time and opportunity he wanted to run down the evidence and work for a new trial, and he is now almost fanatical in his belief in the condemned men. We separated then and I went to the Club to find Margaret with poor dear old Mrs. Evans, who has been on the case since the first day and enlisted Felix Frankfurter in it and all the rest of us. The Frankfurters were there, and Katy Codman and Margaret Shurtleff who has for the last two months taken turns with Katy in visiting the men twice a week and in seeing after Rose Sacco.[b] Katy has been a dear about it and of course the feeling of those three women was not abstract as ours was, but personal, for they have all grown fond of both men. We were together till the end, some others joining from time to time and bringing news from State House and Defense Committee. They were planning how to secure the bodies—of two men living then—it seemed too ghastly. We had meant to go back on an afternoon train but Katy begged me to stay and keep Mrs. Evans from trying to get into the jail for goodby or even going to the Defense Committee, for she was so wrought up no one knew what might happen after the pleasant excitement of signing checks to bail out the pickets was over. So we stayed. We all went up on the roof garden of the Club, looking down on the city, the Charles River basin and Cambridge. Mrs. Evans sank into quiet and we sat there listening to the quarter hours striking from the church below. I shall never forget that strange death watch. At half past twelve we knew both men were dead, and we went down. On the steps of the Club were Paul Kellogg and a group of men, with the two exhausted lawyers, Thompson and [Herbert] Ehrmann. They all greeted Mrs. Evans as if she were the bereaved mother, and then we separated and early the next morning Peg and I came home. It has taken her two full days to get rested but she is glad she went.

There cannot be a short summary of this case, any more than there could be of the Dreyfus case. Lowell's report is that and it can be picked to pieces in every sentence. Such cases are masses of detail, they can't be summarized briefly. The best approach to that is F.F.'s article in the Atlantic. I am going to see if it is in pamphlet form and send it to you.

This is a terribly long letter but I grew so interested in telling it all to you that I could not stop. I am enclosing nice letters from Norah and Quint to take the taste out.

Yours always
AH.

We are taking poor old Simpson to the hospital in Middletown this afternoon. He thinks it will help him but I fear nothing can.

I am sick of "conscientiousness." Torquemada was conscientious, so was Cotton Mather. A conscience may be a terrible thing in a man who has no humility, who can never say "I might be mistaken."

a. Margaret Deland, a Boston novelist and short story writer, had reached the peak of her popularity before World War I.

b. Margaret Homer Shurtleff, a designer of early American furniture, belonged to the Boston reform circle of Katy Codman. (The family name was later changed to Shurcliff.)

HENRY A. CHRISTIAN, who held the Hersey chair of the theory and practice of physic at Harvard and was also physician-in-chief of the Peter Bent Brigham Hospital, wrote Hamilton on August 23, asking "what sort of an investigation" she had made before signing the petition requesting Governor Fuller "to change his action" in the Sacco-Vanzetti case. Noting that he had always respected the "careful investigation and mature thought" she brought to her work as a scientist, he added: "My confidence by [sic] you has been shaken by what seems to me an action, that could not be based on careful, thorough investigation of a nature required by a scientifically trained person. I do not wish to misjudge you and so I have propounded the above question."[45]

83 To Henry A. Christian

Hadlyme, Conn.
August 30, 1927

My dear Dr. Christian:

You ask me what sort of investigation I had made of the Sacco-Vanzetti case before petitioning the Governor to "change his action." First, may I call your attention to the fact that the petition our committee made to Governor Fuller was for commutation of sentence to imprisonment for life, an act which involved no change on his part save that of choosing a merciful but equally legal course, rather than an extreme one.

My acquaintance with the case has lasted for seven years, because I became interested in it from the beginning. I have read the record of the trial and followed with close attention all the appeals made to Judge [Webster] Thayer during these years. So familiar am I with the evidence that when I read the Lowell report I could identify the items there discussed and recall most of those not discussed.

But had I been as unfamiliar with the case as most of the reading public, had I been dependent solely on the pages of a conservative daily, such as the New York Times, I should still have joined in the thousands of petitions which went to the Governor for commutation or stay of sentence "till the sober doubts of the world as to the guilt of these men and the fairness of our treatment of them had been allayed." This was the plea we made to Governor Fuller and we secured five hundred signatures to it in forty-eight hours, from lawyers, professors, clergymen, doctors, teachers, editors.

It was our sense of the wide-spread protest against the execution of men whom multitudes of Europeans as well as Americans believe, rightly or wrongly, to have been condemned, not because they were murderers but because they were Italians and anarchists, that led us to hope the Governor would listen to our plea. When such Conservatives as Mussolini, the Pope, President Masaryk, Louis Loucheur, together with practically the whole European press including Tory, Monarchist and Fascist papers, express this belief, it cannot be ignored.[a] You may say that it is the result of clever propaganda from American radicals, but so long as that propaganda is spread by some of our most eminent men, lawyers of international reputation, you cannot dismiss it as negligible.

The dead men may have been guilty, Judge Thayer may have been fair, the Governor and his committee doubtless were conscientious in their decision and the courts legally correct in their rulings. But something vastly more important than that is the question of world opinion. These men have become a symbol, as was John Brown, as was Francisco Ferrer.[b] Conscientious people who have no sympathy for anarchism believe that the evidence on which they were convicted was insufficient and in such matters what becomes fixed in the minds of a large body of people is more important than the fact itself. If you read the papers you must have seen that Massachusetts has shocked the civilized world, not only by a trial of questionable impartiality, not only by an antiquated legal system which refuses the appeal provided in more modern systems, but most of all by insistence on the death penalty. Dreyfus was condemned by both the military and the civil courts, he was a despised Jew and his vindication meant discrediting not only the courts but the whole military caste, yet France faced it and came out with honor.

I hope that this letter has at least convinced you that I did not go into the matter ignorant of the facts nor of their implications.

Yours sincerely[c]

a. Thomas G. Masaryk was first president of Czechoslovakia. Louis Loucheur was French Minister of Labor and Social Planning.
b. The execution in 1909 of Francisco Ferrer Guardia, a Spanish republican and anticlerical educator, for taking part in an uprising in Barcelona had provoked demonstrations in western Europe.
c. This is an unsigned carbon copy.

IN RESPONSE to her letter, Christian acknowledged that Hamilton's action "would seem to me to have been entirely consistent with the results of your familiarity with the case and with your views of the situation. Any criticism of you that may have been implied in my letter I very gladly retract."[46]

Hamilton's interest in the case persisted. She served on the Sacco-Vanzetti Memorial Committee, which organized a commemorative meeting for the second anniversary of the execution, August 23, 1929.

309

Her speech at that event, which was held in New York because no large public hall in Boston would book it, was a compelling statement of her empathy for the poor and oppressed and her commitment to working for a more just society. She strongly condemned what she considered the backwardness of American legal methods in adhering to ancient practices, a theme to which she returned, in more humorous but still biting fashion, in an article entitled "What About the Lawyers?" published in *Harper's* in October 1931.[47]

VIII

ELDER STATESWOMAN
1928–1935

D URING ALICE HAMILTON'S Harvard years, recognition came more readily from others than from her home institution—she was still an assistant professor when she retired in 1935. In addition to receiving honorary degrees from Mount Holyoke (1926) and Smith (1927), she was honored at a special dinner by the American Public Health Association in 1931. On that occasion, C.-E. A. Winslow, professor of public health at Yale and a leading figure in the field, took note of the unusual dual nature of her achievement—as "crusader and scientist"—and credited her earlier investigations with providing the basis for subsequent progress in industrial medicine. Two years later a less public but no less apt tribute came from the president of a chemical company writing to the technical director of a firm that sold solvents: "I don't know what your Company is feeling as of today about the work of Dr. Alice Hamilton on benzol poisoning. I know that back in the old days some of your boys used to think that she was a plain nuisance and just picking on you for luck. But I have a hunch that as you have learned more about the subject, men like your good self have grown to realize the debt that society owes her for her crusade. I am pretty sure that she has saved the lives of a great many girls in can-making plants and I would hate to think that you didn't agree with me."[1]

In the late twenties and early thirties, Alice Hamilton continued her consulting work and her efforts to advance public and professional knowledge of industrial diseases. Her second book, *Industrial Toxicology* (1934), a volume in Harper's Medical Monograph Series intended for general practitioners, was if anything even more favorably received than her first. Commentators on this "condensed work of a lifetime" made much of Hamilton's eminence and unique position in the field. One reviewer assumed that she was too well known to need any introduction, even to lay people, "at least those who read." In the first editorial it ever devoted to a book, the *Detroit Medical News* proclaimed: "The name Hamilton parallels the building up of the awareness and the study of poisoning in industry."[2]

Sometimes, as in the case of the New Jersey watch dial workers, Hamilton played an important if largely behind-the-scenes role. That case, which had been taken up by Katherine G. T. Wiley of the Consumers' League of New Jersey, continued to make headlines after the release of the Drinker report. In 1926 New Jersey made "radium necrosis" a compensable disease, and the following year Raymond H. Berry, a Newark lawyer, brought suit on behalf of five severely ill women formerly employed by the U.S. Radium Corporation. He eventually won an out-of-court settlement. Berry later wrote Hamilton: "As I need not tell you, you, more than any other person, have gotten positive results in the radium situation and have blasted an opening for me each time I came apparently to an impasse." There remained the larger problem of controlling radium, a substance whose properties were little understood and which often took years to work its harm. In the spring and summer of 1928, Alice Hamilton and Florence Kelley initiated through the National Consumers' League a campaign for a national conference, on the model of the one held three years earlier to discuss tetra-ethyl lead. Twenty-three prominent individuals, mostly physicians, followed Hamilton's lead in endorsing a letter requesting Surgeon General H. S. Cumming to call such a meeting. Through Walter Lippmann, the campaign received excellent coverage in the *New York World*.[3]

On December 20, 1928, nearly fifty individuals, among them representatives of industry, state labor departments, the Public Health Service, and the National Consumers' League, as well as physicians who worked in a variety of capacities, met in Washington to discuss the radium situation. Hamilton, the first speaker at the conference, hoped that several bureaus of the federal government would investigate "whether in using radium in industry we are using something that cannot be rendered free from harm to the worker" or whether proper precautions would prevent injury. Representatives of the radium industry maintained that conditions would be safe if workers ceased pointing radium-dipped brushes in their mouths, but medical experts disagreed, emphasizing the danger of radioactivity to all exposed workers. The hardest-hitting speech came from Ethelbert Stewart, Commissioner of Labor Statistics, who attacked the use of radium—which he did not think could ever be made safe—to manufacture luminous dials, "purely a fad" item in his view. The conference passed two resolutions. The first, proposed by the manufacturers, suggested that the Surgeon General appoint two committees, one to prepare a statement about which individuals were most susceptible to radium poisoning, the other to codify methods of protection. The manufacturers indicated that they were not proposing an investigation of the health of radium workers, but the second resolution, put forward by Stewart, called for a follow-up health study of all individuals who had ever worked with luminous paint in New York City. The plan of the investigations was to be left to the Surgeon General, assuming that he could find the necessary funds.[4]

Shortly after the conference, Katherine Wiley wrote Florence Kelley that she could not understand why Alice Hamilton had been satisfied with the resolutions; another physician had considered them "quite useless," while a representative of the New Jersey Department of Labor thought the meeting had been a "white wash" of the U.S. Radium Corporation. Wiley also complained about Frederick B. Flinn of the Columbia University College of Physicians and Surgeons, a consultant for the corporation who had minimized the dangers of radium; though not a physician, he had apparently examined the corporation's employees.[5]

Responding to Wiley's criticisms, Hamilton's letter to Kelley demonstrates her willingness to accept the good intentions of all parties as well as her faith in publicity as the best weapon against industrial diseases. A short time later she wrote Kelley that, in view of the newspaper coverage: "I should think that any case of obscure illness in a dial painter would be suspected at once of being due to radium, and that it is highly improbable that cases would remain undiscovered." In later years Hamilton still maintained that the conferences on radium and tetra-ethyl lead had effected genuine improvements in the conditions of manufacture of these substances. Given the national preference for voluntary reform, she considered the conference method a constructive and typically American solution to problems that had not yet been brought under public control. To the end of her life, she

313

retained her belief in public exposure as the key to changing attitudes, behavior, and eventually laws. It was an optimistic belief, the quintessential faith of an entire generation of reformers.[6]

In fact, as Hamilton had predicted, the Public Health Service conducted a serious investigation, which included detailed study of the work environments of seven factories, "electroscopic determinations" of radioactivity in workers, laboratory tests, and physical and dental examinations of individuals currently and formerly working with radium as well as of a control group. The study found evidence of radioactivity in the bodies of even those workers who had not used brushes and recommended strict standards of cleanliness as the best prophylactic. The three-part report appeared in the *Journal of Industrial Hygiene* in 1933, by which time the problem of unemployment had eclipsed any nascent interest in industrial disease.

84 To Florence Kelley

Harvard University
School of Public Health
[ca.] January 9, 1929

Dear Mrs. Kelley:

I am sorry to be so long in answering your letter but I only got back last Friday and I have not yet been able to get any time from my secretary so I have had to do my own typing and deal with the accumulation of letters as best I can.

Probably I slipped up on the resolutions in Washington, at least I cannot be at all sure I did not, because I got in after they had been read. It was a great mistake going off to lunch with Mary Anderson and I would never have done it had I realized the distance and the time it would take.[a] But though I was bewildered by having the resolutions read without being able to think them over I was not willing to stand out against them, partly because it seemed to be taking too much on myself, partly because I was told they were only temporary, only to meet the situation as it confronts us now and subject to change as our knowledge increases. I quite sympathize with Miss Wiley's wish that Ethelbert Stewart's suggestions could have been followed, but that seems to me at present impossible. So long as well managed watch works can show that they have had no cases of radium poisoning it will not be possible to prohibit its use, especially as such

314

a prohibition would have to be accompanied by a prohibition of all importations. It may be that unrecognized cases will be discovered even in the best plants, as phossy jaw was discovered in connection with the best match works, and in that case we should be in a strong position. But white phosphorus would never have been forbidden if it had not been shown that poisoning simply could not be prevented, that it appeared under the strictest hygienic control.

I think that if lawyers were present they must have come with the representatives of the watch companies, or possibly as representatives. Mr. Berry might have been invited, but I never thought of suggesting him. What could he have added to the discussion? The other lawyers did not speak and I do not think they belonged there at all, but then I cannot see how their presence did any harm.

As for Dr. Flinn, I give him up. He is impossible to deal with. It is for the doctors in New York to stop his practicing medicine, as he is apparently doing.

The conference struck me as very successful and the manufacturers far meeker and readier to be good than the tetra-ethyl lead men were. It is, after all, the weapon of publicity which we hold up our sleeves that impresses them and makes them ready to do what we tell them to. If the Surgeon-General appoints as well-chosen committees as he did for the study of tetra-ethyl lead you need not be afraid that the matter will not be well and thoroughly handled. Dr. Thompson will be the one who will do this selecting and he is very well oriented in the situation.[b]

I left J.A. Wednesday night, much better but not yet allowed to sit up out of bed, for she still has a rise of temperature every afternoon. But she is back to her old self, interested in things, with fair appetite and much better sleep. It never went on to pneumonia and her heart stood the strain of her constant coughing amazingly well. I think she will go to Tucson with Mary [Rozet Smith] and Mrs. Bowen the last of January.

Yours always
Alice Hamilton

a. Mary Anderson headed the Women's Bureau, for which AH acted as an occasional consultant.

b. Lewis R. Thompson was senior surgeon with the U.S. Public Health Service.

ALICE HAMILTON became an advocate of birth control about a decade after she went to live at Hull House, at a time when even the distribution of information about contraception was illegal. In 1909 she supervised a pioneering and widely cited Hull House study of 1600 working-class families, both immigrant and native-born, which demonstrated that among all ethnic groups "child mortality increases proportionately as the number of children per family increases." She subsequently studied European progress in contraception for the Children's Bureau in 1915 and became a council member of the Citizens' Committee on Family Limitation when it was organized in Chicago in 1917. Participating also at the national level, she served on the council of the American Birth Control League and spoke on poverty and birth control at the Sixth International Neo-Malthusian and Birth Control Conference organized by Margaret Sanger and held in New York in 1925. Alice Hamilton explicitly rejected the eugenic argument for imposing birth control on the working class (a position advanced by conservative members of the movement), arguing instead that large family size was detrimental to the health and well-being of women of the poorer classes and especially to their children. She also pointed out that birth control was already widely practiced in tenements—in the form of abortion. In Boston, Hamilton was a member of the State Council of the Birth Control League of Massachusetts, as was her Cambridge friend the writer Cornelia James Cannon.[7]

85 To Cornelia James Cannon

Harvard University
School of Public Health
January 21, 1932

Dear Mrs. Cannon:

First, I should love to come Sunday night at seven o'clock and surely will.

Secondly, I am of course willing to sign an appeal for money for a birth control clinic but when it comes to accepting membership on a committee of physicians who are supposed to pass on methods used in the clinic and to select and approve physicians to practice for the committee there I cannot see myself doing it conscientiously. In the first place I am quite ignorant of the practical side of birth control. Not being a practicing physician I have never even read it up, let alone seen it in use. I do not know physicians in Boston, outside the laboratories of the Medical School, and could not intelligently pass on their qualifications. And, finally, I expect to go to Germany this Spring, so I could not even grace the meetings of the committee with my presence. You see, I am not trying to get out of it, but I never consent to "lend my name" when it is only an empty gesture and not backed up with any real work.

You will be amused at this. I spoke at a hearing before the Governor's Council yesterday and as I was going out a queer-looking man stopped me and said "When you were speaking a woman told me you were for birth control. Is that true?" The Cardinal's Catholic ladies were there in full force. I said "yes it is quite true." "That's enough" he said. "And then you have the face to call yourself a friend of the poor and of the little children." I was so bewildered I could only say, "But that is just why," and he simply grunted in disgust and left me. I really should like to know just what twist he has in his mind. Do you suppose it is a vision of little souls longing to be born and me trying to keep them out, like an anti-immigration law?

Well, we can talk about the committee on Sunday, but I am sure you can see that signing the appeal and serving in an advisory capacity are quite different and while I am ready to do the one I feel quite incompetent when it comes to the other.

Very sincerely yours
Alice Hamilton

IN HER EARLY SIXTIES Alice Hamilton experienced the first of the inevitable personal losses that befall the long-lived. Katherine Hamilton died on February 5, 1932, the first of the Hamilton cousins. In little more than two months both Florence Kelley and Julia Lathrop were also dead.

86 To Agnes Hamilton

227 Beacon St.
Friday
[February 5, 1932]

Dearest Agnes:

I cannot grasp that it is really true and I wonder if even you can, though you may have known for days that it must come. I never thought of it as really possible and now I feel as if all our lives had been changed of a sudden and the old life could never be again. We are all so closely bound together and now one of us is gone. All these years together in Deep River and Hadlyme have made Katherine so close and intimate a part of life. I can't think of it without her. I have loved her very, very much—more every year. I can't think of you and Jessie, of your mother, of the boys, what the loss is to you. Only, with you and Jessie I know that the unseen world is more real than the seen. But that does not really stop the sense of loss and the acute loneliness which everything around you makes you feel. I can't tell you how I ache for you, all I can say is how much I love you.

Alice

87 To Nicholas Kelley

Harvard University
School of Public Health
February 27, 1932

My dear Ko:

I saw your mother on the Monday before she died. I had had no idea that the end was so near, we had heard good reports about her, but when I saw her it was clear that it must come soon. She knew me and was very sweet, drifting away often but suddenly coming back with flashes of her old self. Once, as I sat there waiting as she dozed, she turned to me and said, in her ringing voice, "Oh but Alice, I have so many compensations." And then for a few minutes she spoke of you and your family, and how you had all been there at Christmas time. I think you must have great comfort in the thought of what you gave her all your life, and especially during these last months.

It is to me as if the first break had been made in our old Hull-House

318

group and one of the few people who started me on my road in life is gone. I was so immature when first I went to Hull-House and your mother, J.A. and Julia Lathrop were the ones who began my education and continued to help in it for many years. Her gallant spirit is what I shall remember most, her courage and honesty of mind and generosity in all her many battles. My mind goes back to those days when she used to come home from the Crerar Library at half past ten and I would stay up and make cocoa for her, as an excuse to sit with her and hear her talk. Always I shall be grateful to the fate that gave her to me as a friend.

My love and sympathy to you, dear Ko.

Alice Hamilton

ALICE HAMILTON and Jane Addams had received from the Oberlaender Trust, an endowed branch of the Carl Schurz Memorial Foundation, traveling fellowships designed to permit distinguished Americans "to become better acquainted with the achievements of the German people in their respective fields," and thus to promote better relations between the two countries. In March Hamilton canceled her trip. Addams had already withdrawn because of illness, and Hamilton wished to complete her second book, which was proceeding more slowly than she had hoped. She decided to spend the spring in Hadlyme, and looked forward, although somewhat apprehensively, to the experiment of living alone there. "Everyone prophesies that I shall not be able to do it," she observed, "but I really will."[8]

88 To Agnes Hamilton

Hadlyme
April 27 [1932]

Dear Agnes:

It is quarter past seven (by this silly new time which all the neighborhood has adopted) but I had my tea rather late and do not feel like supper so I will write to you until I get hungry. Then I will have a shirred egg and toast, some watercress I got from the Cheneys' brook, a glass of milk, some gingerbread Mrs. Babcock brought me and some preserved quince which the wife of my painter sent me.[a] He is an elderly man, Wilcox by name, living up on Town Street and I got him because I find there is a good deal of feeling about not employing the people this side of the River. And he is certainly a good workman and it was nice of his wife to send me the quince, wasn't it?

I went over to the Post Light today. Mr. Markman picked me up at the ferry and I took over fifty little self-sown annual larkspur which Mr. Osaki wanted to get rid of.[b] I have put a lot in my beds too for his are always beautiful. Mr. Markman showed me an empty bed and we put in a double row. I did not have time to explore much but I looked at Jessie's roses which are coming on splendidly, all of them with the possible exception of a little one in the upper bed. The plum trees are in full blossom and the forsythia was so lovely I brought back a bunch. It is up on the mantelpiece in front of me, and so are five spears of blue grape hyacinth, only five, for I sent the rest up to Mrs. Day by Winthrop Durfee.[c] He is doing my work and very well indeed. Think of it, he walks down here after school, works two hours and walks home, for 35 cents an hour. I meant to tell him I would give him 40, but Mr. Markman says 35 cents is the regular wage now for a grown man and the village of Deep River is only paying 30 cents for work on the roads. They are doing one good thing, offering seed and tools and help in ploughing to anybody who will plant a vegetable garden for food for next winter. Winthrop has been accepted by Dartmouth and is keen to earn some money this summer. They are so honest, these Hadlyme boys. I reckoned that I owed him three dollars today but he presented a bill for $2.10. All my boys have leaned over backwards like that, the Italians as well as the New Englanders.

It is really interesting seeing how much it costs to live here. For five dollars a week I could have Mrs. Babcock keep the house clean, and wash the dishes and get breakfast. As for food, it is incredibly little, even if I do not buy it all at the A. and P. in Deep River, where Mrs. Ernest Selden took me last week and Mr. Markman today. The chief trouble I have is miscalculating. I like lentils very much and I cooked some as one cooks pork and beans. But they lasted for seven meals. It will be a long time before I cook another lentil. I got a thickish piece of ham and simmered it in milk a long time, as Mrs. Wilson told me to. It is very nice but I have had it three times and there is still some left.

Life has slipped into a pleasant routine and I am more than keeping up with my schedule for the book. I get up early, seven or half past, breakfast is quickly got, for I put the coffee over at once and it is done by the time I am dressed and there is only bacon to fry and toast to make. Then I go out to view the landscape o'er, see what the day is like and what new things are up. The little Dutchman's breeches and bloodroot are blooming and the ferns are sending up white furry knuckles and the lilies, some of them, have big sprouts. But unless you look groundward you do not see that it is Spring. The elms and the sugar maples are only fuzzy, the oaks and locusts are not even that. I settle to work by nine and keep at it till one, then cook lunch, read my letters and paper till about three and then go out for all the endless gardening chores that are waiting to be done. At half past four I go in for a cup of tea to brace me up for the really heavy work when Winthrop arrives at a quarter to five. This afternoon we have been spreading wonderful hen manure which Mr. Seymour has let me have. Supper comes any time after seven, bed at half past nine. If I don't go to bed early I cannot get up early and that puts out everything. My routine will be broken into by another trip to New York this coming Saturday, for the Social Trends Committee is to have a two-day meeting there.

Do plan to come in May. Tell Aunt Phoebe that else I shall not have more than two week-ends with you, for Margaret will be coming. And I am sure she wants me more than that.

Yours always
Alice—

a. Maude Babcock cooked occasionally for the Hamiltons.

b. Mr. Markman was the caretaker of the Hamilton cousins' house in Deep River. Yukitaka Osaki lived across the road from the Hamiltons on the estate of William Gillette, the noted actor, whose dresser he had been for some forty years.

c. Winthrop Durfee was the son of AH's neighbor Dorothy Day Durfee. Mrs. Day was Dorothy Durfee's mother.

IN MARCH 1930 Alice Hamilton became a member of the President's Research Committee on Social Trends, a group composed mainly of social scientists charged with surveying all aspects of American life. The five male members, who had been appointed by Herbert Hoover several months earlier, selected Hamilton following a request from the president that a woman be added to the group. The study, a landmark of social science research that is still valuable, eventually resulted in two thick volumes, surveying trends in twenty-nine fields, and several supplemental monographs.[9]

Alice Hamilton called the experience "one of the pleasantest things that have happened to me as a reward for having been born a woman." She played a secondary part in the project, which was dominated by social scientists with a passion for quantification she did not share. ("How interesting it would be," she observed at one session, "if in addition to the accurate reports that you are getting out you got out the things that you think and cannot put into figures.") Her contribution consisted principally in evaluating the preliminary reports; she took special responsibility for criticizing those on women, health, and national vitality. Her strongest comments were reserved for the chapter "Labor Groups in the Social Structure," by Leo Wolman and Gustav Peck, which she criticized sharply at the meeting of May 16, 1932. Acting at the suggestion of William F. Ogburn, professor of sociology at the University of Chicago and research director of the committee, she expressed her sentiments directly to Wolman, who was professor of economics at Columbia University, former head of the research department of the Amalgamated Clothing Workers of America, and in 1932 president of an investment trust owned by the union.[10]

89 To Leo Wolman

[Hadlyme]
May 31, 1932

Dear Mr. Wolman,

Your chapter on Labor for the Social Trends Com. is to me the most interesting of all and I have just finished my second reading of it. Now, as you know, I am no student of this subject, all I know is from personal observation during a good many years of life in H.H. and of work in industrial medicine. Maybe I ought not to venture to give my own views, therefore, in this field, but I believe you will not mind if I do and I have really what the Quakers call a "concern of the spirit" on this report of yours, parts of it, that is.

Up to page 42 I should have only enthusiastic comments to make, but from then on your handling of the subject seems to me over-charitable, optimistic to a degree not warranted by the facts as I know them. I admit my comparative ignorance, we will take that for granted at the outset.

For instance, on page 42 you seem to be drawing a comparison between the public measures in European countries and private benefactions in the U.S. but surely the list given on page 43 contains social services which are provided in Germany, Great Britain, Scandinavia, Switzerland, Holland, along with social insurance. You do not wish to give the impression that private philanthropy here makes up for governmental services there. I should like a comment on the amazingly niggardly sum given for mothers' pensions. The general impression is that Americans pension mothers who need it.

Page 45. Here one might quote the Metropolitan Life statistics which show that the industrial policy holders have an average life expectation seven years less than non-industrial policy holders and that this difference is not due to poor living habits for it does not appear in the early age groups, not till over the age of 18 years.

Page 46. May not the working class be buying gasoline instead of meat, radios instead of Sunday clothes? In "Middletown" they did just that.[a] The last paragraph on this page is an instance of the philosophic calm and detachment which so often makes me see red for a minute. Isn't "infinite variety" a rather over-mild word to use for the persistence of the long work-day in some industries? And surely it was only a little while before the panic that that survey of the steel industry came out, showing that the twelve-hour day and the seven-day week still obtained in certain jobs.

Page 50. Ought not the Amalgamated to come in here? I am ready to hand in a minority report if it is quite ignored. Perhaps the second paragraph on page 53 is the better place for it.

Page 64. Here is where I grow really restive. I am willing to accept 64 to 66 as one side of the picture but the other side should be given— the "American plan" with its bitter war against organization and its use of so-called welfare and personnel work as camouflage for espionage and the blacklist.[b] And what about the history of the effort in this field made by the C. F. & I. Co.[c] Think of United States Steel, of Bethlehem Steel, of the Ford Company, of practically every industry in Detroit and in the Pittsburgh region, of the coal regions. I have been in towns where the power wielded by the employer was far greater than that wielded by any noble in Europe since the War.

As for stock-purchase plans, the recent collapse of the Insull companies has shown that all his employees, even in these two last years of partial unemployment and cut wages, were forced to buy stock in Midwest Utilities.[d] Surely you do not wish to leave the impression that such plans are always made in the interest of the employees.

323

Page 70. I think that here you have not made it clear that most American compensation laws cover accidents only, not diseases. There are only eleven states which provide for compensation for occupational diseases and of these only five cover all such diseases. The great industrial state of Pennsylvania gives no compensation even for permanent complete disability—or for death.

As I lay the chapter down I feel that I have had a clear picture given me of the better side of American industry and an able criticism of some of its deficiencies, and the same with regard to American trades unions. But the black side has been shown only by glimpses. The trends as I see them myself do not appear here: the increasing loss of individual freedom and initiative with a consequent loss of self-respect and of pride in work. The efficiency expert decides how the job is to be done. The welfare department takes charge of all else. Labor seems to me to be increasingly helpless and dependent because as mechanization increases the individual is less and less valuable, anyone almost can do the job. We are moving more and more toward industrial feudalism which is benevolent or hard-boiled according as the men at the top desire. And if the benevolent employer goes bankrupt, his serf must enter into the service of the hard-boiled, with nothing of advantage left from his former pleasant estate except its memory.

Now I do not get a glimpse in this report of yours of the contrast between our feudalism and the modern systems in other countries, Germany with her work councils established by law, England with her almost complete union organization. The last time I was in England the storage battery industry, a very dangerous lead trade, was being brought under a new code of hygienic regulations and these had been laid down by experts of the Home Office sitting in conference with representatives of the employers and of the workers. Imagine a storage battery worker in this country having anything to say about his own protection against lead poisoning. To me the opposition of the employing class toward any real organization of the workers is as implacable as it ever was and the absence of revolt as shown by strikes is evidence more of the increased sense of helplessness and loss of initiative and self-respect than it is of contentment with the welfare work of the management.

I suppose this sounds extreme and lacking in a judicial spirit and I admit that I am stating one side only. But it is a side and I think it deserves as complete a statement as does the side with which you have dealt so ably.

A.H.^e

a. *Middletown* (1929) was an important sociological study of Muncie, Indiana, by Robert Lynd and Helen Merrell Lynd.

b. The "American Plan" was the designation for the open shop favored by employers, who during the 1920s waged a strenuous campaign against unionism.

c. The Colorado Fuel and Iron Company, of which John D. Rockefeller, Jr., was the principal stockholder, had established a company union (or "employee representation") plan.

d. The firm of Samuel Insull, who headed a complex utility holding company network, collapsed early in 1932.

e. This is a draft.

In THE SUMMER of 1932, the heads of five companies that manufactured fire extinguishers wrote the Macmillan Company that Alice Hamilton's book *Industrial Poisons in the United States* "contains matter, inaccurate and untrue, which is damaging to us as manufacturers and vendors of fire extinguishers of the carbon tetrachloride type." There ensued lengthy, and to Hamilton extremely irritating, negotiations with Albert F. Jaeckel of the firm of Chadbourne, Hunt, Jaeckel and Brown, counsel for the manufacturers. She brought to this affair a reservoir of resentment (unusual for her) against lawyers, whose concern for technicalities and precedents rather than scientific truth and justice never ceased to mystify and distress her.[11]

Carbon tetrachloride (CCl_4), widely used as a solvent, has toxic properties under certain circumstances. In the case of fire extinguishers, the danger derives from the formation of phosgene, a severe respiratory irritant used in chemical warfare, that is released when carbon tetrachloride decomposes under intense heat. The companies took exception to Hamilton's assertion, based on experiments conducted for the Bureau of Mines, that "the use of carbon tetrachloride fire extinguishers in mines or in other enclosures without abundant air supply" resulted in the production of dangerous gases, including phosgene. They also objected to her references to cases of injury and death resulting from the use of this type of extinguisher. Hinting at legal action, they asked Macmillan to withdraw the book from sale, and, wherever possible, to recall all copies in circulation. Hamilton later learned that the action against her book was one of several initiated by the companies after carbon tetrachloride extinguishers had been rejected as hazardous by the Eighth Avenue Subway in New York City.

In August, at Felix Frankfurter's suggestion, she turned for legal advice to Benjamin V. Cohen, a New York lawyer and a Frankfurter protégé. Cohen, later a prominent New Dealer who played a major part in drafting legislation aimed at curbing big business, took the case without charge. After some preliminary skirmishing, Hamilton decided to publish an article that would bring the subject up to date. Cohen suggested that if the companies approved the article prior to publication, the case against the book would probably be dropped. She agreed to the arrangement, and in October Cohen submitted to Jaeckel a copy of Hamilton's article "Formation of Phosgene in Thermal Decomposition of Carbon Tetrachloride." The letter that prompted Hamilton's of November 5 is missing, but Cohen must have reported the lawyer's objections to her article.[12]

90 To Benjamin V. Cohen

[Boston]
November 5, 1932

My dear Mr. Cohen:

I have read your letter several times and have thought very carefully over the suggestions you make. Probably it is the fact that all the men with whom I discuss this strange case are physicians that explains why I cannot quite see eye to eye with you on the wise way to handle it. To all of us medical people it seems an intolerable impertinence that commercial men should assume to edit what I have written or propose to write in my own field. The only thing that could make me submit to it would be fear, a fact which I can see Mr. Jaeckel realizes and on which he is basing his strategy. But I am not willing to let fear of the expense of litigation drive me as far as that, though I admit readily that such an expense would be very disastrous for me.

I am willing to see Dr. Olsen and to listen to his criticisms, but I cannot promise to do it in the spirit you advise.[a] Dr. Olsen's object is to minimize the danger of carbon tetrachloride extinguishers when used in enclosed places from which there is no easy exit. My object is to emphasize that danger, not to exaggerate it but to make clear

326

beyond all possibility of doubt that all experts are at one on that point. Dr. Olsen's object is to make it possible for the companies for which he works to obtain contracts such as the one they recently lost, namely for subways. If I allow him to tone down my statements to what he would consider acceptable I shall be aiding him to the accomplishment of this object and yet I know that CCl_4 extinguishers should not be used in subways and that all the existing authorities are on record as holding that view. How then can I undertake to meet him half way, to re-write my article so that it becomes unobjectionable to his employers when, in order to do so, I must obscure or minimize a danger which I know exists and about which the public has a right to be informed? For it is clear that nothing else will satisfy them. They lost the subway contract through me and they intend to make me say something that they can successfully use when next they try for a subway contract.

All this effort to alter the text of my seven-year-old book and to have me write an article favorable to them is so futile and foolish. For I am not their only nor even their most formidable adversary, all I have said is based on the word of experts. If my book were annihilated—burnt by the public hangman—the situation would not be changed. Their rivals, the purveyors of other kinds of extinguishers, could cite any number of authorities to show that the use of CCl_4 under certain conditions is dangerous. Of course this attack has set us all—the Bureau of Mines especially—to searching the literature and there is now a complete list of articles ready for all who ask for it. To have me back down, as Mr. Jaeckel demands, would in the end be of no value to them. Yandell Henderson, H. W. Haggard, the Bureau of Mines experts, the German and French authorities, all could be quoted against me, I could be shown to have no first-hand knowledge, to have receded for mysterious reasons from a stand which I had taken publicly and on sound foundations, and my testimony would simply be discredited.[b]

You see, I find it hard to continue to be the only victim of this attack when the Bureau of Mines experts are the ones really responsible and when Dr. Henderson is as much an offender as I. Do you not think there would be some value in joining with Dr. Henderson and his publisher in a concerted defense? I cannot help believing that Mr. Jaeckel is concentrating on me as the weakest of his adversaries and I would like the help of the stronger ones, of the Federal Bureau of Mines and of the American Chemical Society which published Hen-

327

derson's book. I think we might make Mr. Jaeckel see that the matter is bigger than he thinks, that it is not simply a question of frightening one woman into submission but that, even with me quite demolished, the facts would remain as they are and there would be plenty of indignant men to proclaim them.

That appeals to me as the best step. In the early days of the Council of Pharmacy of the American Medical Association any number of suits were threatened by commercial companies whose remedies had been condemned, but when they realized that the strength of the Association was behind the Council, they desisted and now there is no trouble of that kind at all. My friends urge me to write an editorial for the Journal of the A.M.A. outlining the whole case, and perhaps that would be a good thing to do, but I should always wait for your opinion before doing anything so drastic. It seems to me and to my medical friends that the extinguisher companies do not realize the damage that publicity of that sort would do to them.

I will, as I said, consent to see Dr. Olsen and I will do my best to listen patiently to his suggestions but I doubt if I accept any of them, certainly not if they tend to obscure the facts, and I fear that is the only kind he will be interested in. If he will come on the 9th, 10th, 11th or 12th, I will promise to give him two hours, but I think I will not consent to hold up my article longer than that.

Do not come with him, please, for I might lose my temper with you too and that would make me very unhappy. You see, I simply cannot look at this from the lawyer's point of view. Much as I dread a lawsuit I do not wish to go into court with a record of having made all possible concessions to the extinguisher companies. I could not make concessions on any vital point and that is the only kind they ask for.

Very sincerely yours[c]

This is horribly faint—you see I am my own typist.

A.H.

a. J. C. Olsen, professor of chemical engineering at the Polytechnic Institute of Brooklyn, was the technical representative of the Vaporizing Liquid Division of the Chemical Fire Extinguisher Association.

b. Yandell Henderson and Howard Wilcox Haggard, both physiologists teaching at Yale University, were specialists on ventilation and authors of *Noxious Gases and the Principles of Respiration Influencing Their Action* (1927).

c. This is a carbon copy.

Dɪsᴛʀᴇssᴇᴅ ʙʏ Hamilton's letter, Cohen responded two days later: "I hope that you have a better opinion of me than to think that I would be so wicked or so stupid as to suggest that you give up your scientific birthright." On November 14 Hamilton apologized, reporting that Frankfurter had told her that "I had treated you quite outrageously, although he admitted that I did it in ignorance" and had also made her understand, for the first time, "what is meant when I am told that this case will not be dealt with on scientific grounds but on legal."[13]

The matter dragged on. By the beginning of the new year there were three points still at issue. On January 4 Hamilton wrote Cohen that she could accept only one of Jaeckel's proposed changes. She found the conclusion "colorless enough as it is" and feared that she had already conceded too much. Two days later she reiterated her position.[14]

91 To Benjamin V. Cohen

[Boston]
January 6, 1933

My dear Mr. Cohen:

I have been going over our correspondence and the criticisms of Mr. Jaeckel transmitted by you and have come to the conclusion that the summary as I wrote it out in my last letter to you, of January 4, is the best version and the one I would wish to adhere to, unless you have cogent arguments against it. This is, to repeat: "the conclusion to be drawn from the literature is that extinguishers of the CCl_4 type should not be used in enclosures where the ventilation is not sufficient to dilute the products of thermal decomposition below the danger point, especially if escape from the place is not easy."

The points in favor of this are these: It is stated in technical terms and therefore less easy for use by commercial people; it guards against opening the way for rivals of the CCl_4 people, because it makes the conditions under which such extinguishers should not be used a matter for scientific determination; and it also prevents the CCl_4 people from using my statements as an argument for the use of their type under any given conditions unless an analysis of the gases formed by the

329

decomposition of CCl_4 in that particular locality has been made. And of course if such an analysis proved that the amount of phosgene formed was within safe limits, such use should be allowed.

Now I hope this does not mean starting another two-month controversy with Mr. Jaeckel and I do not think it should. In the first place your covering letter of Oct. 21, a copy of which you sent me, gives the conditions under which my paper was submitted to his clients, viz: "As a matter of courtesy, Dr. Hamilton desires me to submit to you a copy of her proposed paper (which is here enclosed). Without wishing in any way to commit you or your clients to any statements therein contained, Dr. Hamilton will consider any criticisms or suggestions of a scientific nature that your clients may care to make. But of course the paper, as published in form and in substance will represent her views on the subject matter and literature uninfluenced by any suggestions that do not commend themselves to her scientific judgment."

That was a very complete and adequate wording of the conditions which you and I thought were necessary for the protection of my independence. The responsibility for the article you state to be mine alone, the manufacturers are in no way involved in what I may state to be the published facts nor in the conclusions drawn from those facts. Only if an inaccuracy on my part were detected, or if the matter were handled by me in a partizan, unfair spirit, would there be any justification for dictation on their part. But I think you will agree with me that we have travelled far from that point during the last ten weeks. As I look at it now I see that Mr. Jaeckel has brought about a situation in which we are assuming that his clients are actually involved in this paper, that they share the responsibility for my statements and are entitled to object to any which do not coincide with their own opinions.

Take some random instances in illustration of this. I must drop the word "verify," because that sounds as if I still believed the statement of the insurance man, but suppose I do? Surely that is my responsibility, not theirs. I must not use the expression "such a case" because that implies that there have been cases caused by CCl_4. But that is just what I do believe. I must not say "the fumes" because that means that carbon tetrachloride fumes were the cause of poisoning. But that also is what I do believe. I cannot use the phrase "closely confined" because that seems to laymen to be indefinite. But suppose I prefer an indefinite phrase? I must not say "consensus of opinion" although

330

the only dissenting voice is that of Dr. Olsen, whose opinion is of no weight.

You see, all those points, all of which I have conceded, are pressed in order to bring my paper into line with what the CCl₄ extinguisher men believe to be safe publicity. Now that was not in your mind nor in mine when the paper was submitted to Mr. Jaeckel. None of the safeguards you provided in your covering letter have been maintained, Mr. Jaeckel has little by little pushed his demands until, with the added pressure produced by dragging out the procedure to the point of exhaustion, he has brought me to the point of conceding everything that is not absolutely absurd or untrue, in the effort to bring matters at last to a conclusion.

I am not proposing to take a bold stand even now. I still face the fact that I am very much afraid of Mr. Jaeckel's power to bankrupt me and that I cannot afford to be bold. But I wonder whether, if you called his attention to the conditions as outlined in your letter of Oct. 21, and to the history of negotiations during these two and a half months, you might not induce him to accept my many concessions, to let this final summary stand, and to bring the whole matter to an end.

Very sincerely yours[a]

a. This is an unsigned carbon copy.

THE FOLLOWING DAY Cohen replied that he thought Hamilton was allowing Jaeckel to agitate her unduly and that her article "fundamentally taken as a whole . . . has been unaffected in any substantial sense by Mr. Jaeckel. If anything it seems more against Mr. Jaeckel now than it did in your first draft."[15]

The companies raised no further objection to the article, which was published with its conclusion intact in the May 1933 issue of *Industrial and Engineering Chemistry*; a discussion by J. C. Olsen appeared in the same issue that sought to minimize the dangers of carbon tetrachloride fire extinguishers and to cast doubt upon Hamilton's scientific credibility. Hamilton's major concession was in a footnote: "I have never been able to verify the statement, quoted in my book, of an industrial insurance man to the effect that six deaths had occurred from the use of this form of fire extinguisher." Macmillan agreed to insert a similar disclaimer in her book.

331

The report of the President's Research Committee on Social Trends was published in January 1933 as *Recent Social Trends in the United States*. To mark the event, President Hoover invited committee members to dine at the White House on January 26.

92 To Edith Hamilton

Baltimore
January 28 [1933]

Dear Edith,

I sent back your "October House."[a] It is first-rate. I love that kind of dialect and it is a good story. I am having a lovely quiet week-end here, but Washington was not really tiring, only the dinner was, in anticipation, not in reality. It was actually a very interesting affair and made no demands on me whatever. We were told to be at the White House at ten minutes to eight and when I got there—in Katy's black bead dress—I found a semi-circle of men in dress suits gathered in the wide hall facing the front door. There were eight of them and no woman, so I was prepared when Mr. Strother murmured to me "The President will take you out."[b] Then in a few minutes a functionary appeared and announced "The President and Mrs. Hoover." Mrs. Hoover was very gracious, said she was glad I was there or she would not have got in, the President gave me his arm and we went in to that beautiful panelled dining room, which he told me was done in the Roosevelt administration. There was a very beautiful table, a wide oval, with the President and Mrs. Hoover sitting opposite each other at the narrow part. There were only eleven of us so, to my enormous relief, conversation was general, mostly between Mr. Mitchell, Mr. Merriam and Mr. Hoover, with others dipping in and out but no tête-à-têtes.[c] I was sure the President did not want to be bothered by me so I hardly ever put in my oar and made no attempt to converse with him directly. So it was really easy and I began to enjoy it. He made a curious impression on me. His first remark was about the attacks during the campaign on his commissions and it was made with real bitterness, and after that, again and again, he referred to his defeat, to the Democrats, to his successor, as if he wanted to show us he did not care, or to say it first so that we should not have a chance to. It was surprising and it was not dignified nor admirable. When we left the table I found I must walk through a long hall and way up the marble stairway, on the President's arm, really an absurdly difficult performance and why? So then I had to speak to him and I told him this was the most exciting thing that had happened to me, except perhaps having an audience with the Pope, and he said "Oh well, why call this exciting? We Presidents come and go so fast it doesn't mean anything." I was taken aback and couldn't think what to say, except that Popes came and went pretty fast too and then hastily asked him how many he had known and he told me about playing bridge with the present Pope when he was Papal Nuncio in Warsaw, and then we reached the Lincoln Study and my efforts for the evening were over and from then on it was very enjoyable. There was an open fire and coffee was brought to us, and Mrs. Hoover took out her knitting and we all sank into comfortable chairs and began to discuss Social Trends, and the President did most of the talking. He would sometimes forget his defeat, but not for long—the bitterness would crop out and he couldn't help digs at Roosevelt—but he talked well and interestingly. He surprised me by saying that before the disaster, production was not excessive and distribution, while faulty, was better managed than ever before, and that it was the financial group that brought on the disaster. And Mr. Mitchell agreed with him. A good bit of what he said the men received in silence, as his insistence on raising the tariff, for bargaining purposes, and his blaming Europe for our plight, but when he discussed government re-organization Mr. Merriam and he were agreed. He is pitifully nervous, his feet are not quiet for a minute, and he wore creaking patent leather pumps so that, as I sat beside him, I heard the undertone of little squeaks all the time. And Mrs. Hoover's foot swayed or tapped continually. Poor things, they have been through a purgatory. Once he said, when we were talking of the thirty-hour week, "Who wants a short day, who wants to work only eight hours? What do you suppose I feel like when I think of going back to California?" That is the hard thing about our system. He ought to pass into the Senate and fight there for the things he believes in. At quarter to eleven he rose and we rose, they both shook hands with all of us and went out. French Strother drove me home and I asked him where the Secret Service men were. He said they were at the entrance when I came in but of course were unobtrusive.

I had two mornings in Washington and went to two Senate Committee hearings, one on the thirty-hour week and one on the agricultural allotment. It so happened that the men heard were opponents

333

chiefly, manufacturers, and so I found it very interesting for of course I know all the arguments pro. I was a good deal impressed by the way it was done and the efforts of the Senators to get at people's opinions.

Margaret has at last got over the bad throat which has lingered on since before Christmas. She does not look well but is, Clara says, much better. I think we are going to Germany, Clara and I. A letter has come from the Oberlaender Trust saying they are willing to send her with me because of her work for the Quakers, and she is very eager to go and freshen up her German. It will leave Margaret alone for only April and May, for I could not sail till April first. I don't really want to go but I think it is silly not to.

Yours always
Alice.

Will you send this on to Norah?

a. A mystery story by Kay (Cleaver) Strahan about a "psychic card reader and palmist."

b. French Strother was Herbert Hoover's administrative assistant.

c. Wesley C. Mitchell, professor of economics at Columbia University, and Charles E. Merriam, professor of political science at the University of Chicago, were chairman and vice-chairman of the committee.

Dᴜʀɪɴɢ ᴛʜᴇ 1920s and early 1930s Hamilton closely followed the frequent strikes in the woolen textile mills in Lawrence, Massachusetts. She agreed to speak at a rally to be held there on February 1, 1933, ostensibly at the invitation of the Women's Club of Lawrence.[16]

93 [To Agnes Hamilton][a]

Codman Hospital
15 Pinckney Street
Boston
[February 1933]

the rain to a cafeteria where I had a supper of Frankfurts and sauerkraut and coffee. She was quiet and intelligent and very discreet but she admitted that she was an organizer and I felt sure she was a Communist. We went to a school house and the big auditorium was filled with men, mostly, about four hundred I thought she said five hundred. I spoke for about forty minutes and the audience was very nice. I recognized some men I saw with Mrs. Ripley during the last strike.[b] When I stopped I asked for a discussion from the floor but June Crowe took the platform and kept it for almost an hour, a real Communist speech attacking the A.F. of L. telling the audience they had been sold out in the last strike, betrayed by their leaders, and reading a violent denunciation from a girl Communist who is to be deported. Well, why the audience let her go on I do not know, for two thirds of them were against her, but they were orderly—thank Heaven—only when she stopped they were on their feet denying and protesting and an old Scotchman I knew pointed his finger at me and demanded to know what I was doing on a platform that was used to vilify good Union men. Of course I assured him that I knew nobody had sold out the strike but I would not turn on June Crowe as he wanted me to, though I was pretty mad at her. It turned out that there is no Women's Club of Lawrence, two reporters told me so afterwards and she simply made up the whole thing, got a textile worker to write me, using the address of Communist headquarters and then captured me and the meeting. She was gaily impervious to anything I had to say to her and assured me she was very grateful to me for coming, she had never before been able to get up a big meeting! Well, the Scotchman and three other labor leaders—A.F. of L.—talked to me after it was over and said they did not blame me, they knew I had been taken advantage of. I really didn't mind much, for I said only what I would have said to a club or a Chamber of Commerce or anybody else. And I was glad to go and to feel there was something I could do. Seven children sat in the front row and only one was fairly normal, the other six were pasty, grey, heavy-eyed, emaciated. They looked like German children in 1919, they made me feel sick.

Susan Kingsbury wants me the 16, 17, and 18 of this month so I will spend Sunday the 19*th* with you.[c]

Yours always
Alice—

a. The beginning of this letter is missing.

b. Ida Davis Ripley, a member of Katy Codman's Boston circle, was active in various labor and social reform causes.

c. Susan M. Kingsbury, director of the graduate department of social economy and social research at Bryn Mawr College, arranged for AH's yearly lectures.

DURING THE DEPRESSION Alice Hamilton often expressed concern over the financial losses of friends and relatives. Her own income was reduced by "about one third or a little more," and in 1933 the cousins were unable to open their house in Deep River.[17]

Her feelings for her family extended to members of the next generation, who were always welcome at Hadlyme. These included Rush Hamilton, second child of Taber and Gail Hamilton, and his siblings, young Taber and Phoebe, as well as Dorian Reid, nephew of Doris Reid and Edith Hamilton's adopted son, and his sister Betsy. Russell Williams, Allen's son, frequently visited the Hamilton sisters after coming east to attend Harvard College and Johns Hopkins Medical School. He sometimes brought with him two friends from the Cate School, Philip and Tudor Owen, who became deeply attached to the Hamiltons; Alice variously referred to them as "my two boys" or "a couple of college lads who have more or less adopted us." Tudor Owen, a polio victim, spent some time in a mental hospital during this period. He later settled in Hadlyme, as for a time did Philip, who became a physician.[18]

94 To Agnes Hamilton

24 Gramercy Park
New York City
April 4, 1933

Dear Agnes,

I was immensely relieved to have your letter yesterday when I got back from Plainfield. Your short note had worried me very much and Edith too. I did not tell Margaret because she is up to her eyes already, but I did show it to Edith and we both thought it meant Deep River might be lost. Sunday evening Rush came in to supper and he was so sure that could not be, but I was not sure till your letter came. If you are planning for Margaret in June and for all Rush's family over Commencement, it must be all right.

I have had two busy days getting through chores, taking my finished book to the medical editor, going out to Plainfield, visiting Tudor in White Plains and talking him over with Philip and Mrs. Wickes. He is so far as I can see perfectly normal and his doctor says he has been so for five months but they seem not to feel safe yet and though they say he may go to Harvard in the Fall they would not tell me when he would be discharged this summer. August was tentatively spoken of. He wants to come to us and I said we would take him. Edith says I am not fair to Margaret, that she does not feel toward the two Owen boys as Norah and I do, and that I am putting on her an extra expense and robbing her summer of a good deal of its peace. All of which is true but Margaret will never say it. Only what can I do? Tudor looks back on Hadlyme as the one place of peace and affection and integrity in his world. He is bitter toward his aunt and Philip and Mrs. Wickes, everyone who was connected with this breakdown, and they say he must be rid of that before he can be really normal, but we seem to be the only people he will trust and so the only ones who can drive the bitterness out. If we refused, I don't know where he would go. But it is hard on Margaret. The money part I don't so much mind for I shall be spending nothing of my own for the next three months. That has had something to do with my going.

Aunt Louise and Uncle Fred were just themselves and very dear. Uncle Fred is going to write you about Buchmanism and I hope you will be able to understand what he means.[a] I couldn't. I spoke of the guidance of the Spirit and he said "My dear, does she not know there are two spirits. Which is this? I fear it is only a subtle disguise of our Adversary. Agnes will point to changed lives but surely the Devil is wise enough to know that he could never tempt a soul like hers with debauched lives. No, he has prepared the net with his deepest subtlety." Well, if you can make that out, you win. Whether the people whose lives are changed are victims of the Devil's subtlety and think they are changed when they are not, or are in league with him, I don't know. Anyway he is going to write you about it.

Margaret and Clara come this evening, late. We sail at midnight tomorrow, to a land that seems far madder than Russia but not so scaring. I will write you, of course, but colorlessly, because letters are opened.

Love to Jessie and Allen, Helen, Andrew, yourself.[b]

Alice—

a. Agnes Hamilton was an enthusiastic participant in a popular evangelical religious movement founded by Frank Buchman. Known as the Oxford Group (and later as Moral Re-Armament), the movement stressed personal reformation and featured public confessions. The Hamiltons had always considered the religious views of their uncle Fred Jennings odd and indecipherable.

b. Allen Hamilton (1874–1961), Agnes' brother, was a physician who practiced in Fort Wayne. His wife was Helen Knight. Their son, Andrew Holman (1910–1980), later became a professional historian and used the name Holman Hamilton.

ACCOMPANIED BY Clara Landsberg, Alice Hamilton finally left for Germany on her Oberlaender Trust fellowship in early April. She arrived less than three months after Hitler had come to power. Fluent in German, she traveled extensively for nine weeks, talking "to every possible person" in an effort to get beneath the surface of the "amazing, preposterous sayings and doings." Always an admirer of Germany's intellectual and scientific life, she had deep ties to that nation, going back to her student days and probably also to the German servants of her childhood. She had friends of long standing there, principally social workers and scientists, many of them Jewish, who talked freely to her. What she saw and heard appalled and bewildered her—the terror and spying, the book burnings, the loss of personal freedom and economic livelihood by those ostracized by the regime, above all the anti-Semitism, to which her response was immediate and unequivocal as it had been in 1895.[19]

A postcard depicting a rathskeller that she sent to Felix Frankfurter sounded a theme to which she returned many times.

95 To Felix Frankfurter

[Berlin]
April 24, 1933

Dear F.F.

This charming place was the scene yesterday of a very tragic interview, of which I have had only too many. I wish to confess to you that I am a convert to a strong constitution with all kinds of guarantees and even to a conservative Supreme Court and to private welfare work rather than Governmental. Also I am now a passionate patriot for the first time. Love to Marion.

A.H.

96 To Agnes Hamilton

Hotel Kaiserhof
Berlin
April 28, 1933

Dear Agnes,

Margaret's letter has just come, with the disastrous news of Ru-rode's, of Helen's losses, of your having to send away maids, give up the house and close Deep River after June first.[1] I suppose if I had read it in safe, pleasant Boston I should have felt overwhelmed, but here, where our hearts are just torn to pieces every day it seems bad, but not dreadful. Nobody is ill, and nobody has to face living squalidly or with people she dislikes. We all are fond of each other and know each other so well that we shall not need to be adaptable even. It will be hard for you to not have your own house, I know. I think we will do it this way. Put pull-back curtains and a dressing table and comfortable cot in the sleeping porch for you. Put Jessie in Norah's room, Norah up opposite me—that is what she would want—Elizabeth in the spare room of the ell. Then we shall still have our three spare rooms and after Miss Addams and Mary are gone we can spread out if we want to. We will list all the house chores and divide them up and then nobody will feel she must offer to do anything. There are three big sitting rooms and porch and arbor, surely we shall not feel on top of each other. As you say, things do not count, only people do. And really it is so little beside the things other people have to bear. It is nobody's fault and nobody is doing it to us out of hatred and nobody wants us to suffer, just the contrary. And some day it may clear up. So long as you can hold on to Deep River the worst has not come.

We are just back from a big meeting of women's organizations where we met some such very fine, wise, clear-headed and patient women. One admires them tremendously. I wish I were home to help talk things over at least but of course we must do justice to the Trust and then it will be such a help if I can write something this summer. I am keeping it all till then.

Love to Jessie, Allen, Helen, Andrew—
Yours always
Alice

339

This is the most beautiful gentian I have ever seen. I am going to try to get some seeds.

a. Rurode's was a Fort Wayne department store housed in a building owned by various members of the Hamilton family.

IN MAY Alice Hamilton and Clara Landsberg toured the Polish Corridor and East Prussia under the auspices of an official organization. Although "most clearly a propaganda affair," this arrangement seemed the most practical way of seeing a remote part of the country. The trip brought them face to face with German eagerness to repossess the German-speaking parts of Poland which had been lost after World War I. During their two-week side trip, Hitler delivered a conciliatory foreign policy speech denouncing war and declaring that Germany was willing to give up its military establishment if other nations did so. His talk was made in response to an appeal for "peace by disarmament" that President Franklin D. Roosevelt addressed to the heads of fifty-four nations on May 16, in an effort to counter Germany's insistence on the right to limited rearmament, which had stalled the international disarmament talks at Geneva.[20]

97 To Edith Hamilton

No. 10
Savoy-Hotel
Breslau
May 16, 1933

Dear Edith,
This will be a diary letter, for we still are being personally conducted and have only brief periods to ourselves. This is one, between a most satisfying mid-day dinner—German food is very much to my taste—and a trip we are to make in the Landeshauptmann's motor car to an old cloister. We have been here three days, bitterly cold and raining days, but we have an electric heater in my vast bedroom and a warm bathroom. The last is a luxury we find we can afford in the provinces and it is certainly delightful. The ten dollar allowance is the outside limit and I do not want to average that sum quite, and it has to cover all travelling expenses and cabs and entertaining, though we do not

340

do much of the last. We are more than comfortable, except for the cold. The thermometer is about forty and I do wonder at these Germans, who dine out of doors and go about so lightly clad. The generation that feared drafts is gone, this one is so vigorous that it puts young Americans to shame. They walk with a beautiful carriage, they bicycle, they have such a glorious color.

The problem they are all absorbed in here is the Polish-Lower-Silesian boundary and we spent all yesterday going over it. But Sunday we had met a colleague of mine, a delightful person, and I was so sick over the fate that has suddenly descended on him that I could not squeeze out one bit of sympathy for the oppressed German minority in Poland. Finally, at the end of an hour's outpouring of eloquence, I blurted out that it was really very difficult for a foreigner at this moment to sympathize with the German minority in Poland when the Jewish minority in Germany had so much worse a fate. It was like a bolt out of the blue—and it completely destroyed all rapport between me and our guide. I am quite sure it was the very first time such an idea had come to him and I took pleasure in assuring him that in America that would be the very first thing we would have said to us [sic]. They seem never to have realized how it looks to the outside world.

May 19
Dresden

The last three days were very full and we reached here last night determined to take three days off—not to present a single letter of introduction, not to do anything except what we darn please. From Breslau—the last day ended with dinner in a grand house and bed not before midnight—we went down to Gleiwitz and Beuthen in what is most misleadingly called Upper Silesia. I had always pictured it as a sort of Pittsburgh, ugly, sooty mining and smelting towns scattered through a wild mountainous region in the northern part of Silesia. The soot and ugliness are there but it is all as flat as Indiana and it is in the southern part of Silesia, they call it "upper" because of the rivers, which flow from south to north. The towns are dreary—not as bad as South Chicago or Homestead for the Germans do manage to have green spots and lovely little gardens on the outskirts of all their cities—and the people much more wretched than any we have seen. It is a mad situation there, as if one drew a scalloped line in and out of one of our big coal and iron centers, breaking up industrial

341

units into two parts with a boundary between, cutting one plant off from its railway junction, another from its supply of sand. Of course I have heard it only from one side and cannot speak impartially but after all, everything there was built with German capital and planned and managed by Germans. A great plant for the production of fertilizer by means of nitrogen from the air has been given to Poland, yet that was a German invention and the plant was German made. The Germans speak always, all along the boundary, of the fairness of the English and sometimes also of the Italians, but never once have they spoken of the French with anything but resentment. Which does seem a mistake on the part of the French.

Our guides—all government people—asked us if we would mind if they stopped for an hour to hear Hitler's foreign policy speech and of course we were delighted to do so. They took us to somebody's apartment and there we sat around a typical German table—Clara and I of course on the sofa—and drank good coffee and ate cakes while we listened in. My impression was absolutely favorable and yesterday I read it and was even more impressed. It is as different as possible from his style when speaking to his own people. I cannot see how even the French can refute it. The people here are much stirred up over Paderewski's spending a night in the White House, and a Polish general going around addressing the American Legion.[a] I told them I did not believe anyone took Paderewski seriously, but maybe they do. Roosevelt's speech was very well received here. We get no real news of America in the papers, only absurd things like all Roosevelt's closest advisers being Jews, and the States of the Middle West being about to split off and found a separate Republic. In Munich I do hope to get some English or American papers.

Yours always
Alice—

a. Ignace Jan Paderewski, Polish pianist, composer, and statesman, was on a concert tour of the United States that included dinner at the White House. During the tour, newspapers publicized rumors about his political future.

98 To Jane Addams

Dearest lady,

We are in a dense fog, crawling into port where we hoped to be at ten this morning. Now we may have to lie outside for hours and perhaps miss the last train for Saybrook.

I am coming back in a mood of passionate patriotism worthy of a D.A.R. The Statue of Liberty will give me a real thrill for the first time. I don't care what faults a democracy has, it is a safe and human and liveable system and nothing else is. As the days go on, Germany gets more and more unreal, nightmarish, and it is hard to believe that eight days ago I was really in Hamburg, talking with people who were living in fear, an increasing fear, who woke every morning with dread of what they might see in the papers, who would buy nothing and could make no plans because any day they might have to fly. You see, it has been growing worse instead of more lenient. About the middle of June there was a Conference of Leaders in Berlin, behind closed doors, and immediately after that the program was stiffened, the great organization of the Steel Helmets—former soldiers and officers, aristocrats, conservatives—was dissolved, the party of Hugenberg and Papen was forbidden, and four violent speeches were made by Hitler and others, proclaiming that what has happened is only a curtain-raiser, the Revolution will sweep on.[a]

We found Frankfurt and Hamburg more terrorized than any cities we were in. Frankfurt is very tragic. Do you remember Tilly Edinger, a spirited young thing we met there in 1919?[b] Her mother was the one who founded the "Air Baths" in the parks, where we saw those pitiful starved children. Tilly's father was a famous neurologist and founded the Neurological Institute of the university. Her mother and her mother's family were known for generosity and there was a street named for her father, while her mother's bust stood in the city's center for social work. Tilly is a gifted young scientist and worked in the Neurological Institute and of course she grew up with the consciousness of belonging to one of Frankfurt's best families and of being the daughter and granddaughter of highly honored people. When we saw her she had been requested to leave the Institute and formally expelled from the City Hospital where she did scientific work. The city authorities telephoned her to come and get her mother's bust or they

343

would throw it out. The night before a beloved old couple, her mother's uncle and his wife, had killed themselves. They had seen everything go, their grandchildren forced out of school, their sons deciding to go to Belgium where life would be possible for them and they could educate their children, they themselves were told that they were not Germans but foreigners and a curse to the country and they decided death was better. He was a great benefactor to Frankfurt and founded the astronomical department, in recognition of which a little planet was named for him "Mauritius."

All the people we met in Frankfurt were Jews and all waiting—to see when the blow would fall—or left suddenly without position or income and with no possible chance of any employment. If only we could open our doors to these people, they are so fine, but of course we cannot.

We went to Amsterdam for two blessed days, for I wanted to see Professor Ripley's wife who is staying there while he recovers from a mental break down.[c] She told me she had been in Geneva and saw Heymann and Augspurg and Gertrude Baer.[d] The two first made a deep impression on her, as of course they would. They are not planning to return to Germany and seem to have grown much interested in astrology. Gertrude Baer was very restless and insisting she must go back but it would be useless. She would simply be sent to a concentration camp for an indefinite time. Just before we left Munich, a young Hindu girl who was on the eve of taking her doctor's degree was arrested because she made a speech on Gandhi to a small group. The British Embassy will doubtless get her out but only on condition that she leave Germany at once—without her degree.

It is much worse than Russia. There the Whites were such a poor lot, one could be terribly sorry for them but nobody could wish them back in power. But in Germany the down-and-outs are the best people they have. So far, Siegmund-Schultze has not been touched personally but his work is hampered in every way.[e] The women of the old suffrage and reform groups, whom you met in Berlin, told us they would never consent to turn Nazi and now they have been dissolved and no German woman will go to the meeting in Stockholm.

Do write to me, you or Mary. I have had no sign of a letter in so long. We hear of sickening heat in Chicago and I am worried. I still count on seeing you in Hadlyme.

Clara sends lots of love. It has been a rather dreadful experience for her but she has been endlessly plucky.

Yours always
Alice

a. Alfred Hugenberg, a member of Hitler's first Cabinet, was head of the monarchist German Nationalist Party, which Hitler dissolved in late June. Franz von Papen, vice-chancellor of the cabinet, had helped bring Hitler to power.

b. Tilly Edinger, a vertebrate paleontologist and daughter of AH's former professor Ludwig Edinger and his wife, Anna, later emigrated to the United States, where she worked at Harvard's Museum of Comparative Zoology.

c. Ida Ripley's husband was William Z. Ripley, professor of political economy at Harvard.

d. Lida Gustava Heymann, a primary school teacher and a leader of the German feminist movement, was a first vice-president of the WILPF. Anita Augspurg, the first woman judge in Germany, was a pioneer in the German woman suffrage movement. Gertrude Baer was active in the German women's rights movement; after Jane Addams retired from the international presidency of the WILPF, she was one of three women who served jointly in that capacity.

e. Friedrich Siegmund-Schultze, a clergyman and pacifist, was head of a Berlin settlement and professor at Berlin University. AH and Addams had met him in 1915.

FOLLOWING HER RETURN from Germany, Alice Hamilton took an active stand against Nazism, becoming a more visible public figure than at any time in her life. She published a series of important articles, including two in the *New York Times Magazine* (the *Times* also built an editorial around an article on German intellectuals she wrote for *Harper's Magazine*), met with President Roosevelt at Hyde Park in late August, and received numerous speaking requests, many from Jewish organizations. Drawing on her reading, observations, and especially her interviews, she depicted the devastating effects of the Nazi regime on Jews, intellectuals, women, and the labor movement. She foresaw no improvement for the Jews as long as Hitler remained in power.[21]

Yet she did not believe that Roosevelt could "make a formal protest any more than Germany could formally protest against the last Maryland lynching," and she saw no immediate remedy other than helping German Jews escape. Both then and later she did her best to assist the many individuals who appealed to her for help in immigrating to the United States. In October she wrote Jane Addams that she found the unaccustomed experience of speaking solely in denunciation some-

345

thing of a strain: "I find I do not like to do it. It is getting distasteful to me to keep on denouncing, accusing, with almost nothing one can say in extenuation. I never have done it before." As a civil libertarian she was also distressed when Jewish groups organized picketing at events where the German ambassador or prominent Nazis were to speak: "And yet these are just the people who protest most when they are not allowed to speak."[22]

Mary Rozet Smith, Jane Addams' most intimate friend since the 1890s, died on February 22, 1934. Addams had recently suffered a heart attack, and friends feared that the shock would impede her recovery. Alice Hamilton subsequently assumed even greater responsibility for Jane Addams' health and well-being, considering her "really my one dependence now." Addams spent the summer of 1934 in Hadlyme, in a wing of the Hamiltons' house that had been remodeled for her comfort.[23]

99 To Edith Hamilton

Hull-House
February 23 [1934]

Dear Edith,

It is before breakfast the day after Mary's death. Do you know, the thirty-six hours before it came I used to say to myself "Mary's death," "the day Mary died" to see if I could bear it and I couldn't seem to see people's lives going on afterwards. And I don't now. Yesterday I was only stunned and kept pushing the reality away from me for fear I would lose my self-control and could not get it back. But last night and now things come moving relentlessly on me. What are we to do? Here is J.A. ordered to bed for four more weeks at the least— and we are in a house that is not Mary's, it is her heirs'—the servants, the heat, the food—it is no longer Mary who gives them to us. The immediate future is bad enough, the remote is worse. I can't see life for J.A. and I have a deep-down fear, a fear that makes me despise myself. But I am sixty-five and I can't take on Hull-House now. Yet probably at the end of next year I shall be free to do it. I feel horribly unworthy pouring all this out and in a few minutes I will push it back and try not to look at it. I can't look at my grief over Mary because

I should lose my grip. When I came out here I told Mary that she must get well, that she could live on without J.A. but J.A. could not live without her.

Mrs. Bowen is coming back, but cannot get here till Sunday. The funeral is Monday. Here comes Eleanor and I must go to breakfast. Norah is lovely in every way.

Yours[a]

a. The letter is unsigned.

100 To Agnes Hamilton

12 West Walton Place
Chicago
February 28 [1934]

Dear Agnes,

It was dear of you to write to me and let me know you were thinking of me all these heavy days. I have needed to feel people caring, it is all that helps. Miss Addams is so endlessly pitiful, it breaks one's heart, so that I can hardly think of my own grief except when I go off upstairs alone and then it comes over me. But grief is different when one is old. All the time the feeling is with me that it is none of it for very long, and that I have had it, had something rich and beautiful for so many years. I cannot really lose Mary, she is part of my memories.

I think Miss Addams knows that it cannot be very long for her. She seems very quiet, very dear, but remote. Her heart is very undependable—this last attack was coronary and that means a much damaged wall. For the present we will stay on here. Mary provided in her will that all should go on for two months after her death. What will happen then I do not know. I think Miss Addams is working it out in her mind but she said to-day that she must not let herself think of the future much, it tired her. She loves this house, but of course it is out of the question for her to keep it up and she would not wish to.

I am leading the disorganized, unplanned life that one leads in a house of sickness. People come and go and I see them or not, and I do very little but it is as if a rope tied me to that sickroom and I am restless if I am not in or near it. But I have long nights and am not tired. Do write me again.

Yours always
Alice—

Norah showed me a *lovely* letter from Jessie.

ALICE HAMILTON, who needed the income as well as the work, did not look forward to retirement. She had been uncertain whether she would be reappointed after her contract expired at the end of the 1933–34 academic year, and had feared she would not be able to find even a lecture series, "with industry in such a bad way and nobody wanting industrial doctors." Having received another three-year contract, she was dismayed when Harvard's new president, James Bryant Conant, who was attempting to institute a more uniform retirement policy, asked for her resignation in January 1935, shortly before her sixty-sixth birthday. Distressed by the threatened loss of work, she spent the next several months looking for paid employment. Yale and the Public Health Service presented the most likely opportunities, she thought, but at the least she hoped for "a local habitation and a name" from which she could continue her consulting work—"Hadlyme will hardly do." Several friends sent out their own inquiries on her behalf. Even the following September, Hamilton seemed to Elizabeth Glendower Evans "a dead woman attempting to function as if she were still alive; surrounded by cousins and cousins' wives full of laughter—and with her work still to find."[24]

101 To Margaret Hamilton

227 Beacon Street
January 10, 1935

Dear Margaret,

I have just been writing to Margaret Shearman. Edith Hilles' note about Mrs. Shearman's death came this afternoon. I wonder if you are going to the funeral. Maybe you will decide to wait till later.

There is a blow which has fallen quite suddenly and I have told nobody as yet and I do hate to tell you, but of course that is silly. President Conant has asked for my resignation. I took the letter over to Dr. Edsall and he was surprised, he had known nothing of it, but he reminded me that when he sent me my three-year appointment he warned me that the President had a right to ask for my resignation at any time after this year. Neither he nor I thought that would ever happen. He said that up to now the matter has been elastic, but he knew that Dr. Conant was in favor of having a definite age limit. Apparently it is in his hands alone. Well, of course I sent in my resignation at once.

Now it means $5500 lost, doesn't it? I am glad it came too late for you to change your mind. Had we known it last Spring you might have refused to resign this year and that would have been dreadful. And you know you and Clara and I can live in Hadlyme for $200 a month, or even less. It is of course quite disintegrating and yet I keep telling myself that two years should not make so much difference. Only I am really so strong and energetic that I am not ready to stop work, and this will stop it. I shall not be available if I am not in the place where people are used to finding me.

Then I am afraid of letting Miss Addams know I am free. She has just found that the person she was counting on to be her successor has refused and she is all at loose ends about the future of Hull-House. I don't mean that she would ever want me to do that but it would be so natural to expect me to help out for a couple of years. And I don't feel that I could do it with any spirit or vim. I couldn't do a new job and that would be new. Maybe something may turn up, but at the worst, if nothing does, it will not be bad to settle down quietly with you and Clara and perhaps do a bit of writing now and then. I have counted up carefully and I shall have enough to last amply through next Summer and till January 1st even if it costs me $100 a month while we three are together and it will not. But I do not think the ocean trip can come in 1936.

I have not told anyone yet, but I mean to tell Katy as soon as I get her alone—and of course Norah and Edith. Norah's last letter sounded happy. I will put it in. I am afraid Dr. Longcope was forgotten this week, but do see him next week, don't wait longer than that.

A.H.

A REVIEW BY Alice Hamilton of Dorothy Dunbar Bromley's *Birth Control: Its Use and Misuse* appeared in the *New York Herald Tribune Books* on January 6, 1935. Hamilton subsequently received letters from

349

several Catholics, mainly clergymen, objecting both to her endorsement of contraception and to her statement about the Church's position on birth control. Having always understood that Catholic women were told that it was sinful to deny their husbands, she had claimed that in endorsing abstinence the Church had changed its position. She exchanged three letters on the subject with Gerald J. McMahon, a Benedictine monk, who wrote her first the day the review appeared.[25]

102 To Gerald J. McMahon

January 11, 1935

Dear Father McMahon:

Thank you for your interesting and informative letter, which I should like to answer in full.

May I explain first that my interest in birth control has come from my contact with poor people altogether. I have never practiced medicine, my specialty is industrial medicine, but I lived for more than twenty years in Hull-House, Chicago, and came in intimate contact with poor Catholic women, Italian and Irish chiefly. With them there was no question of self-indulgence, it was a pretty desperate struggle to keep up some standard of family life and not to exploit the older children too ruthlessly, for the sake of the younger ones.

As to women of the upper classes, undoubtedly selfishness enters in sometimes, but I wonder if a woman selfish enough to hate motherhood would make a good mother. Many intelligent parents whose children would probably be above the average do, as you say, limit their offspring, but it is usually from fear of overburdening the family income so that the children will not have the sort of home and education that the parents think they should have.

My only interest in the birth control movement has been the spread of information among poor women, for I think that it is for them an urgent necessity to learn ways of preventing pregnancy aside from abstinence, since that is not to be expected from men in this class, at least this is the general experience. Your arguments against the method of incomplete copula seem to me to be on my side.

I quite agree with you that the ideal is not to restrict the number of children in a family but to make it possible to have large families without economic disaster, but I fear I shall not live to see such a well-ordered social system, so I must work for a system in which the size of the family can be controlled without disaster to the marriage.

Your statement as to the position of the Catholic Church on birth control is very surprising to me and I thank you for it. I realize now that my understanding of it was founded not on authoritative statements by theologians, but on speeches made by Catholic laymen, lawyers, doctors, and others in legislative hearings, and on conversations with many Catholic married women. I assure you that I have never heard a Catholic say that the Church permitted any form of family limitation, in fact women have told me that their confessors asked them as a routine matter whether they ever refused their husbands and always told them that to do so was a sin. I remember a woman whose husband was a wretched derelict; deserting her for most of the time, but returning often enough to make her pregnant every year or so. Four of her seven children were tuberculous and she was a wreck, but she insisted that while another pregnancy might kill her body, to refuse sexual intercourse with her husband would be to ruin her immortal soul. I heard a Catholic doctor say that the sin of birth control consisted in the refusal to give body to the myriads of souls waiting for incarnation—a doctrine I had supposed to be held by Mormons only.

If I have been misstating the stand of the Church therefore, it is because Catholics themselves misstate it to me. If the Church has really not changed her stand on family limitation ought she not to make that stand very clear, even to the most ignorant of her children?

I hope that, if there is anything in this letter that shows a misunderstanding of Catholic practice and teaching, you will write and tell me so, for it is far from my intention to say or write anything but the truth as I know it.

Very sincerely yours[a]

a. This is an unsigned carbon copy.

103 To Elizabeth Glendower Evans

227 Beacon Street
April 6, 1935

Dear Mrs. Evans,

I have a feeling that you thought me unappreciative and stodgy last Wednesday when Katy Codman and I stopped in and you were so full of exciting plans for my future and I sat by like a bump on a log. But I really was very much touched and no end grateful, only I was embarrassed by all the things you said about me and torn between a desire to protest and a feeling that protesting made me seem more absurd. You see, I simply cannot believe that I am a person of more than ordinary ability, though I know that chance has given me a more than ordinarily interesting life.

But what I really want to say is first that I do feel very deeply your interest in my future and I think it is lovely of you to encourage me to try to write, and I do mean to try, even if my attempt will be more modest than the one you advised. The other thing I want to make sure you understand is that I am not impoverished by losing my salary. I have enough to live on comfortably in Hadlyme and even to afford an occasional visit to Boston. And probably a few paying jobs will come my way. Anyhow, I rather guess I shall be spending more on myself than you are spending *on yourself* just now.

So let me thank you very warmly for caring so much about my future, and please keep on caring and I will promise to write you reports of anything important that happens and I will make Katy and Gertrude keep me posted about what happens to you.

I am putting in a picture of my beloved house, so you can see that a happy home is waiting for me. With love to Anna Bloom.

Yours always
Alice H.

At the beginning of May, Alice Hamilton took part in the twentieth-anniversary celebration of the Women's International League for Peace and Freedom in Washington, D.C. Jane Addams, whose return to public favor had culminated in a Nobel Peace Prize in 1931, figured prominently in the events. Both women spoke, as did Eleanor Roosevelt and Secretary of the Interior Harold Ickes, an old friend of Addams'. Returning to Chicago, Addams worked on a life of Julia

Lathrop. Then on the 18th, her physicians discovered that she had an advanced cancerous condition and gave her at best a few months to live. Hoping to persuade Grace Abbott, a former resident who was now professor of public welfare at the University of Chicago, to become head of Hull House, Alice Hamilton informed her of the situation.

104 To Grace Abbott

1430 Astor Street
Chicago
May 20, 1935

Dear Grace:

This is something I have told only Mrs. Bowen and Weber Linn and nobody is to be told it, for all JA's doctors are agreed that she herself is not to know.[a] She will not get well, she may have a few months of comparative comfort but if she lives on, it can only mean pain, it is quite hopeless. Whether or not I can take her to Hadlyme we do not know, but I hope I can, she wants it and she would begin to suspect if we did not go. I have the feeling that I can carry on to the end if she does not know, but I am not so sure I can if she does.

I am telling you this, as you probably have guessed, because of Hull-House. I want you to be thinking it over while you are on the ocean where there is so much time to think. You are the only person I can think of in this wide world who could do it. I am quite out of the question, I am too old and I have never had charge of big things nor been in authority over people for whose work I was responsible and I have no moral courage when it comes to dealing with things that have to be dealt with firmly. You have had an experience that would make running Hull-House perfectly natural and easy, you could handle the internal fusses and rivalries without taking them seriously, you could lay down the law about keeping within the budget, and you could do all the real part, understand which part of the work must be held on to and which could be sacrificed if need be. I can see that Mrs. Bowen is for cutting rather drastically but she would yield to your judgment as to where it should be cut. And there is nobody the residents would accept as they would you.

It seems to me that you need not live there. You are so used to keeping your hand on things without being there all the time. Ethel

Dewey is an excellent assistant and has taken that place for some years now.[b] She would be even better with a little more responsibility, more than J.A. has ever been willing to give anyone. I do not see how it would interfere seriously with your University work, provided you could cut down on your hours of teaching somewhat. And it is worth doing, surely. Hull-House must not be left to evaporate into a tasteless, colorless institution, going on more and more slowly as its moving power dies away. You could make it grow, change with the changing times, not keep on doing things because that is the way Miss Addams always did. Nothing is more contrary to her ideas than that, "the dead hand," she would call it.

Do think it over prayerfully, as my grandmother used to say. You are my only hope.

Yours always
Alice H.

Don't hurry to answer. Give my love to Emily Balch.

a. James Weber Linn, professor of English at the University of Chicago, was Jane Addams' nephew.

b. Ethel Dewey taught English and citizenship at Hull House.

JANE ADDAMS had dedicated her book *The Excellent Becomes the Permanent* (1932) "To Alice Hamilton whose wisdom and courage have never failed when we have walked together so many times in the very borderland between life and death." That borderland had included Jane Addams' many illnesses, and during Addams' last days Hamilton was again by her side.[26]

Jane Addams' death on May 21, 1935, brought forth a vast outpouring of sentiment throughout the nation and thousands of mourners to her funeral. It left Hull House, over which she had presided for nearly half a century, "aghast and adrift," for no arrangements had been made for a successor. On May 25 forty-seven residents petitioned the trustees of Hull House to ask Alice Hamilton to return for at least two years to take charge of the work: "No one in our opinion is so well fitted . . . both by natural ability and temperament and by understanding of the situation to carry on the work of Hull-House without interruption." Hamilton declined. As she explained to Grace Abbott (who also declined), if she served as a stopgap she would lose touch with her own field and be required to take up "all kinds of civic work" she had never done. Nor could she afford the position, for she would not accept a salary for nonprofessional work. To Agnes she wrote: "The residents are lost and panicky and want one of the old guard whom they can trust to cling to old ways and do nothing new and so they turn to me because that is just the sort of person I am, but it is not what Hull-House needs and I will not." The matter was settled by July, when Adena Miller Rich, a longtime resident and director of the Immigrants' Protective League, became the second head of Hull House.[27]

IX

SEMI-RETIREMENT
1935–1949

OLD AGE CAME SLOWLY to Alice Hamilton. Distressed when people asked why she had resigned from Harvard, she had observed: "I feel them suddenly relegating me to another category, of the aged, but I do not feel aged." In fact, she maintained an active professional life for the next fourteen years. That life centered in Hadlyme, where she now lived year-round with her sister Margaret and Clara Landsberg. She loved the house overlooking the river and the domesticated countryside; after eighteen summers, she felt they had established deep enough roots to cushion the shock of retirement. As in the very different milieu of Hull House, she was a good neighbor. She joined the reading and gardening clubs, enjoyed town meetings, and got along well with her mainly conservative neighbors. But she still wanted her work. And, since she received no pension, she needed to supplement the slender income she had from Fort Wayne real estate.[1]

It was the old-girl network rather than professional colleagues that brought her the work she wanted. Given her choice of two government posts, she turned down a full-time job with the Public Health Service, under the direction of assistant secretary of the treasury Josephine Roche, and chose instead to return to the Department of Labor, headed by Frances Perkins, a longtime member of the women's reform network. Hamilton had rejoiced in Perkins' appointment; it was enough, she claimed, to make her an adherent of Roosevelt. (She had voted for Norman Thomas in 1932.) Always loyal to her old department, in years past she had complained of the Public Health Service's interference with her work. She was also unwilling to tie herself down to a full-time job. She did not want to retire, but neither did she wish to be deprived of the comforts of Hadlyme.[2]

In the fall of 1935, Alice Hamilton took up her work as medical consultant to the Division of Labor Standards, a unit of the Department of Labor established the year before by Frances Perkins. To improve health and safety standards, which varied widely in the forty-eight states, the division promoted uniform labor legislation and administrative practices, provided technical assistance to state labor departments, and sponsored special training programs to raise the competence of state officials. It also focused national attention on serious health problems, most notably, in the 1930s, silicosis. Officials defined the mission of the division as the translation of already available information into more effective programs of control, contrasting this work with that of the Public Health Service, which concentrated on research.[3]

In her new capacity Alice Hamilton conducted surveys, offered advice on occupational poisoning problems, attended conferences, and testified at hearings. Her first assignment was a study of changes in the painters' trade. Updating her 1913 bulletin on the subject, she discussed the dangers of the new constituents of paint, assessed the hazards of the spray gun (a device resisted by skilled painters that she did not find especially harmful in itself), and recommended prophylactic measures. She also brought neglected health problems to the attention of Department of Labor officials. One of these was the high incidence of tuberculosis and silicosis in the Tri-State mining area of Kansas, Missouri, and Oklahoma, which she had first visited in 1912 while studying lead smelting and refining, and which remained dismally poor. Frances Perkins called a conference in Joplin, Missouri, in 1940 to investigate conditions in the region.[4]

Hamilton's most important work during these years was a study of poisons in the manufacture of viscose rayon. This rapidly growing industry used two dangerous substances: carbon disulfide, a poison that acted on the central nervous system and caused manic-depressive insanity, motor paralysis, and loss of vision; and hydrogen sulfide, a powerful asphyxiating poison. Hamilton had been one of the first Americans to call attention to carbon disulfide; she had come across it in 1914 in connection with her survey of the rubber industry and in her 1925 book had reported its use in viscose rayon manufacture. But despite these efforts and a vast European literature, the subject remained virtually taboo in the United States: no systematic investigation had ever been undertaken, and few psychiatrists recognized the substance as a cause of psychosis. During the early 1930s Hamilton became increasingly concerned about the problem as she received reports of serious illnesses among viscose rayon workers. But the industry drew a "veil of secrecy" over its manufacturing processes, and Hamilton had unusual difficulty in gaining access to plants.

Frances Perkins and Verne A. Zimmer, director of the Division of Labor Standards, agreed to investigate the industry, with Alice Hamilton as principal medical consultant. They began by enlisting the cooperation of the Secretary of the Department of Labor and Industry of Pennsylvania, an industrial state with a backward labor record. The multifaceted study included air analyses of workplaces, a search of commitment records to state hospitals for the mentally ill, an autopsy, interviews with 159 workers and their families, and medical examinations of 120 currently employed—and thus presumably "healthy"—viscose rayon workers. (Because the men feared losing their jobs, the examinations were conducted in secret, with unmarked state cars providing transport.) F. H. Lewey, professor of neurology at the University of Pennsylvania, coordinated a team of eight clinical and laboratory specialists who subjected the workers to a battery of sophisticated diagnostic tests, some of which they developed themselves. Among the group's findings: 71 percent of the workers manifested psychiatric symptoms ranging from irritability and severe insomnia to manic-depressive psychosis; 76 percent sustained injury to the peripheral nerves; and 54 percent suffered damage to their eyes. The Pennsylvania study thus conclusively demonstrated the dangers of the viscose rayon industry in the United States. After extending the field investigation to nine other states, Alice Hamilton presented the findings in *Occupational Poisoning in the Viscose Rayon Industry*, a bulletin published in 1940. Although the investigators had encountered considerable resistance to their work—there was even an attempt to switch cadavers, a deceit revealed only by an appendectomy scar—she was satisfied by the rapid changes instituted by the companies following the report. It was her last investigation.[5]

The viscose rayon study made use of techniques that had not been available when Alice Hamilton began her work a quarter of a century earlier, among them more accurate diagnostic tests and technical devices that measured the quantities of dust to which workers were exposed. As long-developing conditions like silicosis became compensable and clinical procedures improved, employers and insurance companies required workers in dangerous industries to undergo physical examinations, with a view to identifying those who showed early symptoms of disease. Workers sometimes objected to this approach, which often went hand-in-hand with neglect of hygienic precautions and an untimely loss of work. Hamilton, who always stressed prophylaxis, favored physical examinations but was disinclined to force workers out of their jobs permanently.

105 To John B. Andrews

Hadlyme Ferry
February 17, 1938

Dear Mr. Andrews:

The question you put up to me, concerning the physical examination of workers, is one of the most puzzling in the whole field of employer-employe relationships. From the point of view of the physician, it is unquestionably true that for a workman's own protection he should be examined, lest he be permitted to continue in work dangerous for him or take on new work of dangerous character. But from his point of view the situation is more complicated. If he suspects that he has already some occupational disease and that examination will reveal it and result in his removal from such work—often the only sort of work for which he has any skill—then he refuses to submit to an examination, he prefers to keep his skilled job and die at an earlier age, rather than to join the unskilled or the unemployed and lower his and his family's social standing and standard of living. And his objections cannot be really met until we can assure him that what we advocate as a protective measure for him does not carry with it such a menace.

When it comes to the detection of an infectious disease which is a danger to his fellow-workers, such as tuberculosis or syphilis, he has no right to refuse and I believe even the trade unions would admit that. I quite agree with your statement that the examination should be made by an impartial doctor and that the information should not be used to the detriment of the worker examined, but of course that last proviso means that it must not be used to exclude him from any

359

sort of employment and that brings us back to the same impasse. I hope you will write me if you have any solution for this difficulty.

Sincerely yours
Alice Hamilton

ANOTHER NEW and controversial approach to industrial hygiene was the setting of "allowable concentrations" or "toxic limits" of dangerous substances to which workers might, with presumed safety, be exposed. Hamilton supported this endeavor and represented the Division of Labor Standards on the Committee on Permissible Limits of Dusts, Gases, and Vapors, established by the American Standards Association in 1938. But she cautioned that the figures were "often based on something not far from guess work" and should be "advisory only." Still, explaining that she was "always willing to accept half a loaf rather than no bread," she agreed to a standard for benzene that was considered unacceptably high by two colleagues, on the grounds that the standard could be "realized practically while a more rigid standard would be rejected, and we should be left with no standard at all." After the war she warned that the "so-called maximum allowable concentrations" were based on the quantity of poison inhaled by a worker breathing at a normal rate for eight hours and thus must be lowered for overtime or for heavy, hot work that resulted in deeper or more rapid breathing.[6]

Alice Hamilton returned to Nazi Germany in 1938, this time as the Department of Labor's representative at the Eighth International Congress on Occupational Accidents and Diseases, which met in Frankfurt September 26–30. There she delivered a paper on viscose rayon that she had coauthored with F. H. Lewey, formerly director of the Neurological Institute of the University of Berlin, now a refugee. During the summer she had received a letter from "the head of a Jewish organization, begging me not to go, but to make a public refusal on the ground that this Congress cannot be international because Jews are excluded." Though troubled by the situation, she felt it was too late to withdraw, noting that her Jewish friends had understood her earlier visit to Nazi Germany and that the trip would "enable me to get fresh material on the situation."[7]

The congress was overshadowed by the threat of imminent war. Hitler had announced his intention of invading Czechoslovakia unless that country returned by October 1 the Sudeten area, territory inhabited by Germans that had been ceded to Czechoslovakia after the war. Since Czechoslovakia had defensive treaties with France and the Soviet Union, a full-scale European war seemed likely. To forestall this possibility, British prime minister Neville Chamberlain and French premier Edouard Daladier, meeting with Hitler and Mussolini in Munich on September 29 and 30, acceded to the dismemberment of Czechoslovakia. Eduard Beneš, president of the Czech Republic, resigned and went into exile after learning of the agreement.

Like many other delegates, Hamilton made a hasty exit from the congress; she presented her paper a day early so she could drive with American colleagues to Holland. After inspecting some viscose rayon factories there, she joined her sister Norah in France for a month's vacation, which they spent mainly on the Riviera. The visible signs of war preparations in Holland and in France brought home the narrow margin by which tragedy had been averted. Although Munich later became a symbol of abject appeasement of Hitler's expansionist policies, Hamilton's sense of relief at the resolution of the crisis was widely shared at the time by many mainstream politicians, including President Roosevelt, as well as by pacifists. By the time she wrote her autobiography in 1942, Hamilton had an uneasy conscience about her position and noted the one-sided presentation of news, not only in Germany but also in France, where she had found no questioning of the Munich agreement.[8]

106 To Edith Hamilton

Menton
October 23 [1938]

Dear Edith,

Just after I had mailed letters and cards saying that we were perishing for letters a whole batch appeared this afternoon, written between Oct. 3rd and 10th, yours with enclosures from Margaret, one from her written on the back of Mrs. Bowen's, and one from Agnes to me and from Jessie to Norah. We were so hungry for them. I wish you could have had better news, about Doris [Reid]. It is so wretched for her to have that illness and it takes so long. Amory Codman went through it last summer, a whole month in bed, then another month with a steamer chair on the porch added to bed. But he really did get amazingly better at the end. Give her loads of love and sympathy from me.

From all you people we get the impression that you think Chamberlain and Daladier did the cowardly thing and that they should have stood by Beneš and defied Germany. But that is because there are three thousand miles between America and Europe. Over here one can't help feeling what war really means and feeling that nothing can possibly be worse. I don't think Chamberlain did take back "peace with honor," there could be no peaceful and honorable way of clearing up a mess made by the three men at Versailles, but suppose he had let war break out, there would have been no Czecho-Slovakia left to enjoy the honor. England and France could not conceivably have saved Prague from bombardment nor the whole countryside, in fact France could have done nothing without deliberately invading Germany. And, as both French and English statesmen are now confessing frankly, their people would hardly have consented to a war *against* the principle of self-determination. For however ugly and dangerous Hitler's methods were, his cause was just. The Sudeten Germans should never have been given to the Czechs, and England had nothing to say when Hitler said that he was asking only for what England granted to Ulster. The crime of France has been, not her backing out at the end but her encouraging the Czechs to rely on her and on Russia and not try to make friends with the Germans and Poles and Hungarians nor to meet reasonable demands. Beneš has not been really the wise statesman we thought him. As for Chamberlain, truly I do believe there would have been war except for him. Hitler wanted it, he had mobilized his great army and he was itching to try it out. He was plainly disappointed when Chamberlain at Godesberg told him Czecho-Slovakia would yield and at once he stiffened his demands—without ever letting his people know he had—and he felt sure that he would be able to march in on October first. I think his Sarrebruck speech showed his irritation and balked ambition, when he hit out at the English so childishly. Had Chamberlain not been able to bring in Mussolini and force another conference, war would have come. And what a war! If you could hear how people talk over here, of the mobilization, of the desperate attempt to evacuate Paris and to protect those who could not go, everyone knowing that no protection would amount to much. Down here the people fled by the thousand, for everyone expected the Riviera to be bombed by Italian planes and the frontier is right here. In Tunisia they waited for the same thing and with not one anti-aircraft gun or even a gas mask. In Holland they had mobilized, but they were waiting in grim helplessness to see what would happen to them. You see, aircraft has made war so swiftly deadly. In former times people could calculate how far off the enemy was and how long it would take to march to a certain point. Now it is a question only of hours before the airplanes are over the city, sometimes of less than an hour.

It is true that Hitler grows bolder all the time, as his demands are yielded to, but in justice one must admit that so far he has demanded nothing that should not have been yielded long ago to the peaceful requests of the Republic. It was mad for France to insist on refusing Anschluss with Austria when Stresemann asked it.[a] When Clara and I were in Germany a young officer said to us "Why do you say we should not believe in the doctrine of force? I tell you Hitler will gain by the threat of force far more than Stresemann ever won by concil-iation." And it is true. When Hitler says that the Sudeten question was settled only because of the threat of the German army, he is telling the truth. The victorious countries have blundered and have sinned against justice and it would have been a terrible thing if they had gone to war in the effort to perpetuate injustices. And if you could have felt the long breath of relief that went up when it was known that war had been averted. Chamberlain was cheered in Germany as heart-ily as in his own country. Probably that is what made Hitler so cross. But don't think that if France and England and Russia had stood firm, Hitler would have drawn back and yielded. He was bent on invasion and Ribbentrop, Goebbels and Himmler were egging him on. Cham-berlain saw that at Godesberg and that is why he worked so desperately for time. I don't know if he is a great man but he did a great deed once in his life anyway.

The letters told of Jessie's eyes, of her learning Braille. Agnes said that her next appointment with her oculist was not till October 17*th*, a week ago. We ought to hear about that before I sail. It is too sad even to talk about, but it can't be hopeless yet. Agnes says Jessie thinks one eye is better, the other worse. I wonder if they would not like to come and live with us until they go west with Taber and Gail. Mar-garet says they seem so alone.

We are in a shabby little hotel, with very pleasant bedrooms and delicious food and a charming young Italian woman to look after us. Menton is right on the sea which is only a[b]

a. In March 1938 Hitler had sent toops to Vienna, thereby attaining Anschluss, the incorporation of Austria into Germany. Gustav Stresemann, German foreign minister from 1923 to 1929, had followed a conciliatory foreign policy in the hope of securing an end to the harsh provisions of the 1919 peace treaty.

b. The rest of the letter is lost.

H‌AMILTON'S FRIENDSHIP with Felix Frankfurter had ripened during her Boston years. Frankfurter, then a professor at the Harvard Law School, was at the peak of his career as a fighter against injustice and defender of the underprivileged. His public image as a radical had begun with his report, as secretary and counsel to the President's Mediation Commission during World War I, that the conviction of labor organizer Tom Mooney for planting a bomb that led to several deaths had been obtained by perjured testimony (the ensuing publicity prompted the governor of California to commute the sentence of execution to life imprisonment). That reputation had been consolidated by Frankfurter's pivotal role in the Sacco-Vanzetti case, and had been further intensified by his position as an influential adviser to President Roosevelt, who appointed many Frankfurter protégés to important government positions. Early in January 1939 Roosevelt nominated Frankfurter as associate justice of the Supreme Court. Although Frankfurter's alien birth, religion, and alleged radicalism all came up at the public hearings, the Senate quickly and unanimously confirmed the nomination. The *New York Times* reported on January 29 that 82 percent of the American public approved the choice.[9]

107 To Felix Frankfurter

Hadlyme Ferry
January 29, 1939

Dear Felix:

I know you have had felicitations galore and I sent my own long ago, but I feel more inclined today to send them than when I first heard the good news, because I have just read the last Gallup poll, on your appointment. It pleases me even more than the appointment did and I wonder if Marian does not agree with me. First, because it shows your popularity, second because it shows a so much more intelligent and liberal outlook on the part of ordinary people than I usually expect, and third because the rate of anti–Semitism is so small. Also I am wickedly glad to know that [A. Lawrence] Lowell is still alive and in his right mind and able to read.

When one thinks of Sacco and Vanzetti and of Mooney and of your filling Washington with young radicals it is quite wonderful, isn't it?

Yours always
Alice H.

Like Alice Hamilton, many of Frankfurter's friends expected him to support liberal values on the court as he had off the bench. Consequently they were dismayed by his opinion in *Minersville School District* v. *Gobitis* (1940), a case that raised the question of the court's role in protecting the religious freedom of a minority against the action of the majority. Frankfurter, speaking for the court, upheld the right of a school district to require the children of Jehovah's Witnesses to salute the flag, an action that violated the sect's religious scruples. He further denied that the compulsory flag salute was an infringement on religious freedom and declared that local authorities had the right to determine appropriate ways of evoking the social unity necessary in a pluralistic society. Writing at a time when Hitler was overrunning the nations of western Europe, Frankfurter, always an ardent patriot, further maintained that "national unity is the basis of national security" and that the flag was one of the foremost symbols transcending internal differences. When Hamilton expressed disagreement, Frankfurter sent her a copy of his opinion and reproached her for dismissing "other people's conception of their duty, simply because it runs counter to your beliefs." Hamilton's response reveals her commitment to the primacy of personal freedoms, a position taken by Associate Justice Harlan Fiske Stone, who had written an impassioned dissent in the case. It was a debate Hamilton and Frankfurter were to have many times.[10]

108 To Felix Frankfurter

Dear Felix:

Your letter deserves a very well thought-out answer and I will try to write one, though I know that it is not easy to argue with you and that I shall probably be far from convincing. But I do want you to understand why I feel as I do about this most important decision of yours.

The Jehovah Witnesses cases have interested me from the beginning and I have followed their course through the courts eagerly, welcoming with joy the decision of the Quaker, Judge Maris, and never doubting for a moment that it would be confirmed by the present Supreme Court. I have read through this decision of yours twice, but I find it no easier to understand how or why you came to write it than I did when I had only the excerpts in the Times to go by. It seems to me at variance with your principles, as I thought I knew them but plainly I did not know them.

I do not see how you can make the distinction you do between freedom of religious belief and freedom to control the education of one's children. In the first place, what is to most of us not a matter of religious belief may be one to others. Judge Maris' decision, as I remember it, stressed that point. It was not a question of whether the Court held the flag salute to be a religious gesture, the point was that Jehovah's Witnesses did.

In the second place, you imply that while compulsion in religious matters is wrong, compulsion for the sake of national unity is right. But all the religious persecutions of the past and the present have been motivated by a passion for national unity, from the days of Ferdinand and Isabella and the Revocation of the Edict of Nantes, to Stalin's Russia and Hitler's Germany. In the past, the Church was the symbol of unity, in our day, we, knowing that we have no longer the Cross or the Bible have turned to the flag, with the same religious devotion as was given to those older symbols. It seems to me that all your arguments in defense of a compulsory flag salute would apply quite as well to Catholic Spain, where the adoration of the Host takes its place. The Spanish clergy would agree with you, that to permit a school child to absent himself from Mass "might cast doubts in the mind of the other school children." But do you really think that would

366

be worse than to have the children say: "You'd better. You'll be expelled if you don't."

I cannot see how you could write the second paragraph of page 8 in defense of your opinion, for it seems to me a direct challenge to it. "So long as men's right to believe as they please . . . is fully respected." But if you forbid action which a certain belief enjoins, surely you are not fully respecting that belief.

You call this an "issue of educational policy." Surely it is more than that. For an educational policy has not the force of law. I should say that it was exactly the province of the Supreme Court to "choose among competing considerations in the subtle process of securing effective loyalty to the traditional ideals of democracy, while respecting at the same time individual idiosyncrasies among a people so diversified in religious allegiances."

"The ultimate foundation of a free society is the binding tie of cohesive sentiment." But sentiment cannot be forced. Will children forced to do what is contrary to the religion taught them by their parents ever be imbued with a genuine love for a country which thus forces them?

You tell us that the religious liberty which the Constitution protects has never excluded legislation "not directed against doctrinal loyalties of particular sects." But, Felix dear, what is the protest of Jehovah's Witnesses but the doctrinal loyalty of a particular sect?

This is the first time I have ever differed from you on a subject on which I feel deeply, much more deeply that I did on a big Navy or any other of the suspect actions of the President. Because this decision opens the door to so much and it seems based on a difference in fundamentals which I did not think existed. It makes me wonder what will happen to conscientious objectors, to anti-War people like [Eugene] Debs and Kate O'Hare, to the many laws that are looming up already, against Fifth Columns and "subversive elements." If national unity is of first importance, where are the dissenters to go? And in the past, dissenters have so often been the salt of the earth.

After I had written this I discovered the opinion of Justice Stone, following yours and now I see I need not have written these pages, for he has said it all so much better. However, here it is and I am sending it on, for I do want you to know just how and why I feel as I do.

With just as much love as if I did not think you altogether wrong.

Yours always[a]

a. This is an unsigned carbon copy.

Aᴛᴇʀ ʜᴇʀ ʀᴇᴛᴜʀɴ from Germany, Alice Hamilton had continued to oppose efforts to enlarge United States military capacity (increased naval appropriations and the fortification of Guam) as well as military training in the schools. For many years, in talks to peace groups, she had stressed the analogy between the prevention of disease and war, arguing that war would cease to exist only when statesmen opposed it as uncompromisingly as physicians did disease. She later wrote in her autobiography that Hitler's invasion of Poland in September 1939 (the act that officially launched World War II) had left her "bewildered," as it had most of her pacifist friends: "My clean-cut principles no longer seemed to apply." Her first reaction to the event was to agree "heart and soul" with Colonel Charles A. Lindbergh who, in a radio address on September 15, urged that the United States keep out of war: "I would let Hitler go on like a raging lion swallowing country after country rather than try to stop him by war, because I do not believe that anything he can do will be as disastrous for Europe and the rest of the world as will war." This uncharacteristic outburst reflected her instinctive pacifism as well as her belief that Germany (which had gained Japanese support and had neutralized the Soviet Union) could not be defeated; nor did she believe that the German empire would long survive.[11]

Events at home and in Europe gradually prompted her to reassess her position. She found the "anti-war movement" (by which she presumably meant the conservative and isolationist America First Committee that predominated after 1940) "narrow and nationalistic." Moreover, as Hitler conquered one European nation after another, and it became clear that two of her deepest principles—maintaining peace and protecting human rights—came into direct conflict, she concluded that to stand aside was a form of selfishness that would "stifle some of the finest qualities in our nation." By December 1940, in an article in *The Nation*, she endorsed sending food to western Europe, a move opposed by some who thought it would undermine the British blockade of the continent. Her response to starvation was hardly surprising, but her argument suggests that her position on the war had already changed: "This war is being fought to save the values for which civilized people stand—not only for freedom and justice but for mercy and a belief in the supreme worth of each human being. If we choose as our weapon starvation, not of the fighting forces but of the most helpless, are we not destroying the very values we seek to save?" And in her autobiography, written early in 1942, she observed: "It is no defense of war as a means of settling disputes to say that when once war has been started by greed for power and helped on by blindness and selfishness we cannot save the world by saving our own selves, we must get down into the arena and throw our strength on the side we think the right one."[12]

United States entry into the war had dramatic consequences throughout the nation. Even rural Hadlyme felt its impact. Within two weeks after the Japanese bombing of Pearl Harbor on December 7, 1941, residents made a collection for the Red Cross, gathered information about empty rooms and spare blankets and cots in case of an accident in a nearby war plant, and planned to open the schoolhouse to encourage women to mend old clothes that the Red Cross had rejected. Alice Hamilton signed up for a first aid course because she thought everyone would expect her, as a doctor, to know about it. "There is none of the enthusiasm or the silliness of 1917," she reported, "nobody rushing to wear uniforms or drive ambulances, just a dull determination to do what each of us can." Accommodating themselves to the consequences of war, the Hamiltons and Clara Landsberg husbanded gasoline and, after losing the black couple who had lived with them to war work, meticulously divided up the cooking and other chores so that no one would worry about not doing her share. Throughout the war Alice and Margaret knitted and sewed for the Red Cross, and Margaret also served on the rationing board.[13]

During the first winter of the war there was a series of spy scares in Hadlyme after a man from the State Defense Council addressed the town meeting on the subject of fifth columnists. Alice and Margaret Hamilton feared the possible consequences for their Japanese neighbor Yukitaka Osaki, a colorful local figure who shared their interest in gardening and sometimes brought their mail from the post office on his donkey. Osaki was distraught over Pearl Harbor and told Margaret Hamilton that he did not expect to survive the winter (he died in October 1942). Believing that silence was the best policy, the Hamiltons were alarmed when Allen H. Gates, minister of the Congregational Church in nearby East Haddam and an indefatigable writer of letters to the *Hartford Courant*—there were five in January 1942 alone—publicized Osaki's plight in a melodramatic appeal for tolerance.

369

109 To Agnes Hamilton

Dear Agnes:

I have no business to be writing you this morning, just after breakfast, with letters to people who are a chore waiting for me, but your letter which came yesterday makes me feel like it, so here I am doing it. You must think of me in front of the library door, looking out on sun-lit ice, with a narrow strip of deep blue water where a great tanker has just passed through. There is snow everywhere, and the thermometer is twelve above, but mercifully there is no wind as there was yesterday. Today our usual routine is broken into, we go to lunch at the Old Lyme Inn with Mrs. Stutz and then on to meet Clara on the Boston train, then for fish and the week's marketing, for we go over to Deep River only once a week now, to save tires. Clara must make up her mind between Arizona and Alexander, and that is what she is coming for, to talk it over with Margaret. I suppose you had better not say anything to Miss Barthelow, but if she decides on Boston, it will be rather hard on Miss Barthelow.[a] Only it probably will not mean that she has refused somebody in order to hold on to Clara's room, for you have so few there this year. It is strange that more people do not go to Arizona, it is so much safer than the Coast.

We are in the midst of spy scares, the most absurd thing for a little country village. I wonder how many thousands of denunciations the F.B.I. gets, if this is an average place. Margaret says she told you about Osaki, how we felt quite safe about him and how nice Mr. Gates was with him. But, three days later Mr. Gates wrote me enclosing the carbon copy of a letter he had written to the Hartford Courant, recounting all about his conversation with Osaki and even adding what Margaret had repeated, Osaki's saying that he would never live through another winter. Mr. Gates said that if we objected to it, he would "countermand its publication." We drove up at once to his house and I told him it must be stopped, that it was a violation of confidence and that we had been warned to keep Osaki's presence as quiet as possible, whereupon he said that he was sorry, but the letter had appeared that morning. There had been a slip-up in the mailing of his letter to me. "But you could have telephoned us," I said. We did leave him in a pulp of apologies, but I can't help suspecting that he did not want us to get the letter in time to stop the Courant.

As I got in the car, he came out and said, "The paper is going to send a reporter down to interview you and Mr. Osaki." My experience with reporters has been that they are the most ruthless and unscrupulous lot, so I mounted guard that afternoon and caught the man before he went over to Osaki's. He proved to be a wonder, though he did not look it, he was a Pole, Comiskey, uncouth and chewing gum. But I made him see how dangerous it was to make Osaki conspicuous and while we talked in came Osaki himself and put the last touch. Comiskey went away promising to write nothing at all about it. Mr. Gates has written me a letter of abject repentance and of course I won't be nasty to him but I suspect he is pretty self-righteous and gets most of his satisfaction in denouncing the less righteous who just now are those with intolerance toward our enemies. That is all right, to work for tolerance, but it is really a quality he has very little of.

We must have written you about Mr. Peterson, who has the house Mrs. Largen built up on the hill above the village.[b] The F.B.I. came to investigate him, at least so we are told, and there are all sorts of stories about their waiting around for weeks to find out if he had a powerful sending set. It seems that he was reported to Washington by a dreadful man whom he had been trying to pull up, a drunkard who, left alone in the house, drank up all the liquor in the cellar and then shot himself. I am sure that if the story about the F.B.I. is true, they found nothing, for Mr. Peterson was at our Town Meeting.

Now it is Miss Shoe. Mrs. Hanna called up the other evening to say that she had been summoned by an agitated East Haddam lady to come to a meeting of the Women's Auxiliary of St. Stephens Church to consult what to do about Miss Shoe, who is a German spy with a short wave transmission set. It seems that the informer this time is Leavenworth, Hazel's husband, who works occasionally for the Kihns. What will come of it, we do not yet know, but Mrs. Hanna said she would go to call on her at once, and do what she could with the Church women. Only unfortunately Miss Shoe has offended Mr. Page, by saying something tactless about inefficiency in some church affairs, and he, being a Canadian, is especially fervent in his pro-war feelings. We are going to have her over for tea on Monday when the Reading Club meets here. As Margaret says, such things force one to be so much more attentive than one wants to be.[c]

I am so sorry you could not give us any news of Helen, but I know that this has been a terrible blow to her. Clara says that Bessie Knight

371

was one of the warmest and most vivid women she ever met. I remember her only as a very pretty Sunday School girl and I never saw Carole except once in a movie and then I did not care much for her but all that the papers say about her make her seem an unusually human and kind and generous actress, very much loved. And there could be no more tragic form of death than that.[d]

Now I must stop. This letter has been interrupted by household chores and news radio and now it is time to start for Old Lyme.

Love to you both
Alice—

a. Mildred Barthelow managed Adobe House, the inn in Scottsdale, Arizona where Agnes and Jessie Hamilton and sometimes Clara Landsberg spent their winters in later years.

b. Paul Peterson kept the part of his basement where he made hard cider locked.

c. Miss Shoe, a nurse, lived across the road from the Hamiltons; Grace Hanna was the wife of the first selectman; W. Langdon Kihn, a painter, and his wife, Helen, lived in the house next to the Hamiltons; Mr. Page, an Episcopal priest, lived in East Haddam.

d. Elizabeth (Bessie) Knight Peters was Helen Knight Hamilton's sister. Carole Lombard, daughter of Elizabeth Peters, died in a plane crash on January 16.

Ever since Alice Hamilton's retirement, Edward A. Weeks, editor of the *Atlantic Monthly* and an admiring cousin on her mother's side, had been trying to persuade her to write her autobiography. As her investment in work and the demand for her services diminished, she bowed to his campaign, and in early May 1942 she submitted a first draft. The act of self-revelation demanded by the genre proved difficult for one who valued her privacy as much as Alice Hamilton did. "When you sit down to write about yourself," Weeks wrote, "it is just as if you threw half the keys to your life into the Connecticut: half the doors to you—the real Alice Hamilton are locked throughout,—your likes and dislikes, the impressions people made on you, the exhilarations and disappointments you suffered in your work, the grief, the humor, and the love of people." Although she informed him that she was over seventy and could not "of a sudden become frank and outspoken and able to confide personal intimacies to the public," she took well to his editorial guidance. A "fast worker," according to Weeks, she completed her revisions in a few months.[14]

More detached and less immediate than her letters, *Exploring the*

372

Dangerous Trades is the quietly understated work of a woman who refused to see either her life or herself as unusual. She told the truth, but not all of it: she revealed no family secrets, dwelled on no personal slights, and said nothing that might hurt anyone living, at least no one she mentioned by name. Felix Frankfurter, for one, wanted "more personalia" and missed in the book some of the qualities that delighted her friends, including the skill in argument that enabled her to best even Bertrand Russell in debate. But if *Exploring the Dangerous Trades* lacks intimacy, it is nevertheless deeply personal, for Alice Hamilton wrote entirely out of her own experience. She made excellent use of the personal vignette, whether as poignant reminder of the hardships faced by others (an aspect highlighted as well by Norah Hamilton's vivid illustrations), or as humorous puncturing of her own enthusiasms or the absurdities of people and situations.[15]

If her book did not tell all, it was, as she intended, a record of the changes that had occurred in the field of industrial medicine since she had entered it three decades before. Taking the long view, she concluded that the lives of most men and women had improved immeasurably. In her own field, she believed that workers would never again be subject to such grossly hazardous conditions as she had witnessed in the early years. She recognized the existence of callowness and selfishness, but by both temperament and experience she preferred to dwell on generous motives and positive changes. Today, when the dangers of industrial and environmental diseases have forced themselves anew on the attention of workers and public alike, Hamilton's optimism seems complaisant if not misguided. But her trust in the good faith of others had been amply repaid over the years. In her seventies, she could well look back on what had been accomplished rather than focus on what remained to be done. She later noted that in writing her autobiography she had striven for impartiality: "I knew that I was prejudiced in the workman's favor and I must try to be fair and unemotional." Nevertheless, she managed to convey the passion and determination that had prompted her to spend her life confronting poverty and suffering and to do what she could to eliminate one potent source of misery. Alice Hamilton's autobiography was, as she once wrote of another's work, "fair but not too fair."[16]

The reviews of *Exploring the Dangerous Trades* (published first in an abbreviated four-part serialization in the *Atlantic Monthly*) were uniformly favorable. Characteristically, Alice Hamilton disliked a glowing tribute in the *New York Times* that she thought "makes the book sound like a muck-raker which it is not" and considered Emily Greene Balch's review in the *Social Service Review* far more discerning. She must have been moved by the tributes that poured in from all manner of friends and acquaintances, from Frances Perkins and former professors who had followed her career to the man who mended the ferry slip. She also heard from strangers whose lives she had somehow touched: a man from Fort Wayne who had known of the Hamiltons as "quality folk, [a] real 'first family,' " alumnae of Miss Porter's School, prospective physicians. People were as impressed by the manner of the book as by its substance: "My God, how you and your sister Edith can write!" one admirer exclaimed. Perhaps one of the most revealing statements came from Marion Frankfurter, who noted: "You are one of the rare people who write as they speak." But because Alice Hamilton talked so rarely about herself, even friends of many years were surprised at the range of her work and the dangers to which she had repeatedly subjected herself.[17]

110 To Emily Greene Balch

Hadlyme Ferry
May 10, 1943

Dear Emily Balch:

This is the best review my book has had, the only one that brings out the parts I care most about, and shows you really read it, which most reviewers did not. There is nothing in it I should want changed. It pleases me enormously to have you say such nice things about it, for you are one of those who know what I was writing about. And you are merciful not to point out a ghastly blunder, which Gertrude Winslow has written me about, the way I have mixed up two distinct Lawrence strikes.[a] If ever there is a second printing, I shall have to make five corrections.

I am so sorry to hear about your sister's relapse. That is so hard an illness for others to watch, so impossible to help and yet so sure one ought somehow to be able to. I hope it clears up soon.

The Hull-House residents and trustees have begged me to go to the annual meeting of the H.H. Associates on May 19th, and I am going, though I feel quite inadequate to cope with the situation that has arisen there during late years, when there is a paid staff, unionized and "bargaining" with the trustees about wages and hours, the residents being apparently only side-issue, the neighbors demanding equal or majority representation on the board of trustees, and the Boys' Club becoming ungovernable under the war time atmosphere. It is all so different from settlement life as I knew it. I saw Dick Winslow, of South End House in Boston and Mrs. Simkhovitch in New York, but neither of those settlements has undergone such radical changes as those and they could tell me little.[b] Of course I know that our neighborhood is no longer filled with newly-arrived foreigners, helpless and defenceless, but with products of our public schools, who do not want charity, who want to have things as they decide. But how to bring about a more democratic state of things, when there is no responsibility for providing the money, I do not see. Also I cannot accept the change from a volunteer group to a professional staff with all the psychology of a fighting trade union. However, if I am to do any good I must try to see it from the new point of view.

Don't think I have stolen your "Women at The Hague." I really do mean to mail it back to you.

Thank you very much for this review.

Yours always
Alice H.

a. Gertrude Winslow, secretary of the Community Church in Boston, belonged to AH's circle of liberal and radical friends there. She had been active in the Sacco-Vanzetti case.

b. Richard Winslow, Gertrude Winslow's son, was head worker of the South End Settlement House in Boston. Mary Kingsbury Simkhovitch was head of Greenwich House, a New York settlement.

DURING THE EARLY YEARS of her retirement, Alice Hamilton had continued her yearly talks at Bryn Mawr College (through 1943), adding annual lectures on industrial toxicology at the Woman's Medical College of Pennsylvania (1937–1942) and for some years also at Tufts Medical School in Boston. She lectured at medical societies on special occasions as well as at Harvard and at Connecticut College for Women. She also served as a trustee of Hull House. And, after many years as a vice-president of the National Consumers' League, she became president of the organization in August 1944, thereby breaking her "life-long rule, never to take a position simply as a figurehead, lending my name without doing any real work."[18]

As her work for the Department of Labor tapered off after 1940,

Hamilton was drawn away from the domestic scene less frequently. Sometimes the sisters left Hadlyme during the coldest part of the winter; they visited Morocco and Spain in 1936, Mexico in 1937, New York and Boston during the war years, and Guatemala in 1948. Alice, who noted her "absurd clinging to my own place," was eager to return home after any trip. Edith and Norah evidently thought that their Hadlyme sisters vegetated in retirement, and Alice sometimes felt like a "country mouse" when she visited them in New York, which was fairly often before Edith's move to Washington in 1943 and Norah's death in February 1945. Only one surviving letter alludes to Norah's death, but a letter Alice had sent Edith Abbott on the death of her sister Grace suggests something of what the loss must have been: "The deepest and closest relation I have ever known is that between sisters and I need no imagination to know what this means to you."[19]

111 To Jessie Hamilton

Hotel Statler
Detroit
[1945]

Dear Jessie,

I have been watching for more than half an hour the most wonderful parade past the hotel, tens of thousands it seems to me, chiefly Italians, Poles, Bohemians and negroes. The negro girls have just gone past in the pattern of an American flag. In between come khaki soldiers and innumerable bands of music. I wish Holman [Hamilton] were here, though I think you are really the one who would like it best and believe in it most. Well, I believe in it too, some of it anyway. Just at this minute they all mean what they are doing and the crowd on the sidewalks means it too and that is a lot even if it does not go over into all the facts of every day life.

You were good to me, wonderfully. You were so charming over at Caroline's and it was so specially nice of you to know that I would not want to be alone sewing up there in Norah's room. You are very understanding. This check is your hat and the telegram and a dollar and a half for the man to cut the grass in the yard.

Yours ever
Alice

376

As a consultant for the Department of Labor, Alice Hamilton had again chosen to work for the institution doing the most innovative work in industrial hygiene. After the war, when she had no institutional affiliation, she seemed at times more critical of trends in her field. In 1945, for example, she maintained that under the influence of compensation laws and insurance companies the profession had adopted "over-narrow rules . . . for the diagnosis of occupational poisoning." And in the Cummings Memorial Lecture to the American Industrial Hygiene Association in 1948, she delivered a masterly summary of developments in the field, warning against such pitfalls as applying the results of animal experiments too closely to human beings.[20]

In semiretirement, Alice Hamilton continued her efforts to warn workers, the general public, and the medical profession of the dangers of industrial diseases, tailoring her remarks to her particular audience. She published in her field throughout the 1940s. For *The Oxford Medicine*, she wrote the section on industrial toxicology with Rutherford Johnstone; this appeared as a separate volume in 1945. Her last major project was the revision of her 1934 textbook, *Industrial Toxicology*. For this task, she chose as her coauthor Harriet L. Hardy, a physician then employed by the Massachusetts Division of Occupational Hygiene, whose work on beryllium poisoning in the manufacture of fluorescent lamps had received considerable acclaim. At their first meeting, Hamilton ignored Hardy's disclaimer that she was too new to the field to undertake the assignment and proceeded to divide up the work. Hamilton wrote her chapters with customary dispatch, as did Hardy. They submitted the manuscript to their publisher in August 1947, but the volume was held up by a printers' strike for almost a year. The revised edition of *Industrial Toxicology*, which appeared in 1949 when Alice Hamilton was eighty, was the capstone of her life's work. She shrewdly informed her coauthor, who had written the chapters on beryllium and radiant energy among others, that without the collaboration the volume might have been dismissed because "everyone would think it all out-of-date."[21]

Life at Hadlyme followed the well-ordered routine that Alice Hamilton found so congenial. She put in several hours each morning on lectures or books or her extensive correspondence and then usually helped with the main midday meal. (An enthusiastic cook, she was absent-minded and sometimes burned the pots.) She spent much of the afternoon outdoors. At seventy-nine, she regretted that she was up to no more than "an hour of gentle gardening." Sometimes she sketched. Often she walked, even in bracing weather. In season she was an irrepressible collector of wild flowers, which she placed liberally about the house. Guests often joined the Hamiltons, including old friends and colleagues as well as cousins of various generations. Neighbors had an open invitation to tea, an invariable ritual, served in winter in the sitting room by the open fire that Alice alone tended. News and music usually preceded supper, and reading aloud—mainly by Alice—occupied a good portion of the evening. Punctually at eleven, Alice, the last to retire, ascended the steep steps to her third-floor room overlooking the river.[22]

As she approached her eightieth birthday, Alice Hamilton began to cut back on her public activities. During the war she had observed that the growing number of specialists in her field enabled her to look forward to a quiet old age without any feeling of "shirking." At the end of 1947 she urged the general secretary of the National Consumers' League to find a new and younger president: "I shall be eighty years old in 1949 and must not pretend any longer to be an active person." A year later she wrote a friend: "It really is not decent to keep on to such an age." The "old-old age" for which she had once stored up memories was finally at hand.[23]

112 To Rose Haas[a]

Hadlyme Ferry
February 28, 1949

Dear Miss Haas:

Thank you for your nice birthday letter. It came just as it should, on the very day, for our neighbor always goes for mail on Sunday and the office is open for an hour. I find it hard to believe that I am really eighty years old, but it is true. Old age has come to me so slowly that I have been able to grow used, little by little, to giving up the active life I love so much, and settling down to just being old and physically idle. Fortunately that does not mean mentally idle so far and I fervently hope I shall die before real senility comes on.

As it is, my sister and I enjoy the quiet life here, even in winter, and this winter has been most merciful, the River has never been frozen over and the thermometer touched zero only twice. Our Austrian friends are still with us, which means both comfort and companionship.[b]

Thank you again for your sweet letter.

Faithfully yours
Alice Hamilton

a. Rose Haas and her sister Clara Haas were refugees from Germany, whom AH had met through Tilly Edinger. Rose Haas, who had been a teacher in Frankfurt, studied home economics at Simmons College and became a dietician and house director at Tufts University.

b. The Austrian friends were Nora and George Szarvasy, anti-Nazi expatriots who had settled in the United States. They lived in the wing of the Hamilton house for a number of years, helping with housekeeping and other tasks.

X

"OLD-OLD AGE"
1950–1970

$\mathbf{A}$FTER THE PUBLICATION OF *Industrial Toxicology*, Alice Hamilton settled still deeper into the routine of Hadlyme. With advancing age, she hoped that each day would bring "no unexpected happenings." In fact, the three women of Hadlyme remained remarkably healthy, although Margaret suffered from coronary heart disease and Clara from cataracts. They continued to live together in unusual harmony. Alice and Margaret respected Clara's wishes in household matters, while Clara contributed her share—and probably more—to the expenses. Only in the matter of the dark meat that both the others preferred did Alice, who carved, display a sisterly preference. In 1950 Jarmila and Bohus Petrzelka and their daughter Jane, a Czech refugee family whom the Hamiltons sponsored in the United States, joined the household. They lived in the ell and saw that the three old women were well looked after for the rest of their lives.[1]

Though they honored the civilities of an earlier era, the women of Hadlyme lived fully in the present. Rarely did they discuss the past, and then without nostalgia. Alice especially prided herself on her knowledge of current affairs. A close reader of the *New York Times*, as was Margaret (whose ultimate reproach was to suggest that someone had not read the *Times*), she subscribed to numerous journals, among them the *Manchester Guardian*, *The Nation*, *The New Republic*, the *Reporter*, and the *Saturday Review*. For divergent opinion, she took *The Worker* ("Elizabeth Gurley Flynn's Communist weekly"), *The National Guardian*, and William Buckley's conservative *National Review*. She also received the *Washington Post*, read the latest works on politics, the Supreme Court, and foreign affairs, and listened avidly to the news. In later years, she enjoyed the first-hand view of politicians that television afforded.[2]

In the years immediately after the war, Alice Hamilton had feared that many New Deal gains would be undone, a concern doubtless fueled by talks with Edith's conservative Washington friends, many of them Taft Republicans. To the end she favored expanded federal benefits for the minimum wage, health care, and migrant farm workers. But such issues paled before two alarming tendencies of the postwar era: the preoccupation with communism abroad and the concomitant encroachment on personal freedoms at home. These developments concerned her more than did trends in occupational health, more even than atomic weapons, although she recognized the "staggering problems" presented by atomic power and "its radiation perils" and wished that "those workshops of the Devil [at Los Alamos] could be pulled down."[3]

Though less active than in the past, she was just as eager to make her life count. Quick to write a congressman or a newspaper, or to sign an appeal, she protested violations of basic freedoms and a foreign policy based on military force rather than diplomacy. Many of the petitions were sponsored by organizations deemed to be subversive, and reports of her activities sometimes found their way into the file the FBI maintained on her during these years. The file also included the charge by a "confidential informant" that she was a "concealed communist" and a copy of a letter she had written a prominent civil liberties lawyer criticizing the FBI and the government loyalty program.[4]

From the start of the Cold War, Alice Hamilton deplored the dualistic thinking that prompted the United States to proclaim itself the defender of freedom while shoring up authoritarian (but anticommunist) regimes: on these grounds she questioned the Truman Doctrine, which provided military aid to Greece and Turkey, as well as assistance to Chiang Kai-shek, Syngman Rhee of Korea, and Emperor Bao Dai of Indochina. She favored recognition of the People's Republic of China and its admission to the United Nations, even at the height of the Korean War when that nation had been branded an aggressor; only when its "pariah" status had ended, she argued, could China be expected to act responsibly. A letter on this subject to the New York Times in December 1950—asking that Americans try to look at the situation from China's point of view—brought twenty-four responses, all but one of them favorable. The following year

Alice Hamilton supported a cease-fire and the withdrawal of all foreign troops from Korea and subsequently opposed the rearming of Germany and Japan. Some, like Senator Paul Douglas, rejected her view that disagreements among nations could be resolved by discussion and persuasion and thought she "underestimate[d] the toughness and the deep-seated character of communism's aim for world conquest." In fact, though she always concerned herself more with American deficiencies than with those of other nations, she recognized that the obstacles to peace were not all on one side.[5]

To Alice Hamilton probably the worst consequence of the Cold War was its undermining of basic American values. An "old-fashioned" civil libertarian, she rejected virtually all restraints on freedom of speech or association, even countenancing public expression of a creed as repugnant to her as anti-Semitism. She gave little credence to the danger of internal subversion and once claimed that the communists could do no more to undermine American life than their opponents had done already. With growing alarm she watched the domestic rituals that became familiar during the Cold War years—the succession of congressional investigations into alleged subversion, the loyalty hearings for civil servants, most of all the activities of the Department of Justice—"Department of Prosecution" she considered an apter name—with its deportations of suspect aliens and lists of "subversive" organizations. Her own values remained essentially what they had been. As she had urged clemency for Sacco and Vanzetti, so she signed an appeal to President Truman to commute the death sentences of Julius and Ethel Rosenberg, who had been convicted of conspiracy to commit espionage. Still deeply solicitous of the foreign-born, she protested the practice of deporting aliens or incarcerating them without due process and urged repeal of the McCarran-Walter Act which sanctioned such actions.[6]

Alice Hamilton elaborated her political views in letters to two exceptionally well-informed men of affairs: one was Felix Frankfurter, the other Charles C. Burlingham (1858–1959), a prominent lawyer and former president of the New York City bar. Her passionate defense of individual freedom and fair play reveals the continued emotional relevance of issues that had engaged her for half a century. Indeed, Alice Hamilton's late letters constitute an extraordinary testament to the values by which she had lived. Their spirited intensity also conveys the qualities of mind and heart that had long delighted her friends. Frankfurter, frequently the target of her reproaches, re-

382

ported that at eighty-seven she was perhaps "a little more fragilely exquisite," but otherwise showed no sign of her years; certainly he saw no lessening of "her pugnacity of views regarding the things of which she disapproves."[7]

Evidently responding to a remark of Alice Hamilton's, Frankfurter wrote her on February 22, 1951, expressing dismay that "people who are committed to the Christian ethic pass adverse judgment" on Secretary of State Dean Acheson (an old and close friend with whom he often walked to work). He implied that Hamilton lost her scientific objectivity when she stepped outside her own field and claimed that she would be "surprised with how much I agree on foreign affairs, subject only to the reservation that I am not so sure of my predispositions in these matters as you are of yours."[8]

113 To Felix Frankfurter

Hadlyme Ferry
March 2, 1951

Dear Felix:

Your "sizzling" letter was here when I got home. Of course I should not have assumed that you discussed foreign affairs with Dean Acheson on your daily walks, but I could not imagine two men engaged in such interesting affairs as you and he are confining yourselves to talk about the weather. I must say it seems to me a bit dull, but maybe it is right.

Also I must plead guilty to not being entirely open-minded about the Korean situation or our foreign policy in general, but neither is our Government, which has to make decisions one way or the other. So long as different roads are open to us and a choice must be made, I think the rest of us ought not to simply sit by and watch the outcome. We ought to use what judgment we have to back up or oppose the Government policy. To do anything else is to trust blindly to the wisdom of a few men, maybe to a single man. Time was when I thought that the decisions of our State Department were based on knowledge which was not revealed to the public, and must be taken on trust, but the First World War destroyed that belief.

After all, it is not as if my opposition to the State Department policies were based just on my own ideas. As far as my reading goes the majority of English, European, Indian thinkers are of the same opinion. Really you must not think this is all a matter of impulsive, emotional feeling on my part. Ever since the First World War our foreign policy has been one of my chief interests and I have read and listened to everything I could in that field.

Don't blame either Ethel Moors or me for lack of a scientific approach.[a] Several first-class scientists are with us, Harlow Shapley, Kirtley Mather, Anton Carlson, others.[b] In times of emergency, a stand must be taken, for or against, a neutral attitude is equivalent to approval of what is being done.

I am ready to accept your characterization of Dean Acheson as "essentially right-minded, fearless and dedicated," but that does not mean that his doctrine of "containment of Russia," "situations of strength" is wise. I believe he is leading us into war, he is sowing the wind and we shall reap the whirlwind.

Anyway, I have C. C. Burlingham with me.

Yours always
Alice H.

a. Ethel Paine Moors, member of a prominent Boston family, was active in several pacifist organizations. Like AH, she opposed the Korean war and placed a large measure of blame on Acheson.

b. Shapley was professor of astronomy and director of the Harvard University Observatory; Mather, professor of geology at Harvard; Carlson, emeritus professor of physiology at the University of Chicago. Shapley and Carlson were past presidents, Mather the current president of the American Association for the Advancement of Science.

DURING ALICE HAMILTON's presidency of the National Consumers' League, she had been pressed into service in the League's campaign against the Equal Rights Amendment, which had been given a new lease on life during the election year of 1944 when both major parties endorsed it. Writing in the *Ladies' Home Journal* in July 1945, Hamilton had emphasized the ERA's potential threat to state labor laws that regulated women's hours and working conditions. Consequently she was distressed when Florence L. C. Kitchelt, a former opponent of the amendment who had changed her mind, claimed that many supporters of the ERA also favored protective legislation and did not believe the amendment would interfere with these laws. Fearing that her article had been written "without up-to-date information," Hamilton declared that if Kitchelt's contention was true, "a large part of my opposition fades away." Even earlier, she had expressed dissatisfaction with the laws prohibiting women's work in the poisonous trades. When examined closely, she thought, they made "a very poor showing . . . there is an absurd stress laid on unimportant ones and complete ignoring of really dangerous ones." In 1948 she went even further in questioning the "over-susceptibility" of women to hazardous work since, with some exceptions, "careful observation of large groups in the poisonous trades has not given proof of it."[9]

Hamilton and Kitchelt continued to exchange occasional letters, and on May 7, 1952, Hamilton wrote that she would no longer take a stand against the Equal Rights Amendment; soon after, she prepared a short statement that she knew would be widely publicized. Kitchelt, who considered Hamilton "the top card of our opposition" for nearly thirty years, quoted her in full in a letter published in the *New York Herald Tribune* on Sunday, May 22.[10]

114 To Florence L. C. Kitchelt

[Hadlyme]
May 13, 1952

Dear Mrs. Kitchelt:

If, as I surmise, you intend to quote my statement on the Equal Rights Amendment, I had better make it out carefully. I think this will do.

My long opposition to the Equal Rights Amendment has lost much of its force during the thirty years since the movement for it started. The health of women in industry is now a matter of concern to health authorities, both State and Federal, to employers' associations, insurance companies, trade unions. I do not believe that this situation would be changed by the passage of the Amendment now. Moreover, it seems best for our country to join in the effort of the United Nations to adopt such a principle internationally. I still believe that the legal disabilities of American women could have been lifted years ago had the effort been made State by State.

Please quote it in its entirety.

Sincerely yours
Alice Hamilton[a]

a. This is an unsigned carbon copy; AH's name is typed.

Hᴀᴍɪʟᴛᴏɴ's political frustration reached something of a peak in the early 1950s, when McCarthyism was at its height and charges of communist subversion in government and other American institutions dominated the headlines. During these years her acquaintanceship with C. C. Burlingham ripened into affectionate friendship. In her eighties and his nineties, they progressed from Mr. Burlingham and Dr. Hamilton to C.C.B. and A.H., and ultimately to Alice. In letters, overnight stays at his Park Avenue apartment, and visits to Black Hall, his summer home in Niantic, Connecticut, Hamilton found Burlingham the perfect companion. At a time when so many of her friends were "dead or crippled or just a shadow of themselves," he was also a source of inspiration. Frankfurter considered Burlingham, whose intellect and zest for life were undimmed at 100 despite total blindness and partial deafness, "the finest physiological specimen" he had ever known, surpassing even his idol Oliver Wendell Holmes.[11]

Considered by some New York's "first citizen," for decades Burlingham had been a force for good government, his special cause the creation of an honest judiciary. A man of broad interests and exceptional charm, he agreed with Hamilton on many controversial points of foreign and domestic policy: he favored recognition of Red China, sought early withdrawal of troops from Korea, was outraged by the jailing of Elizabeth Gurley Flynn, protested the indictment of Owen Lattimore, and urged clemency for the Rosenbergs. Though cognizant of the danger of being "triangulated" between friends, in the early 1950s, when Hamilton seemed most undone by Frankfurter's court decisions, Burlingham urged the justice to keep on explaining himself to her. As the relationship between his elders progressed, Frankfurter wrote Burlingham wistfully about "your Alice—she was mine long, long ago."[12]

115 To Charles Culp Burlingham

Hadlyme Ferry
January 14, 1953

Dear Mr. Burlingham:

My typewriter is old and decrepit, but it is, at its worst, better than my handwriting, so whoever reads your letters to you will be thankful that I use it.

I had hoped to write you that I must go to New York for a medical meeting late in January but I have had to give it up because my sister Margaret, who lives with me and is younger than I but has a bad heart, had an attack last week and I cannot leave her alone just now. So I must write instead of, as I hoped, seeing you and having a long talk.

Every year at this time I pay a visit to my sister Edith in Washington and I went off this year as usual but was called back by Margaret's illness. Still, I saw my old friends there, including of course Felix and Marian. Felix invited me to the Supreme Court, which I had never seen, to listen to a case on the deportation of a man after some twenty-five years of life here and of citizenship. This was under the McCarran Act.[a] So I went and was taken sight-seeing by Felix' very intelligent Negro chauffeur around that wonderful and impressive building. Felix himself ushered me into the reserved section of the Court room and then went to robe himself. Then came the august summons of the clerk or marshal and then the Justices marched in. But, sadly enough, it was an anticlimax. I do think our Justices made a mistake when they gave up the English wigs. Because the contrast is too great between the lofty setting, the truly palatial surroundings, and the nine elderly—not to say aged—gentlemen, who file in on their lofty dais and then sink into their chairs, showing only a shrunken, balding head. Wigs would really help a lot. Only Jackson and Vinson are tall, only Douglas is even middle-aged and in possession of all his hair. The case was that of a man the Department of Justice wants to deport, why I did not gather, evidently that was not the question before the Court, but as he was a Russian and Russia will not accept him, the question is, can the Department keep him indefinitely in confinement at Ellis Island. I could not hear the Justices well, but I gathered a favorable attitude toward the victim and I hope Felix will send me the decisions. His secretary told me I may have to wait a long time for them.[b]

While I was in Washington I saw a friend whom I had known as a very able secretary of the Consumers' League, the youngest we ever had, Mary Keyserling, the wife of Truman's economic adviser, Leon Keyserling. She had been suspended from her job in the Commerce Department because, last February McCarthy had denounced her on the Senate floor as a Communist. It seemed this was based on the statement of one [Paul] Crouch, who is said to be second to Budenz as an ex-Communist denouncer, that he had known her to be a Communist in 1928, when she was a Freshman at Columbia.[c] She has just

387

been vindicated, for he described her as five feet, seven and blonde, while she is five feet, two and brunette, but it took till January to clear her. The loyalty board at first insisted she had dyed her hair, and she told me it is very hard to prove you have not. Apparently they did not have to prove anything positively, the burden of disproof was on her. It is all so changed from the time before World War II that it makes me wonder, and makes me feel lonely. Then I have to remind myself of Elijah on Mount Horeb. Were you brought up on the Old Testament? I was, in a stanch Presbyterian family. Elijah is a good cure for self-righteousness. You see he has fled from the wrath of Jezebel to Mount Horeb and there the voice of the Lord comes to him. "What doest thou here?" and Elijah answers "I have been very jealous for the Lord my God. They have destroyed Thine altars and slain Thy priests and I alone am left to serve Thee and they seek my life to take it." And the Lord retorts "Behold I have left to myself three thousand in Israel who have not bowed the knee unto Baal." I am sure there are many thousand in our country who are as revolted as I over the way things are going.

We are having the mildest of winter, snow and ice and wind all around us, but we are in a sheltered little pocket and escape it all. I should love to have a few words to tell me how you do and how the eyes are.

Faithfully yours
Alice Hamilton

a. Among its several stringent anticommunist provisions, the McCarran Internal Security Act of 1950 authorized the Department of Justice to deport "subversive" aliens or to detain them indefinitely.

b. The case AH heard was *Shaughnessy* v. *U.S.* ex rel *Mezei*. Ignatz Mezei had been born in Gibraltar, not Russia, and was an "alien resident" rather than a citizen.

c. After leaving the Communist Party in 1945, Louis Budenz, former managing editor of the *Daily Worker*, became a principal witness against ex-associates at many trials and congressional hearings.

To HAMILTON's relief, in the Mezei case Frankfurter dissented from the court majority, which upheld the right of the executive branch to detain indefinitely an alien whose deportation had been ordered (in this instance on the basis of an anonymous and unspecified allegation and without a jury trial) and whom no country would admit. He allowed for judicial intervention in this and other law enforcement cases on the ground that government officials were bound to adhere to fundamental constitutional guarantees. But for the most part Frankfurter was the court's leading proponent of judicial restraint, an Olympian view of judging that he tried for years to explain—and to justify—to Alice Hamilton. As an individual, he might agree with her, he observed, but as a judge he had no right to pass on the wisdom of specific laws—that was the task of Congress and the state legislatures. Frankfurter thus extended the principle of restraint, applied earlier in the economic sphere by Oliver Wendell Holmes, to the field of civil liberties.[13]

For her part, Hamilton looked to the courts to protect individual rights (as did Frankfurter's colleagues Hugo Black and William O. Douglas), believing that the Supreme Court should act as "a balance wheel" and "an impartial voice in times of popular upheaval." She was dismayed when Frankfurter joined the majority of justices who upheld the Smith Act of 1940, which made it a crime to advocate forceful overthrow of the government or to belong to any group adhering to such a doctrine. Shortly after this decision in June 1951, Elizabeth Gurley Flynn (along with other communist leaders) was indicted on charges of conspiracy. When Flynn received a jail sentence for contempt of court in November 1952, Hamilton wrote a letter to the *New York Times* praising Flynn's sense of honor in refusing to give evidence that might incriminate others. Despite several hostile replies and a "fierce" answering letter in the *Times*, the response was generally favorable—several correspondents commended her courage—and included a prized endorsement from Albert Einstein.[14]

116 To Charles Culp Burlingham

Hadlyme Ferry
January 23, 195[3][a]

Dear Mr. Burlingham:

I am terribly sorry about your eyes. That is just what happened to Mark Howe, as I hear from one of my friends in Boston.[b] It is tragic and there is nothing I can say about it that is comforting, since we both know that none of it would be true. It is just a disaster, one of those that fall into the class of "the mystery of undeserved suffering." If only we were neighbors and I could run in every day and read to you. I should love it and we could have long talks about the parlous state of the country and I could pour out my pessimism, for I am pessimistic for the first time since Mitchell Palmer.ᶜ But how about getting the "talking books" that the Federal Government provides (if that be Socialism, make the most of it!) and does it in such a way as to make it as easy as possible to use. They send you the Victrola and a catalogue of books that have been read aloud by people like your reader. All you have to do is to send in the list of the books you want. The address is "Library for the Blind," 137 West 25th Street. A friend who lives with us, a Bryn Mawr classmate of my sister, has double cataract and she uses the books with much pleasure.

It is a relief to have you say that Brownell's appointment is "excellent."ᵈ Felix could only tell me that he had heard well of him. Certainly it would be dreadful if he were not an improvement on the last three. To me, he is the most important of the members of the Cabinet. I am feeling depressed over the conviction of Elizabeth Gurley Flynn, whom I have known for many years and always respected as a dedicated idealist. I met her first way back in the days when Kenesaw Mountain Landis was trying some hundred and twenty IWWs and other radicals for opposing conscription. She was the only woman and Landis dismissed her case at once, fearing that the attractive young Irish girl might influence the jury. She used to come over to Hull-House to pour out her indignation over this and to talk her particular brand of radicalism. Then I saw her again in one of those depressing Lawrence strikes. She has always harbored "dangerous thoughts," but only now does she have to suffer for them. I cannot forgive Felix— and Learned Hand too—for holding the Smith law constitutional.ᵉ

I am sure Conant has been a good President for Harvard, though when he was proposed I agreed with those who preferred Kenneth Murdock, just because Conant is a scientist and many even of the scientists felt he would not give enough importance to the humanities.ᶠ But I have never heard any criticism of him since then, on that score. There are several men on the Faculty or Emeritus whom I should like to see succeed him, but they are old now, old enough to be mellow and wise—like you—but those do not seem to be the qualities looked for in a President. Unfortunately youth does not guarantee broadmindedness, patience and wisdom.

I have just been reading a circular letter about the proposal to raise the salary of members of Congress, in the hope that more money will keep them from questionable practices. The letter contrasts the $2500 salary of the British M.P. with our $25000, to which may be added

390

(in Nixon's case was added) $65000 as expense account. The writer doubts that we can buy honesty. I wonder why the English should have a higher sense of honor than we. Is it because we have had our English blood diluted by so much blood from countries that have never had decent, honest governments? The Poles, the Sicilians, the Serbs, Bulgarians, Hungarians and above all the Irish, think of Government as something to be suspected and thwarted. I remember when John Morley came to Hull-House and heard us deplore Steffens' "Shame of the Cities."ᵍ He said that the English were partly to blame for it, since the rotten politicians were then mostly Irish and the Irish had had English rule for some eight centuries and had learned their ways from the English.

This letter does not sound cheerful, but I know you do not want Pollyanna sentiments. I can be cheerful about the weather we are having, sunny, windless, beautiful. The River is like a mirror reflecting every tiny branch of the trees on the farther bank. Our little valley is still warmed by the unfrozen River so that we escaped the ice storm that ruined the electric wires of our neighbors just a mile inland. And I can be cheerful about my sister, who is pulling out of her recent attack very satisfactorily.

I forgot to speak of your letter to the Times, which I was so glad to see. Felix too spoke of it when I saw him in Washington.

Faithfully yours
Alice Hamilton

a. This letter is incorrectly dated 1952.

b. Mark A. De Wolfe Howe, a Boston man of letters, was best known for his biographies of prominent New Englanders.

c. Under A. Mitchell Palmer, Woodrow Wilson's attorney general, the Department of Justice conducted raids against radicals and aliens. State governments prosecuted many of the radicals, while many of the aliens were deported.

d. Herbert Brownell was Dwight D. Eisenhower's attorney general.

e. Learned Hand, chief judge of the U.S. Court of Appeals, Second Circuit, and one of the most respected jurists of his era, had upheld the conviction of the Communist Party leaders in Dennis v. United States, the case that first tested the validity of the Smith Act. He was a close friend of Frankfurter's.

f. Kenneth Murdock was professor of English at Harvard and, from 1931–1936, dean of the faculty of arts and sciences.

g. John Morley, a distinguished man of letters and Liberal member of the House of Commons, supported Irish Home Rule. Lincoln Steffens' The Shame of the Cities (1904) was a classic of muckraking literature.

THE *AMERICAN MERCURY* of May 1953 carried an article by J. B. Matthews entitled "Communism and the Colleges," which charged that the Communist Party had enlisted the support of at least 3500 professors. Matthews, a Methodist minister who had been active in pacifist and radical organizations in the early 1930s, had appeared in 1938 as a witness against his former associates before the House Committee on Un-American Activities; he later became its chief investigator. In the article Matthews included Hamilton among the "top academic collaborators with the Communist-front apparatus" and mentioned the American Committee for Protection of Foreign Born, of which she was a sponsor, as one of the "more important units" of the network. The committee, which protested the forced deportation of aliens, had been designated a Communist-front organization by Attorney General Brownell only the month before.

117 To Herbert Brownell

[Hadlyme]
May 16, 1953

Dear Mr. Brownell:

As one who for some years has contributed to the funds of the American Committee for Foreign Born and whose name is on the list of sponsors, I have just been sent a copy of your "Petition" which I have read carefully.

If you will read the list of names on the notepaper of that organization I think you cannot fail to see that many of the men and women who have backed this movement are people of high standing, respected in their communities, and not likely to be carried away by hysterical emotionalism or by Communist propaganda. It would seem that such a group deserved respectful treatment, a careful inquiry, before it is branded as "subversive."

Have you thought at all about the motives that led us to back this organization? Do you know the histories of the men and women your office under the last Administration took from their homes and families and subjected to deportation, not because of any offense on their part but only because of a former adherence to the Communist Party, an adherence often dating back many years. It is hard to see how any pitiful or justice-loving person could read these histories without an indignant resolve to do what he could to help.

In the last two paragraphs of your "Petition" you accuse the Committee for the Protection of Foreign Born of "opposing the enactment of certain legislation regarded by the Party as inimical to its interests, such as the Mundt-Nixon Bill, the Hobbs Bill and the Internal Security Act of 1950"[a] also of consistently supporting "the views and policies of the Communist Party toward the Committee on Un-American Activities of the United States House of Representatives."

Surely you do not think that it is only Communists who hold such views. You must know that many Americans who have no slightest connection with Communism oppose what they consider un-American legislation to nullify the Bill of Rights, and the attacks on individual liberty of thought and of expression and of association made by various Congressional Committees. Is it your rule that a cause in which one believes, an effort one would wish to help, must be avoided because the Commun[ists] also believe in it? I remember when that argument was used against the Child Labor laws. It was said that the program for the protection of children against premature labor, which Julia Lathrop of our Children's Bureau advocated, was identical with that proposed by Kollontai of Red Russia, therefore it was Communist-inspired and un-American.[b] You are placing us in a dilemma when you condemn everything that Communists approve of as "subversive."

I hope that you will take a second look at this group that you are condemning.

We who have watched with deep apprehension the increasing interference with freedom of speech, thought, association, practiced by the last Administration are hoping for a return to our old ideals, now that we have new faces in the seats of power.

Sincerely yours
Alice Hamilton[c]

a. The Mundt-Nixon Bill, which required that all "Communist political organizations" register the names of their officers, became part of the omnibus Internal Security Act of 1950. There were two Hobbs bills in the late 1940s, neither of which passed Congress. One concerned the deportation of aliens, the other the employment of "subversive" persons by government agencies.

b. Aleksandra Kollontai, a Marxist revolutionary, was appointed by Lenin to head the Commissariat of Social Welfare.

c. This is an unsigned carbon copy; AH's name is typed.

118 To Felix Frankfurter

Hadlyme Ferry
May 16, 1953

Dear Felix:

Here is the copy of a letter I have brashly written to the Attorney General. I am sorry it is so very faint, but I have to do my typing myself and I had a new ribbon and was afraid to press too hard. Of course it is perfectly futile but if I am publicly denounced as a Red I don't see why I should not live up to my reputation. Just now I am feeling particularly cocky because I have the May Mercury and my old adversary, J. B. Matthews, who five years ago denounced me before a Legislative Committee in Boston, now puts me in such grand company as to make me thrill with pride. But I am wondering why you and C.C.B. both spoke well of Brownell. Nothing seems ever to change the Department of Justice. This organization he is now attacking is one I have belonged to for years and I am one of the sponsors, together with Bishop Edward Parsons, Rabbi Robert Goldburg, Professors Kirtley Mather, Albert Guerard, A. J. Carlson, Philip Morrison, Robert Lovett, Vida Scudder, Ellen Talbot, and others less well known.[a] I admit that Dashiell Hammett is also there, and John Kingsbury, so I can be accused of associating with fellow travellers, as they certainly are.[b] The latest move of the Attorney-General is to put on the subversive list a perfectly innocent and ineffective group of "grass-roots" pacifists, to which also I belong, which brings my number up to seven, *not* over ten, as Matthews claims.

I heard a story the other day which will amuse you and Marian. An old radical, who calls herself a follower of the "Clara Zetkin ideal" writes me that a grandson of Judge Thayer, of Sacco-Vanzetti fame, is now a Trotskyite Communist.[c] He is Thayer Burbank. Pity his grandfather did not live to see it.

Love to you both
Alice

394

a. Edward L. Parsons was Protestant Episcopal Bishop (retired) of California; Robert Goldburg was a reform rabbi from New Haven; Albert L. Guerard was professor of French and general literature at Stanford University before his retirement; Philip Morrison, a physicist who had worked at Los Alamos, taught at Cornell University; Vida Scudder was professor emerita of literature at Wellesley College and a prominent Christian Socialist; Ellen B. Talbot had been professor of philosophy and head of the department at Mount Holyoke College.

b. Dashiell Hammett, popular writer of detective fiction and a supporter of many left-wing causes, is generally believed to have been a member of the Communist Party. John A. Kingsbury, former secretary of the Milbank Memorial Fund, was chairman of the National Council of American-Soviet Friendship.

c. Clara Zetkin was a founder of the German Communist Party.

ALICE HAMILTON's pride in being labeled a subversive was tempered at times by alarm at the dogmatism—not to say fancifulness—of some of her associates. For the most part she accepted this aspect of the radical temperament. As she once wrote a friend who had joined a radical organization shortly after taking out American citizenship: "You will find your Radical colleagues sometimes unreasonable, wrong-headed even, but you can always look at the alternative and ask yourself if they are not, at their worst, better than the Republicans . . . I am so glad you are starting on the right side."[15]

Over the years, Alice Hamilton supported many appeals (often for legal expenses) from organizations whose policies she did not entirely agree with. One of them was the National Council of American-Soviet Friendship, which had early been labeled "subversive" by the attorney general. The organization was frequently in the news—for its refusal to turn over membership lists and minutes to federal authorities and for its politics (it even sponsored a memorial meeting for Joseph Stalin shortly after his death in March 1953). As early as 1950, Alice Hamilton had informed the executive director of the council that she found herself "at total variance" with its recent expression of confidence in the desire of Soviet leaders for peace, a view she thought was based on "wishful thinking, not the result of a wide-open-eyed search for the truth." Some time before March 1955 she ceased to be a member of the organization. It was a period when she seemed unusually hesitant about endorsing some of the requests from left-wing groups that came her way.[16]

119 To Charles Culp Burlingham

Hadlyme Ferry
March 1, 1955

Dear C.C.B.

Thank you for the Radcliffe letters. I know I shall find them interesting. Just now I am struggling with a pile of letters which accumulated during my very brief absence. Some of them fill me with acute depression and doubts. I hate to feel myself a coward, yet what else can I feel when I read a most reasonable plea from the Jefferson School in New York, which the Department of Justice is closing down because it teaches Marxism? I am sure I shall not sign their protest, yet I do believe in Jefferson's verdict that any doctrine should be permitted expression, provided truth is permitted to reply. So it is probably lack of courage that holds me back.

Then there is an impassioned plea for money to fight the case of an American woman who was "kidnapped" from Ceylon and brought back to the U.S. because she wrote a book defending the innocence of the Rosenbergs. It sounds outrageous, but all I can think of doing is to send it to Patrick Malin and ask him what about it and what will the Civil Liberties do?[a]

And then there is your friend John Kingsbury, with a long letter, wanting to know why I left his organization and why I do not come back, now they are in trouble, and now that he is faced with a five year sentence and $10,000 fine. I am going to send him the poem from Pilgrim's Progress that Peggy Radcliffe sent you.[b] And I shall have to say that I left the Soviet-American Friendship because I could not continue to back the sort of publicity it was getting out, completely one-sided and naively credulous. But maybe I will send him a small sum to pay legal expenses.

All that in the mail of three days, and more of the same kind, only not so troubling. And so this letter, which was meant to thank you for a delight[ful] if short visit, has turned into a wail from an unwilling radical, who does not like the company she is in, but is ashamed to turn her back on it.

I did enjoy my visit very much. It was so pleasant to find you quite as young as ever, and to meet Mrs. Hirst again, and to come to know Mrs. Jay and start an acquaintance with Abraham Flexner.[c] Thank you for all of it. I do hope something will take me to New York again before the summer, but if not, then I will see you in Black Hall.

Yours always
Alice

Warm greetings to Mrs. Hirst.

a. Patrick Murphy Malin was executive director of the American Civil Liberties Union.

b. AH meant Peggy Ratcliff, daughter of Burlingham's friends S. K. Ratcliff, a British journalist and writer, and Katie Ratcliff.

c. Louisa Jay was the widow of Pierre Jay, former chairman of the board of the Federal Reserve Bank of New York. Helena Hirst was the widow of Francis W. Hirst, English editor and writer and a close friend of Burlingham's. Abraham Flexner, author of the influential report that had prompted reforms in medical education, had also founded the Institute for Advanced Study at Princeton, New Jersey. Of this dinner, Burlingham wrote that the age of the five diners added up to 421 years.

THROUGHOUT her eighties, Alice Hamilton remained in singularly good health, better health than either Margaret or Clara, both younger than she, as she could not resist pointing out on occasion. She complained of diminished strength, but others found her physical and mental stamina inspiring. Harriet Hardy recalled observing Alice Hamilton in denims and sneakers, attacking a pile of manure, while she, many years younger, prepared for an afternoon nap. "We can't all be Hamiltons," another friend remarked of someone who seemed to be losing his grip. Alice recovered quickly from a broken hip caused by a fall in 1954. "We Hamiltons are a tough and long-lived lot," she reported, after she, Margaret, and Edith, all in their eighties, snapped back from their respective orthopedic injuries as if they had been in their twenties. Her gravest problem was loss of hearing. She had difficulty coming to terms with a series of hearing aids and wondered "whether the wear and tear on my temper that my 'hearing aid' makes is not worse than the thing that it is supposed to cure." Characteristically she added that deafness is "really hard on the people who have to live with you . . . and of course I shall some day get used to the wretched thing."[17]

Alice Hamilton lived long enough to become something of a wonder once more. In 1947 she was the first woman to receive the prestigious

Lasker Award for her contributions to workers' health. The following year her alma mater, the University of Michigan, awarded her an honorary doctorate, and she gave the Cummings Memorial Lecture of the American Industrial Hygiene Association. The Knudsen Award of the Industrial Medical Association and the Elizabeth Blackwell Citation of the New York Infirmary followed in 1953 and 1954 respectively. She was amused and puzzled at being named New England Medical Woman of the Year in 1956, "when I have done absolutely nothing for more than ten years." But although she complained about the fuss, including a *Time* article she considered vulgar, she welcomed the bulging mail from acquaintances who had learned she was alive, including a letter from an old lawyer who was "an ardent anti-woman-suffrage fighter and loved to ignore my wrong-headedness on the subject and imagine me a perfect lady." Her attitude toward the ceremonial invitations that came her way as an aging dignitary was characteristically crisp. Preparing to attend a function of the United Mine Workers in Washington in 1958, she noted that it had seemed "tempting when it was not imminent, my last chance to dip down for a bit into the old life." But once the event was at hand, she feared it would be a repetition of a recent ceremony in Boston, with "grey-haired, bald, arthritic old men greeting me and asking if I remember meeting them back in the nineties."[18]

As Alice Hamilton had earlier conveyed the uncertainties of youth, so her late letters record the ambiguities of growing old. Without self-pity or false optimism, she reported the inevitable, but still unlooked for, incidents of age: the loss of friends, their sudden or lingering illnesses, and the decline of her own physical and mental powers. Most of all she dreaded surviving as a shadow of her former self, and prayed for a quick end. "I am afraid I shall outlive all my generation," she wrote the day after her eighty-fifth birthday. Nevertheless, her hold on life remained strong: "Life is still as interesting as ever," she wrote at eighty-eight. "I should hate to leave it and not know what will happen next. Maybe we can sit on a cloud and watch." Two years later she claimed: "My being able to reach the age of ninety is more important than anything in my life."[19]

120 To Katherine Bowditch Codman

Hadlyme Ferry
January 13, 1959

Dear Katy:

Your letter just came. I had no idea that it was so long since I had written you. It must be because there were so many duty letters after Christmas and also belated Xmas cards, that I put off personal ones.

That is a sad letter. I never realized before, as I do now, that the saddest thing about old age is the way one after the other of old friends die. It makes death come very near. Of course no day passes that I do not think of Margaret's death, for I know I shall outlive her.

Somehow, I do not know how or when, I lost my belief in individual immortality. I don't mean I disbelieve it, I simply do not know. When death comes, I shall know. "Now we see through a glass darkly, but then face to face. Now we know in part but then shall we know even as we are known." That is from St. Paul. My whole relation to God and to Christ has become cloudy, and I think will be till I die and then really know.

I am so sorry to hear of Katherine Homans' death. Marian I never knew but somehow I met Katherine quite often. I remember her driving me out to the Medical School in my early days, and when she told me that we were on Avenue Louis Pasteur, I said, how fine of Boston to name an important street for a foreigner. She said "Well I always thought it should have been named for Katy Codman's uncle." I had a vision of the street signs carrying that name, before I realized she meant Henry Bowditch.[a]

It is exciting to know that John Codman's daughter, with husband and children, are going to Portuguese Africa. I think I will send you a booklet that a member of the—to me quite unknown—American Committee on Africa, sent me on the eastern Portuguese colony. Mozambique is more civilized than Angola, where your young people go, but I gather that the governing principles are the same, and quite dreadful they are. But probably they, as foreigners, will not be allowed to see any of it.

As for us three we go along a very unchanging, but pleasant routine. Margaret has slight heart attacks which make Dr. Ely urge quiet, but he wants her to be up and about, not bedfast.[b] Clara has less sight all the time, as her cataracts thicken, but she listens to her "talking books" by the hour and enjoys it very much. We have the radio a lot, or rather we did have till just now. Suddenly it has been cut down, to our great dismay. I read aloud in the evening, just now it is Galbraith's book on Poland and Jugoslavia, pretty thin, but we are waiting for Ashmore's book on Little Rock.[c] Of course old age brings me many maddening handicaps, I no longer walk with ease and surety, I use a cane, I have what my grandmother called "butterfingers," which means I spill and drop and fumble. Worst of all, I tire after the least exertion. But I may easily live to a hundred. I am sure Browning was still young when he wrote that about "the best is yet to be."

I have just heard a story from Washington which I will pass on. A Senator is reported to have said: "Roosevelt showed us that a man can be President as long as he likes. Then Truman came and showed us that any man can be President. And now Eisenhower shows us that we do not need a President."

Yours always
Alice

a. Henry Pickering Bowditch, a distinguished physiologist, had taught at Harvard.

b. Julian Ely was the Hamiltons' physician for many years.

c. The books are John Kenneth Galbraith, *Journey to Poland and Yugoslavia,* and Harry S. Ashmore, *An Epitaph for Dixie.*

PROFESSIONAL ASSOCIATES, led by Harriet Hardy, looked for a fitting tribute for Alice Hamilton's ninetieth birthday, February 27, 1959. Knowing her disdain for empty gestures and public display, they passed up a formal event and established the Alice Hamilton Fund for Occupational Medicine at the Harvard School of Public Health. The fund has been used principally to sponsor special lectures, of which Harriet Hardy gave the first. Harvard itself was less forthcoming. Despite efforts by influential friends, the university never awarded Hamilton an honorary degree. Margaret, who had a long memory for slights, was indignant, especially after Edith in June 1959 received one from Yale, an institution with which she had no prior connection.[20]

Two events that June left Alice feeling bereft. C. C. Burlingham died on the sixth, and two days later Frankfurter joined the court majority in upholding, by a five-to-four vote, the conviction of Willard Uphaus, a Christian pacifist, for refusing to turn over to New

Hampshire's attorney general information about individuals who had attended a summer camp of the World Fellowship, an organization he headed. Ordinarily, she explained to Katy Codman, she would be writing Frankfurter to mourn the death of their friend, or pouring out her feelings about Frankfurter's conduct to Burlingham: "As it is, I cannot even find an outlet for my grief or my indignation." The Uphaus case provided the occasion for a sustained exchange between Hamilton and Frankfurter on philosophical issues.[21]

121 To Felix Frankfurter

Hadlyme Ferry
July 3, 1959

Dear Felix:

Ever since C.C.B.'s death I have [been] thinking I must write you, for it seemed strange not to be telling anyone how much I miss him and how much I have loved him. I do not know anyone in his family well enough to write to them, so Nora Brennan was the only one I could think of (she sent me a really lovely answer) except you and I felt sure you were too deep in majority and minority opinions to want letters that were personal only.[a] Now at last the Court has entered on its summer vacation and so you are again accessible.

I had hoped to see you this summer when you came up for your yearly visit at Black Point, for I missed my yearly winter visit to Washington this year. Instead, Edith came to Hadlyme twice, that last time to get her Yale degree, which I was able to witness and which went off with an efficiency and perfection that was purely American. Equally American was the fact that we, the close friends and relatives of the recipients, were provided with a delicious and certainly expensive lunch, but had to wait on ourselves, no servants present.

Katy Codman sent me a copy of "Wisdom," devoted to you. I had never heard of that magazine but of course I read all the part about you and was deeply impressed, especially by the "from the wisdom of." It was nice to find Marian included. I wonder how she is and if you are spending still another summer in Washington. Still, you do have air conditioning, don't you?

Willard Uphaus is an old friend of mine. I wonder whether the people who appear before the Court do really appear, or only their lawyers. Because if you saw him I think your only impression of him could be that of a typical religious pacifist, fervently sure he is following Christ and ready to face any sentence, rather than give way before authority, when that authority is wrong. He will probably come up before you again, for, as I understand it, he is sentenced to prison until he submits and surrenders his list of fellow campers. Of course he will never do that, New Hampshire's Attorney General is not much of a psychologist if he does not know that there are people who welcome martyrdom, but I suppose his lawyer could complain that indefinite imprisonment constitutes "cruel and unusual punishment" forbidden by the Constitution.

My friends seem bent on getting into trouble with the Government. Here is A. J. Muste, whom I knew first years ago during one of the Lawrence strikes, when he was a minister in one of Lawrence's Presbyterian churches.[b] He took the side of the strikers and had to leave his church. Then he went through a very Radical period, but later came back to religion and was one of the founders of the Fellowship of Reconciliation, to which I have belonged for many years. Now I hear he is in jail for trespassing on the site of an Atlas missile plant, I suppose as a protest against war. He is 73 years old. Somehow I would think such an action was natural for the 22 year-old lad who was with him, but not for an oldster.

I have wandered on without realizing that I had turned the paper and written too much. It is a cool and lovely evening here, and we have had only two hot days so far this summer. For aged women we are well and even a bit active. I hope Marian, so much younger, can say the same.

Love to her and to you
Alice

a. Nora Brennan was Burlingham's longtime housekeeper.

b. A. J. Muste, a Christian pacifist and executive secretary of the Fellowship of Reconciliation, was arrested on July 1, 1959, with other members of the Committee for Nonviolent Action for trespassing on a missile base near Omaha, Nebraska.

FRANKFURTER wrote that on receiving her letter he had informed his wife: "Alice is getting mellower. She has evidently given me up as hopeless and no longer even scolds me for judicial conduct of mine of which she thoroughly disapproves." Contrasting the freedom she enjoyed with the restraints his judicial office imposed on him, he supposed that even she would agree "since you are not an anarchist, that we are not free to choose what laws we will obey and what we won't obey." He added that he believed the state of New Hampshire had a right to exact the information from Uphaus and wondered "if you sat on the Supreme Court and could not escape the conclusion that Muste clearly violated an unquestionably constitutional statute whether you would reverse his conviction." Finally, enjoining her to secrecy, he confided that he had written the citation for Edith's Yale degree.[22]

122 To Felix Frankfurter

Hadlyme Ferry
July 15, 1959

Dear Felix:

That was such a specially good letter from you that it made me feel waiting for an answer to mine was much more than worth while. Anyway I knew you were terribly busy till adjournment came. I will try to write for you at least a sentence that you can use, about C.C.B. It is curious how I really miss him, keep thinking all of a sudden that it is about time I arranged for a drive over to Black Point.

Maybe I am some sort of an anarchist, at least I am a bit staggered by your assuming that "we are not free to choose what laws we will obey and what we won't obey, no matter how religiously rooted our reasons for disobedience." But that disposes of Foxe's Book of Martyrs, of the Quakers' Underground for runaway slaves, and a lot more, including Peter's reply in the Book of Acts: "We ought to obey God rather than men."[a] Of course I recognize that it is difficult to know just what God is commanding and what our own obstinacy and self-will is doing, but I can't subscribe to that sentence of yours, and I wonder what future generations are going to think of our acceptance of the dicta of our Attorney-Generals—have we had a single respectable one since Francis Biddle?—ever since about 1940.

That is just the sort of thing I would pour out to C.C.B. and he would laugh and pat my hand and tell me to go slow, but give me the impression that he was not ridiculing me at all. As for A. J. Muste, I cannot escape the impression that he acted as he did for the sake of publicity. But of course all rebels want publicity. I met a young professor in Hitler's Germany in 1933, who told me he had been ordered to change his lectures on history to accord with Nazi principles. He said "It means submission or starvation for my family. But there is not much inspiration in choosing the last, when one's martyrdom is known to nobody. Nowadays it is forbidden to publish dismissals from the universities, it makes a bad impression abroad." I suppose that the desire for publicity is what inspires Dorothy Day to refuse to take shelter during a staged air-raid, and young Bigelow to sail into the prohibited zone in the Pacific.[b] By the way, did you know he was the husband of Josephine Roche?[c] He seemed to all Boston then nothing but a play-boy. I wonder what made the change in him. His second wife was by way of being an actress, but not a successful one. It could not have been her influence.

Another thing I would be discussing with C.C.B. is, why are we the only western country that lives in terror of native Communists. All the European countries have open and above-board political Communist parties some even have members of Parliament or whatever, and they do not have Un-Dutch Activities Committee. Look at the contrast between the English treatment of Klaus Fuchs and our treatment of the Rosenbergs.[d] Fuchs is a scientist (which Rosenberg was not) he gave valuable atomic secrets to the Russians (Urey testified that Rosenberg did not know enough to do that) he confessed (the Rosenbergs refused to, though offered their lives as reward) Fuchs acted during the war, the Rosenbergs during peace.[e]

Well, I will not put you through the sort of talk that C.C.B. encouraged because it amused him. I do think you wrote a wonderful citation for Edith. I could hear every word of it and could bring back a printed copy for Margaret and of course Edith had one. She thought it was the most successful thing that had ever been written and showed so much insight, it could not have been the work of Griswold, who did not know her at all.[f] I wish you would let me tell her— why not?

I am so sorry to hear of Marion's set-back, and this heavy, humid weather does not help. We have had it, with rain, for three days and are promised three more. Somehow air-conditioning does not satisfy, it seems not the same as a fresh breeze. I do hope for her sake that Washington has a change soon. Much love and sympathy to her, and to you.

Yours always
Alice

a. Foxe's *Book of Martyrs,* a compelling work of Protestant hagiography written in the late sixteenth century, had remained a household staple in Britain and America for three centuries. AH cited it frequently in her late years.

b. In April, Dorothy Day, a pacifist and leader of the Catholic Worker movement, had led a small group of men and women who protested a civil defense drill by refusing to take cover; with four others she went to jail rather than pay the fine. Albert Bigelow was captain of the Golden Rule, a ketch that members of the Committee for Nonviolent Action planned to sail into the Eniwetok nuclear testing area in the Marshall Islands in 1958. The ship reached Hawaii, where the passengers were arrested for violating a restraining order.

c. AH meant Josephine Rotch, a niece of Kitty Ludington, who died in 1929 at the hand of Harry Crosby in a bizarre murder-suicide that left Boston's first families reeling.

d. Klaus Fuchs, a British physicist, had been arrested and convicted in 1950 for giving British and American atomic secrets to the Soviet government. He was released from prison in 1959.

e. Harold C. Urey was a Nobel Prize–winning chemist who had worked on the development of the atomic bomb.

f. A. Whitney Griswold was president of Yale.

FRANKFURTER reproached Hamilton for seeming to imply that he would "ridicule, anymore than C.C.B. would, *any* opinion of yours. Least of all, any view of yours regarding man's relation to God and the State." He reminded her that he had taken an oath to " 'administer justice . . . agreeably to the Constitution and laws of the United States,' not 'agreeably to the Command of God' "; nor was he appointed to enforce his "*private* views of what is right and just." Declaring that he respected martyrs "almost to the point of veneration," he insisted that "martyrdom, like every kind of freedom, must be paid for."[23]

123 To Felix Frankfurter

Hadlyme Ferry
August 25, 1959

Dear Felix:

I am just back from Edith's place on Mount Desert where I took your last letter, thinking to answer it in the quiet up there. But they have no typewriter and my handwriting has grown so poor that I hated to send it to you and so I waited to get home to my own machine.

Your letter has given me much food for thought. I wish I could talk it over with you. You do seem to me to take an attitude toward the Constitution that is like the attitude of the orthodox Jew to the Old Testament, and that of the Catholic to the Church, and the Fundamentalist Protestant to the Bible. But they all believe in a supernatural authority whose word needs no defense. You cannot put our Founding Fathers in such a place. Yet you feel that their wisdom, their rules, must be followed, not your own sense of right and justice and honor. You "search the scriptures," for authority and accept it.

Of course all human experience has shown the deep desire of us human beings for an authority which we can accept without controversy, without inner doubt. But can we who have rejected the supreme authority of the Bible, the Church, Moses, tradition, accept that of the Founding Fathers, who lived in a country so different from the one we now live in.

You will say we must have some principles that are accepted by the great majority, some code of laws to which we can appeal when in doubt, not questioning their rightness. Of course I know that is practically desirable, but we can have it only by sacrificing some of our liberties, among them our own individual standards of justice and truth. Maybe there is no way of avoiding this if we are to have an established rule of law, which everyone except Anarchists want, I do not know. But you have not dispelled my reluctance to agree with you about the over-riding authority of the Constitution when it comes to matters of right and wrong. You have not even the comfort of believing that it is Divine.

All the time I have been writing this I have been thinking how I would have laid it before C.C.B. some afternoon over in Black Hall, and he would have listened patiently and then shown me where I am, let us say confused, he would have made me feel that I was taking a wrong start, from premises not well thought out, that I must re-think it all over again. So you must take his place and try to make the muddle clear.

It was beautiful up in Maine but I stayed only eight days, being always a bit uneasy at leaving Margaret so long. Clara, who still lives with us, is almost blind and very uncertain in her walk, and Margaret herself has a damaged heart and is partially lame. So though I am ninety years old I am far younger than both of them. Our Czech couple is still with us and housekeeping is no problem, but I do like to be on hand for emergencies which do happen. I have thought often about Marian when I read in the Washington Post about the heat

406

down there. I do hope it has not pulled her down. Give her much love from me.

Yours always
Alice

In his lengthy response, Frankfurter further elaborated his views. He also sent copies of two opinions, one of which spelled out "the kind of torture" he went through when deciding whether procedural safeguards had been properly observed (the case involved a death sentence, a practice he considered especially "repellent"). It was precisely because the constitution left such matters for future interpretation, he explained, that reaching a decision was for him quite unlike a search through scripture. After reading Frankfurter's letter and the two opinions, Hamilton wrote: "I can see your stand clearly at last. I do understand what your attitude is toward such questions, and I do respect it." To which she added: "Whether in your place I could follow it, I am not sure, but then several of your colleagues cannot, so my attitude is not really law-despising or sentimental. But never again shall I wonder what you can possibly think, when you vote against a measure that I believe is vitally necessary."[24]

Alice Hamilton, who as a young woman had longed to leave behind some "definite achievement, something really lasting . . . to make the world better," looked back over her long life with a sense of fulfillment. "I wouldn't change my life a bit," she said in a 1957 interview. "Taking part in a new and expanding discipline brings out the best in one. For me the satisfaction is that things are better now, and I had some part in it."[25]

If for Alice old age was a time for reflection rather than for scaling new heights, for Edith Hamilton it was a period of continued productivity and public acclaim. Her books had long sold well, but in the late 1950s she skyrocketed to fame: in 1957, when she was ninety, she traveled to Greece to become an honorary citizen of Athens, and was also elected to the American Academy of Arts and Letters; in the next two years there were television interviews; and in 1959 the Yale degree. Long sought out by prominent literary and political figures, during the Kennedy years Edith Hamilton received homage from the first family, who were great admirers of her work. She also kept on writing. Next to her older sister, Alice in later years felt decidedly

unambitious. "But really," she wrote, "one in a family is enough."
At ninety Alice wrote of Edith: "She was, of course, far more gifted
than I and I always knew it and admired her and we were as deeply
intimate as sisters so near of age usually are." For her part, Edith,
who in earlier years had been high-strung and subject to bouts of
depression, said of Alice: "She never irritated me. She was balm to
my soul."[26]

124 To Katherine Bowditch Codman

Hadlyme Ferry
January 23, [19]61

Dear Katy:

Just scrawl a few words on this self-addressed postcard and have
somebody drop it in the mail. I know mails are queer lately not only
because of the storm but still more because of the strike but I am
worried because I have had no news from you for a long time. Here
there is little news except about snow, the blizzard of last Friday,
which has left drifts as high as six feet on our hillside and the zero
weather, all of which I am sure Boston has had too. But it is in the
country that one can really see the beauty and the fierceness of a storm
like that and we watched it all through the daylight except when we
broke off to see the high points of the Inaugural. Kennedy's acceptance
speech was excellent, Mrs. Kennedy's coiffure and hat were deplorable.

I do hope and long for a word from you.

Lovingly always
Alice

125 To Rosemary Park[a]

Hadlyme Ferry
December 21, 1961

Dear Dr. Park:

Your letter, suggesting that a new dormitory at Connecticut College
be named for my sister, Edith, and me offers us a very gratifying

honor, giving us a place among the outstanding women already so honored. Of course I accept my part with pleasure and gratitude, but with one slight reservation, Edith's name must come first. It is not only that her writings on Greece and Rome will always be of lasting value, while mine on dangerous trades are already out-moded, but she is the elder sister, her name belongs in the first place.

Please give the Board of Trustees the assurance of my gratitude for the honor they propose to do me.

Sincerely yours
Alice Hamilton

a. Rosemary Park was president of Connecticut College for Women.

126 To Francesca Molinaro[a]

Hadlyme Ferry
December 28, 1961

Dear Francesca:

That was a lovely box of presents you sent us, the charming hand-kerchiefs and those soft, warm gloves, which are just what I need in this cold winter weather. I am sure you understand why I am so slow in thanking you. When one is really so very old, over 90 years, one gets tired quickly and Christmas time seems to bring so many extra things to do.

I am sending you—when I get it wrapped up and posted—an old breast pin which I know you will value because of its history. It belonged to Mary Rozet Smith's mother, when she died Mary had it. She left it to Miss Addams and when Miss Addams died it came to me. So it is a memorial of all of us three.

There is no use saying how sad I am over Hull-House and for you more than almost anyone except perhaps Jessie Binford.[b]

Affectionately always
Alice Hamilton

a. Francesca Molinaro was a child of the Hull House neighborhood who later lived and worked at the settlement.

b. Jessie Binford was for many years superintendent of the Juvenile Protective Association, a Hull House affiliate. A longtime resident of Hull House, she led an unsuccessful campaign to preserve it as a working settlement and was the last person to leave. In 1963 the property was sold to the city of Chicago, which subsequently sold it to the University of Illinois; two of the buildings are still standing.

IN HER NINETIES, age began to overtake Alice Hamilton. She was dismayed by the infirmities that became increasingly insistent: clumsiness, forgetfulness, procrastination, and always her increasing deafness. "Fortunately I have not the habit of swearing," she wrote, "or I should be doing it all the time." Then in May 1961, a severe hemorrhage from "a quite unsuspected gastric ulcer" nearly took her life. Subsequently she suffered from occasional small strokes that incapacitated her for a time.[27]

127 To Clara Haas[a]

Hadlyme Ferry
October 10, 1962

Dear Doctor Clare:

It has been a long time since we exchanged letters but we here have even better excuses than you, for in addition to the usual afflictions of ordinary senility we, Margaret and I, are now over 90 years old. This means that, at the best, letter writing is a great effort and often impossible. With me it takes the form of what seem to be tiny cerebral hemorrhages, which put me flat on my back in bed, ordered not to lift my head, fed through a straw and forbidden all excitement. That means that Margaret, whom we have always guarded because of her coronary, has had to take on all the household responsibilities. Clara is much as she has been for several years, increasingly blind and uncertain in walking.

This is not a cheerful picture, nor is the one you give us of yourself and your sister, but of course your picture will probably clear up, even if you never get back your full strength.

I sympathize with your and Rose's longing to get at your garden again.

This is just a note to tell you why you have not heard from us for so long. Love from us three to you two.

Alice H—

a. Clara (or Cläre) Haas, a psychiatrist, worked at the state hospital in Howard, Rhode Island.

ALICE HAMILTON was fortunate in the longevity of those closest to her. Of the circle of Hamilton sisters and cousins, only Norah and Katherine died at relatively early ages (seventy-one and sixty-nine respectively). Then in the early 1960s the losses followed fast upon one another. Jessie died in 1960, Agnes the following year. In 1963 Edith died at the age of ninety-five; she was buried in Hadlyme, near her mother and Norah. Alice and Margaret sat through the funeral without visible emotion but expressed concern for the sobbing Doris Reid, much younger than they, and consequently in Alice's view much more vulnerable.

128 To Arthur Hamilton

Hadlyme
March 26, [19]64

Dear Quint:

I have just subtracted 17 from 95. The first time I made it 88—my usual blunder when I try the simplest arithmetic—then I tried again and it came out 78 which must be right. You see I was 17 when you were born, so you are now approaching 80. All my neighbors are congratulating me on my 95 years. But wait till you do the same and you will find there are some drawbacks. My greatest is increasing deafness, which does cut one off from humankind a good deal.

This birthday check I am sending you is part of an anticipated one from The Atlantic which is from an article they asked me to write about Edith's and my year of study in Germany in 1895–96. First I thought it was out of the question, then I found I remembered it pretty well, militarism and women-despising. So it is now being typed. I am sure Ted Weeks will take it. Did you know he is a cousin of ours? A grandson or great-grandson of Grandma Pond?

March came in like a lamb and is slinking out like a despised dog, chilly rain, no buds anywhere, no Spring in the air. We are all right, only a bit older all the time.

Love to you and Mary. May you both keep on having birthdays.

Alice

ON LEARNING that Alice Hamilton was writing a biographical piece for the *Atlantic,* Felix Frankfurter recalled that he had once heard her talk "fascinatingly" about the household in which she had grown up— "all of you were talkers"—and urged her to write more personally about her early life than she had in *Exploring the Dangerous Trades.*[28]

129 To Felix Frankfurter

Hadlyme Ferry
April 27, 1964

Dear Felix:

That was a most lovely letter which came from you this morning. I wish I could answer it on my old typewriter instead of with a pencil in uncertain fingers. My grandmother called it "butterfingers" and it is one of the many maddening features of 95 years. Somebody sent me a model prayer for the aged to use and one of the petitions is "Lord make me patient with mine own infirmities," a most appropriate one. But I must not wander on, I must answer your very flattering urge for me to write a "Tischreden" which would reveal the intellectual atmosphere in which I grew up. But I can't possibly, for I don't remember any. I think our family talk must have been just the ordinary, day-by-day discussion of things that happened or didn't happen. I wrote in my book more than enough about my childhood and youth.

Do you know, I have forgotten Bertrand Russell, meeting him and talking with him? All I remember is the impression he made on me, of a man who might be a great mathematician but had a curious adolescent love of shocking people, as when he said "Monogamy, that darling superstition of the non-conformist mind." It is all wrong for my memory to hold that absurd remark and forget all the wise ones.

No, I am not planning to do any more writing, certainly not about the Hamilton family. I have raked up what I remember of Edith and me studying in Germany and it has gone to Ted Weeks, but I don't yet know if he will want it.

But what about you? Why are not you writing about a life far more varied and interesting than most men ever have. You could write a whole chapter, vivid and amusing and fair, about your year in England and one about the "House of Truth" in Washington and—oh so many

I can think of. Of course I have read—was it in The Atlantic or Harper's?—the article written by a man and wife who paid you visits and talked to you. It made me envious. Instead of urging an ordinary person like me to take up writing at 95, you should do it seriously before you reach that appalling age.

You don't tell me how Marion is and how you are yourself, nor where. Are you still in the house with the rather formidable stoop steps? And has Marion still her devoted Negro nurses, and where do you spend your summers,—though air-conditioning means that Washington is no longer unbearable in summer. Give her and take yourself much love and best wishes.

Yours always
Alice

130 To Clara Haas

Hadlyme Ferry
November 25, 1964

Dear Dr. Haas:

The "Historic Events in Occupational Medicine" lies on my desk but I had quite forgotten how it got there.[a] Loss of memory is one of my most deplored deprivations. I enjoyed it, especially the illustrations, but was disgusted to find myself coupled with Hippocrates and Ramazzini.[b] We are having wonderful Fall weather. Love to you both from us three.

A.H.

a. *Man, Medicine, and Work: Historic Events in Occupational Medicine* was published by the Department of Health, Education, and Welfare in 1964.

b. Hippocrates, the Greek physician, is considered the father of medicine. Bernardino Ramazzini, often called the father of industrial medicine, wrote the first systematic treatise on occupational diseases in the seventeenth century.

DESPITE HER SENSE of increasing limitation, Alice Hamilton's interest in public affairs remained keen. In 1961 she protested the efforts of the Senate Subcommittee on Internal Security, led by her own senator, Thomas J. Dodd, to secure from Nobel Prize–winning chemist Linus Pauling the names of individuals who had helped him collect signatures for an international petition calling for a nuclear test ban. To Dodd she wrote that the committee's effort "to prove that a prominent scientist has Communist sympathies seems to me ominous for the future" and a "revival of McCarthyism." In 1963 she signed an open appeal to President Kennedy asking for an early withdrawal of troops from Vietnam and two years later protested the use of poison gas there to her other senator, Abraham Ribicoff. As late as July 1965, when she was ninety-six, she wrote Attorney General Nicholas Katzenbach protesting the latest Department of Justice action against the American Committee for Protection of Foreign Born.[29]

During her late years she also wrote occasional articles for the *Atlantic Monthly*. Two humorous pieces on the English language in 1954 and 1959 reflected on the incongruities of the language, principally as observed by the Petrzelka family. In "A Woman of Ninety Looks at Her World," published in 1961, she reviewed some of her experiences and reminded her readers that whatever the deficiencies of organized labor, they were "inescapably out in the open," unlike the "well-concealed and outwardly well-mannered conduct of the employing class." At ninety-five, she wrote about the year she and Edith had spent in Germany; though the article covered similar ground to the account in her autobiography twenty-two years earlier, it was in some respects more vivid. When the article appeared in March 1965, a reader pointed out an error.[30]

131 To Edward A. Weeks

Hadlyme Ferry
March 29, 1965

Dear Ted:

I am really distressed over that blunder I made in the article about Germany, giving the Atlantic the trouble of answering all those letters from indignant German-Americans. Of course in this little village there is no library I could consult, but I should have had sense enough not to use something that needed checking. Unfortunately Margaret's memory was as faulty as mine, she accepted Hauptmann as I did.

Well, this is my last writing effort, and high time! Seeing I am already within four years of 100.

Yours always
Alice

BY THE END of 1966, the strokes had taken their toll. Denied the quick fading she had hoped would be hers, Alice Hamilton lived on for another four years. Bedfast much of this time, her strong body survived excruciating bedsores and a Job's plague of boils. She sometimes dwelled on the events of her childhood and youth in Fort Wayne: the Presbyterian Church, a defeat of her father's, and especially the activities of "the three As." Margaret, alone with her sister after Clara Landsberg's death in April 1966 and Quint's in 1967, sat with Alice for several hours each day, for the most part in silence.[31]

132 Margaret Hamilton to Francesca Molinaro

Hadlyme Ferry
January 10*th* [1968]

Dear Frances

I am grieved over the delicious box of cheese. It is all gone—eaten up eagerly—and you have not heard from me of its arrival. I am sure I wrote—it is crossed out on my list—but mailed, I do not know. At 96 years old my memory is poor. I remember the box—how good every[one] thought it was how it vanished, but mailing a letter is a hard thing—every one who comes in takes my letters. It may be in some man's pocket. However my thoughts have gone to you. Alice is not mentally here any more. She sleeps almost all the time. She needs two nurses. I am not strong enough to even turn her in bed. After all she will be 100 in a year. She is still her lovely self—never complains—never grumbles—just helpful. But she does not know me often and never talks to me. It takes all my energy to do what I must do. I neglect every thing else.

You know that Clara Landsberg died 18 months ago and my brother Quint six months ago, so I am all alone with her now.

Thank you for thinking of me and your lovely gift.

Yours always
Margaret Hamilton

ALICE HAMILTON had correctly predicted that she would live to be 100. Her centenary in February 1969 received considerable attention. The many communications that arrived included a telegram from

President Nixon, praising her achievements in industrial medicine, which he believed had paved the way for workmen's compensation, as well as her work at Hull House, at Harvard, and for the federal government. There was also a query from a researcher on centenarians inquiring whether she was a vegetarian. The *New York Times* carried a long article and an editorial.[32]

Margaret remained alert to the end, which came in July 1969, when she was ninety-eight. Alice lived on for another fifteen months before her strength finally gave out. Her death came on September 22, 1970. She was 101. Three months later Congress passed the Occupational Safety and Health Act, the law that first empowered the federal government to enforce healthier conditions in the workplaces of America over which Alice Hamilton had stood watch for so long.

Abbreviations
Notes
Sources
Acknowledgments
Index

ABBREVIATIONS

AALL	American Association for Labor Legislation
AH	Alice Hamilton
AHC	Alice Hamilton Collection, Connecticut College Library
AHP	Alice Hamilton Papers, Schlesinger Library, Radcliffe College
BLS	Bureau of Labor Statistics
CCB	Charles Culp Burlingham
CLM	Francis A. Countway Library of Medicine, Harvard Medical School
EDT	*Exploring the Dangerous Trades*
EGE	Elizabeth Glendower Evans
EH	Edith Hamilton
FF	Felix Frankfurter
FFP	Felix Frankfurter Papers, Manuscript Division, Library of Congress
HFP	Hamilton Family Papers, Schlesinger Library, Radcliffe College
HSPH	Harvard School of Public Health
JA	Jane Addams
JH	Jessie Hamilton
KBC	Katherine Bowditch Codman
MH	Margaret Hamilton
NCL	National Consumers' League
NH	Norah Hamilton

NOTES

All citations not otherwise identified are from the Hamilton Family Papers.

Introduction

1. Clipping, *The Gazette, Montreal,* Sept. 16, 1931, AHP, no. 4.

2. There is no satisfactory overview of this generation, but see the influential interpretations of Christopher Lasch, *The New Radicalism in America, 1889–1963: The Intellectual as a Social Type* (New York, 1965), pp. 3–37; Jill Conway, "Women Reformers and American Culture, 1870–1930," *Journal of Social History,* 5 (Winter 1971); and Estelle Freedman, "Separatism as Strategy: Female Institution Building and American Feminism, 1870–1930" *Feminist Studies,* 5 (Fall 1979). Brief accounts of many women of the Progressive generation appear in *Notable American Women, 1607–1950,* I–III, ed. Edward T. James, Janet Wilson James, and Paul S. Boyer (Cambridge, Mass., 1971) and *Notable American Women: The Modern Period,* ed. Barbara Sicherman and Carol Hurd Green (Cambridge, Mass., 1980).

3. EDT, pp. 190–191; AH taped interview with Jean Alonzo Curran, Nov. 29, 1963, Rare Book Room, CLM.

4. Lippmann quoted in FF to Katharine Ludington, Friday [Jan. 21, 1916], FFP; AH to CCB, April 5, 1957, CCB Papers; AH to KBC, Dec. 3, 1958.

5. Notes for Speech, "Anne Morgan's Club," AHC; transcript, meeting of May 15, 1932, p. 45, Recent Social Trends, in William F. Ogburn Papers, box 9. See also AH to Mary O'Malley, Sept. 27, 1922, National Woman's Party Papers, microfilm reel 17, and AH to Mary Beard, June 18, 1938, Mary Ritter Beard Papers.

6. AH interview with Jean Curran, Nov. 29, 1963; "My Day" (1936), clipping in AHP, no. 2.

7. EDT, p. 252.

8. "To the President and Fellows of Harvard University," A. Lawrence

Lowell Papers, no. 1604; Lowell to Henry P. Walcott, Dec. 21, 1918, Lowell Papers, no. 311; EDT, pp. 252–253.

9. See letters 59 and 61 in this volume.

10. Paul Reznikoff, "The Grandmother of Industrial Medicine," *Journal of Occupational Medicine,* 14 (Feb. 1972), 111; FF to Leo Mayer, Dec. 20, 1960, FFP. Jarmila and Jane Petrzelka interviews with author, May 20, 1977, and Sept. 4, 1981, and Dorothy Detzer Denny, taped reminiscences, March 1976, comment on AH's self-sufficiency and reserve.

11. EGE, "People I Have Known: Alice Hamilton, M.D., Pioneer in a New Kind of Human Service," *The Progressive,* 2 (Dec. 20, 1930).

12. Interview with Harriet L. Hardy, Nov. 7, 1974.

13. Recent research makes it necessary to modify Christopher Lasch's interpretation of JA (and, by extension, others of her generation) in *The New Radicalism,* which emphasizes her rejection of "the family claim." On this point, see also Joyce Antler, " 'After College, What?': New Graduates and the Family Claim," *American Quarterly,* 32 (Fall 1980). I am indebted to David Riesman for the term "family culture."

14. AH to EGE, Jan. 27, 1931, and April 6, 1935, EGE Papers; AH to Harriet L. Hardy, Aug. 2, 1949, Hardy Papers.

I. The Hamiltons of Fort Wayne

1. Information about the Hamilton family was obtained from the Allen County Public Library (including its Indiana and newspaper collections), the Allen County–Fort Wayne Historical Society Museum, Clerk of Allen County, First Presbyterian Church, Trinity English Lutheran Church, and Lindenwood Cemetery, all of Fort Wayne; the Allen Hamilton Papers, Indiana Division, Indiana State Library, Indianapolis; U.S. census records; city directories; birth and death records; obituaries; the Presbyterian Historical Society, Philadelphia; and the Mackinac Island State Park Commission. W. Rush G. Hamilton, Phoebe Hamilton Soule, Russell Williams, the late Holman Hamilton, the late Dorothy Detzer Denny, and the late Hildegarde Wagenhals Bowen provided important information about the Hamilton family.

2. EDT, pp. 18–19; AH to Madeleine P. Grant, Dec. 4, 1959, Grant Papers; EH to JH [Dec. 18, 1896]; interviews with Dorian Reid, Dec. 29, 1978, and with Jarmila and Jane Petrzelka, May 20, 1977; Dorothy Detzer Denny to author, June 1976. See also Helen H. Bacon, "Edith Hamilton," *Notable American Women: The Modern Period,* ed. Barbara Sicherman and Carol Hurd Green (Cambridge, Mass., 1980); John Mason Brown, "The Heritage of Edith Hamilton," *Saturday Review of Literature,* 46 (June 22, 1963); and Doris Fielding Reid, *Edith Hamilton: An Intimate Portrait* (New York,

1967). The last, a reminiscence by EH's close companion, is not accurate in all details.

3. EDT, pp. 19–20. Information about MH from the Bryn Mawr School; Bryn Mawr College; Johns Hopkins University School of Medicine; *The New Era* (Deep River–Old Saybrook, Conn.), June 13, 1968; and Russell Williams to author, ca. Jan. 19, 1977.

4. EDT, p. 20; NH, "Creative Childhood," *The Survey*, 49 (Feb. 1, 1923). Information on NH also from *The News-Sentinel* (Fort Wayne), Feb. 7, 1951; *Art News*, 44 (Dec. 1–14, 1945), pp. 28, 33; Art Students' League.

5. EDT, p. 18. Information on Arthur Hamilton from University of Illinois and Johns Hopkins University Archives.

6. EDT, pp. 25–29; Allen Hamilton Williams, "Official Document," Nov. 23, 1882. There were also six much younger cousins, some a full generation younger than the oldest.

7. Agnes Hamilton Diary, Jan. 17 [1895]; Dorothy Detzer Denny, taped reminiscences, March 1976.

8. See Allyn C. Wetmore, "Allen Hamilton: The Evolution of a Frontier Capitalist" (Ph.D. diss., Ball State University, 1974; a copy is deposited in the Allen County Public Library), and Charles R. Poinsatte, *Fort Wayne during the Canal Era, 1828–1855* (Indianapolis, 1969). Most sources, including EDT, p. 22, portray Hamilton as a friend of the Indians. Certainly he had the trust of Richardville and the Miamis, who approved his appointment as Indian subagent in 1841. However, a composite document in the Allen County–Fort Wayne Historical Society Museum reveals that a lawsuit was brought against Hamilton's heirs in 1867 claiming that he had duped Richardville's daughters out of valuable property. The court found for Richardville's heirs, and the land was restored to them. See also Wetmore, "Allen Hamilton," esp. pp. 249–253, and Poinsatte, *Fort Wayne*, esp. pp. 96–100.

9. Information on the Holman family from Israel George Blake, *The Holmans of Veraestau* (Oxford, Ohio, 1943); Holman Hamilton to author, July 28, 1974 and Feb. 15, 1976; *Dictionary of American Biography*, IX, ed. Dumas Malone (New York, 1946), pp. 158–159.

10. Agnes Hamilton Diary, Aug. 18 [1889]; obituary, *Fort Wayne News*, Aug. 17, 1889; Last Will and Testament of Emerine J. Hamilton, Allen County Court; Holman Hamilton to author, July 28, 1974.

11. Agnes Hamilton Diary, Aug. 18 [1889]; EDT, pp. 23–24.

12. AH to Agnes, March 30, 1895, Feb. 11, 1894, and undated [fall 1896]; Margaret V. Hamilton to Katherine Hamilton, April 18, 1895.

13. On the estate, see Allen Hamilton Papers; Allen County Circuit Court; Wetmore, "Allen Hamilton," esp. pp. 321–322. Information on A. H. Hamilton from Allen County Public Library; Scrapbook, HFP; *Biographical Di-*

rectory of the American Congress, 1774–1961 (Washington, D.C., 1961), p. 996; *Fort Wayne Journal,* May 10, 1895; Dorothy Detzer Denny to author, May 1976.

14. Emerine J. Hamilton to A. H. Hamilton, Aug. 18, 1866; Mary Hamilton (Williams) to A. H. Hamilton, July 30 [1866]; Montgomery Hamilton to A. H. Hamilton, Nov. 14, 1863, June 4, 1864, July 30, 1864.

Gertrude Hamilton was older than her husband and apparently dissembled about her age; census records of her age vary. The 1840 birth year is from Parish Register, Christ Episcopal Church, Tarrytown, New York.

15. Allen County Circuit Court; Holman Hamilton to author, July 28, 1974.

16. Obituaries in *Fort Wayne Sentinel,* June 9, 1909, and *Fort Wayne Journal-Gazette,* June 10, 1909; Dorothy Detzer Denny to author, April 7, 1976; Manuscript Library, Princeton University.

17. EH to JH, Tuesday evening [1882?]; EDT, pp. 29–31.

18. AH to EH, postcard, May 20 [1933], EH Papers; EH to JH, Sunday [Aug. 18, 1929]. The bank presidency is mentioned in Phoebe Taber Hamilton to Agnes, Jan. 10, 1887. Information about Montgomery Hamilton's drinking comes from Gertrude Hamilton to Family, April 13 [1900], Dorothy Detzer Denny, taped reminiscences, March 1976, and letter of May 1976, and Hildegarde Wagenhals Bowen, interview, Dec. 30, 1978. The suggested separation is mentioned in NH to EH, n.d. [1893?].

19. Emerine J. Hamilton to Phoebe Taber Hamilton, Aug. [1866]; Phoebe Taber Hamilton to Emerine J. Hamilton, Oct. 26, 1866. The family silver is mentioned in Emily Johnston de Forest, *James Colles, 1788–1883, Life and Letters* (New York, 1926); the portraits in Jarmila and Jane Petrzelka, interview, May 20, 1977 (some of them appear in the photograph of Alice and Margaret Hamilton in this volume); the presence of the Ponds in Fort Wayne in Holman Hamilton, interview, April 6, 1977; the strain between families in Hildegarde Wagenhals Bowen, interview, Dec. 30, 1978; Gertrude Hamilton's independence in EDT, p. 31. Other information was obtained from the Historical Society of the Tarrytowns; Parish Register, Christ Episcopal Church, Tarrytown, N.Y.; Collegiate Reformed Protestant Dutch Church, New York City; Union County [N.J.] Surrogate's Court; directories of New York City; wills; obituaries; the Historical Society of Plainfield and North Plainfield, N.J.; and from Alice Canoune Coates and Holman Hamilton. See also Edward Doubleday Harris, *A Genealogical Record of Daniel Pond and His Descendants* (Boston, 1873).

20. AH to Grace Abbott, July 19, 1935, Edith and Grace Abbott Papers, Box 55, Folder 1; Dorothy Detzer Denny, taped reminiscences, March 1976.

21. EDT, p. 32; Dorothy Detzer Denny, taped reminiscences, March 1976.

22. Dorothy Detzer Denny to author, May 1976; Holman Hamilton interview, April 6, 1977; Phoebe Hamilton Soule interview, May 21, 1982;

W. Rush G. Hamilton; Hildegarde Wagenhals Bowen interview, Dec. 30, 1978.

23. Agnes Hamilton Diary, Dec. 13 [1896].

24. Ibid., Aug. 19 [1886].

25. Ibid., Feb. 10 [1884]; EH to JH, undated [July 1890]; AH to Agnes, July 5 [1923].

26. Sarah Porter to Allen Hamilton, Aug. 13, 1863, Allen Hamilton Papers. On Miss Porter and her school, see letters of the Hamilton women and Agnes Hamilton Diary, 1886–1888; Miss Porter's School Archives; Louise L. Stevenson, "Sarah Porter Educates Useful Ladies, 1847–1900," *Winterthur Portfolio,* 18 (Spring 1983); and Amy K. Johnson, "Miss Sarah Porter and Her School: Bastions of Conservatism or Precursors of Feminism?" (paper, Trinity College, Hartford, 1983).

27. EDT, pp. 35–37.

28. EDT, pp. 34–35.

29. AH to Agnes, Aug. 12, 1887.

30. July 21, 1889.

31. AH to Agnes, July 21, 1889; EH to JH, undated [ca. July 22, 1890]. See also AH to Agnes, Sept. 13, 1891, and May 22, 1892.

32. AH to EH, March 15, 1890; AH to NH, March 18, 1890; AH to Agnes, March 23, 1890; EH to JH, undated [July 20, 1890].

33. AH to Agnes, Aug. 1891. See also Agnes to Allen Hamilton [Feb. 1, 1891, incorrectly dated Jan. 31, 1891]; Agnes to JH, June 20, 1892; AH to Agnes, June 3, 1892.

34. Elizabeth Failing Conner Diary, Monday, Feb. [20, 1888], Miss Porter's School Archives; Agnes Hamilton Diary, April 8 [1890] and Sept. 22 [1895]; Hildegarde Wagenhals Bowen, interview, Dec. 30, 1978; AH to JH, Saturday afternoon [June 1926]; Allen Hamilton Williams to AH, Sept. 10, 1896.

II. Medical Training, 1890–1894

1. EDT, p. 26; Elizabeth Failing Conner Diary, Monday, Feb. [20, 1888], Miss Porter's School Archives.

2. EDT, p. 38.

3. See Mary Roth Walsh, *"Doctors Wanted: No Women Need Apply": Sexual Barriers in the Medical Profession, 1835–1975* (New Haven, 1977); and Regina Markell Morantz, *Natural Guardians of the Race: Women Physicians in American Medicine, 1840–1980,* forthcoming.

4. EDT, p. 38; Agnes Hamilton Diary, Nov. 23 [1889]; AH to Agnes, Nov. 21 [1896]; AH to Gertrude Hamilton, June 5, 1891.

5. Agnes Hamilton Diary, Dec. 7 [1890].

6. AH to Agnes [August 1891].

7. Material pertaining to medical education at Michigan and AH's career there (including her transcript and minutes of faculty meetings) may be found in the Michigan Historical Collections, Bentley Historical Library, University of Michigan. See also Wilfred B. Shaw, ed., *The University of Michigan: An Encyclopedic Survey*, II, IV (Ann Arbor, 1951, 1958); Dorothy Gies McGuigan, *A Dangerous Experiment: 100 Years of Women at the University of Michigan* (Ann Arbor, 1970); Calendar of the University of Michigan; Victor C. Vaughan, *A Doctor's Memories* (Indianapolis, 1926). On medical education, see Ronald L. Numbers, ed., *The Education of American Physicians: Historical Essays* (Berkeley and Los Angeles, 1980).

8. Prescott to Sarah Palmer, April 3, 1893, New England Hospital Papers, Sophia Smith Collection.

9. AH to Agnes, May 22, 1892.

10. AH to Agnes, March 27, 1892, fragment [1892], and May 22, 1892; Janet Howell Clark to AH [1943].

11. Minutes, Department of Medicine and Surgery, May 20, 1892, University of Michigan; Agnes Hamilton Diary, June 21 [1892]; AH interview with Jean Alonzo Curran, Nov. 29, 1963.

12. Phoebe Taber Hamilton to JH, May 24, 1892; AH to Agnes, Dec. 4, 1892.

13. AH to Agnes, Oct. 9, 1892, Sunday evening [Nov. 6, 1892], Dec. 11, 1892.

14. AH to Agnes, Feb. 4, 1893.

15. Agnes Hamilton Diary, Dec. 18 [1892], Feb. 5 [1893]. See also Mina J. Carson, "Agnes Hamilton of Fort Wayne: The Education of a Christian Settlement Worker," forthcoming in *Indiana Magazine of History*, 1984.

16. AH to Agnes, Dec. 11, 1892; Agnes Hamilton Diary, Nov. 22 [1891].

17. AH to Agnes, May 20, 1893.

18. AH to Agnes, May 15, 1892, Jan. 10, 1893, and fragment [spring 1893].

19. AH to [Agnes], fragment [summer 1893]; EDT, p. 42. The patient's death is mentioned in Madeleine P. Grant, *Alice Hamilton: Pioneer Doctor in Industrial Medicine* (New York, 1967), p. 47, and "Interview with Harriet Hardy, M.D.," October 13 and 14, 1977, Oral History Project on Women in Medicine, Medical College of Pennsylvania, p. 23. Both sources place the episode in Ann Arbor, but it is more likely to have occurred during AH's internship. See letter 17 in this volume.

20. AH to Agnes, Aug. 13, 1893, Aug. 31, 1893.

21. Elizabeth Shepley Sergeant, "Alice Hamilton, M.D.: Crusader for Health in Industry," *Harper's*, 152 (May 1926), p. 766.

22. AH to Agnes, Nov. 27, 1893.

23. AH to Agnes, Nov. 5, 1893, Nov. 19, 1893.

24. Bertha Van Hoosen, Explanation of Resignation, Aug. 30, 1891; Special Meeting of Board of Physicians, New England Hospital, Oct. 17, 1894,

both in New England Hospital Papers, Sophia Smith Collection. See also Virginia G. Drachman, "Female Solidarity and Professional Success: The Dilemma of Women Doctors in Late Nineteenth-Century America," *Journal of Social History,* 15 (Summer 1982).

25. Margaret V. Hamilton to Katherine Hamilton, April 18, 1895; Hildegarde Wagenhals Bowen interview, Dec. 28, 1978; Phoebe Taber Hamilton to Katherine Hamilton, Dec. 30, 1904.

26. AH to Agnes, Nov. 27, 1893.

27. AH to Agnes, April 20 [1895, incorrectly dated 1894], and Feb. 11, 1894.

28. AH to Agnes, Feb. 11, 1894, March 8 [1894].

29. AH to Agnes, Nov. 19, 1893, April 1, 1894; AH to Emily Pope, April 26, 1894, and Minutes, Board of Physicians, New England Hospital, May 4, 1894, IV, pp. 75–76, New England Hospital Papers.

III. "I Shall Know, Being Old": Career and Family, 1895–1897

1. An excellent discussion of the subject of women scientists is Margaret W. Rossiter, *Women Scientists in America: Struggles and Strategies to 1940* (Baltimore and London, 1982), esp. pp. 29–50.

2. AH to Agnes, undated [ca. Feb. 24, 1895], March 30, 1895, March 7 [1895], April 30 [1895] (fragment), and Thursday [May 23, 1895].

3. AH to Agnes, June 11, 1895 and Monday [July 7, 1895].

4. AH wrote about her year in EDT, pp. 43–51 and in "Edith and Alice Hamilton: Students in Germany," *Atlantic Monthly,* 215 (March 1965).

5. EDT, p. 45; AH to Agnes, Feb. 16, 1896.

6. AH interview with Jean Alonzo Curran, Nov. 29, 1963, CLM; EDT, p. 47; AH to Gertrude Hamilton, July 16 [1896].

7. AH to Agnes, undated [Nov. 1895], and March 28, 1896.

8. AH to Agnes, undated [July 1896].

9. Agnes Hamilton Diary, Sept. 19, Sept. 22 [1895]; AH to Agnes, undated fragment [Dec. 1895 or Jan. 1896]; Agnes Hamilton Diary [after July 23, 1895]; Agnes to AH, Jan. 27, 1896.

10. Allen Hamilton Williams to [AH], Sept. 10, 1896, Sept. 17, 1896. Information on Allen Williams from Harvard University Archives and Russell Williams. Information on Marian Walker Williams from Radcliffe College Archives, Johns Hopkins University Archives, and Russell Williams.

11. EDT, pp. 51–52.

12. AH to Agnes, Nov. 21 [1896].

13. AH to Agnes, Thursday [Dec. 1896 or Jan. 1897], May 18, 1897.

14. AH to Agnes, Aug. 10, 1897.

IV. Hull House, 1897–1907

1. Typescript, "Memorial Services Held for Mrs. Florence Kelley . . . April 8, 1932," EGE Papers; EDT, p. 94.

2. AH, "Jane Addams of Hull-House, Chicago" [1952], AHP, no. 6.

3. The classic description of settlement life is JA's *Twenty Years at Hull-House* (New York, 1960; orig. pub. 1910). See also Allen F. Davis, *Spearheads for Reform: The Social Settlements and the Progressive Movement, 1890–1914* (New York, 1967) and Dolores Hayden, *The Grand Domestic Revolution: A History of Feminist Designs for American Homes, Neighborhoods, and Cities* (Cambridge, Mass., 1981); the latter emphasizes communal living at Hull House, a point that has been generally neglected.

4. See *Hull-House Maps and Papers,* by Residents of Hull-House (New York, 1895).

5. *Hull-House Bulletin* (1896–1905/1906) and *Hull-House Year Book* (1906/07–1929) are invaluable sources of information about the settlement's activities and residents. Revealing primary sources by early residents that include recollections of AH are Nicholas Kelley, "Early Days at Hull-House," *Social Service Review,* 28 (Dec. 1954) and Francis Hackett, "Hull-House—A Souvenir," *The Survey,* 54 (June 1, 1925). Allen F. Davis and Mary Lynn McCree, eds., *Eighty Years at Hull-House* (Chicago, 1969) is a useful collection of primary sources.

6. EDT, p. 61; AH to Agnes, Wednesday evening [March 30?, 1898].

7. EDT, p. 64.

8. The literature on Jane Addams is vast. See especially Allen F. Davis, *American Heroine: The Life and Legend of Jane Addams* (New York, 1973); John C. Farrell, *Beloved Lady: A History of Jane Addams' Ideas on Reform and Peace* (Baltimore, 1967); Daniel Levine, *Jane Addams and the Liberal Tradition* (Madison, Wis., 1971); and James Weber Linn, *Jane Addams: A Biography* (New York, 1935); the last is by JA's nephew.

A complete microfilm edition of The Papers of Jane Addams, prepared under the direction of Mary Lynn McCree at the University of Illinois at Chicago, should soon be available. It will include correspondence to and from JA, clipping and reference files, published and unpublished writings, and the papers of Hull House.

9. AH to Agnes, May 18, 1897. See also EDT, pp. 64–67, and EH to JH and Agnes, Jan. 6, 1899.

10. EDT, p. 69.

11. EDT, p. 68; AH, "Speech at H.H. 50th Anniversary," AHP, no. 6; AH, "As One Woman Sees the Issues," *The New Republic,* 8 (Oct. 7, 1916); Agnes Hamilton Diary, Nov. 3, 4, 5 [1896].

12. EDT, p. 60; "Speech at H.H. 50th Anniversary"; AH to Agnes, Aug. 24 [1898] and Aug. 9, 1898.

13. Linn, *Jane Addams,* p. 145; AH to Agnes, "Some time in February," continues April 6 [1902].

14. EDT, pp. 69–70; AH to Agnes, Nov. 5 [1899].

15. AH to [Agnes], fragment [March? 1898]. See also Davis, *American Heroine,* pp. 120–125.

16. AH to Agnes, Aug. 9, 1898, Oct. 11, 1898.

17. AH to Agnes, fragment, Nov. 7, 1898, Aug. 24 [1898].

18. AH to Agnes, Aug. 28 [1902]; EH to JH and Agnes, Jan. 6, 1899; AH to Agnes, April 23, 1900. See also Elizabeth Shepley Sergeant, "Alice Hamilton, M.D.: Crusader for Health in Industry," *Harper's,* 152 (May 1926), 767.

19. AH to Agnes, June 18, 1899.

20. AH, "New Ideals and Old Ideals, Baldwin School," undated commencement address, AHP, no. 27. Cf. *The Alumnae Bulletin of Miss Porter's School* (Spring 1951).

21. EDT, pp. 30–31.

22. AH to Agnes, Oct. 25 [1900], Dec. 1, 1906 (letter 34 in this volume).

23. EDT, p. 95; AH to Agnes, Sept. 15, 1899.

24. Gertrude Hamilton to "At Home People," April 4 and 5 [1900], enclosed in AH to Agnes, April 23, 1900. See also Gertrude Hamilton to AH, Jan. 31, 1900, and to "family," April 13 [1900].

25. AH to Agnes, Nov. 5 [1899], April 30 [1900], and May 21 [1900].

26. AH to Agnes, Sept. 2, 1900, Aug. 28 [1902].

27. See, for example, AH to Agnes, Oct. 24 [1938].

28. AH to Agnes, "Some time in February," continues April 6 [1902]; JH to Agnes, June 1, 1902.

29. AH to Agnes, July 28 [1902].

30. EDT, p. 96. See also Morris Fishbein, "Ludvig Hektoen: A Biography and an Appreciation," *Archives of Pathology,* 26 (July 1938), and Frederick Stenn, "The Life and Times of Ludvig Hektoen: Pathologist, Investigator, Administrator," manuscript on deposit, Northwestern University Medical School.

31. EDT, pp. 97–100. See also "An Inquiry into the Causes of the Recent Epidemic of Typhoid Fever in Chicago," *The Commons,* 8 (May 1903), 3–7; "Public Indebtedness to Hull-House," ibid., p. 19; "Improvement Needed in the Health Department," *The Commons,* 8 (August 1903); *Chicago Medical Recorder,* 25 (July 15, 1903), 58.

32. AH to Agnes, Oct. 17 [1903]. See also "Lax Methods of the Chicago Sanitary Bureau," *Charities,* 11 (Aug. 1, 1903), 100.

33. EDT, pp. 100–104. See also "Conference on the Illegal Sale of Cocaine," *Hull-House Bulletin,* 6 (Autumn 1904), p. 21; "The Hull-House War on Cocaine," *Charities and The Commons,* 17 (March 9, 1907), 1034–1035; and "A New Weapon against Cocaine," ibid., 19 (Nov. 16, 1907), 1045.

34. See publications of the Committee on the Prevention of Tuberculosis of the Visiting Nurse Association [of Chicago], 1903–1904; AH, "The Industrial Viewpoint: Occupational Conditions of Tuberculosis," *Charities and The Commons,* 16 (May 5, 1906), 205–207; *A Study of Tuberculosis in Chicago* (Chicago, 1905); Addams, *Twenty Years at Hull-House,* p. 213; JA and AH, "The 'Piece-Work' System as a Factor in the Tuberculosis of Wage-Workers," *Transactions of the Sixth International Congress on Tuberculosis, Sept. 28 to Oct. 5, 1908,* 3 (Philadelphia, 1908).

35. "The Midwives of Chicago," *Journal of the American Medical Association,* 50 (April 25, 1908); AH, "Excessive Child-Bearing as a Factor in Infant Mortality," *Bulletin of the American Academy of Medicine,* 11 (Feb. 1910); "The Social Settlement and Public Health," *Charities and The Commons,* 17 (March 9, 1907).

36. AH to Agnes, "Some time in February," continues April 6 [1902], March 12, 1906.

37. AH to Agnes, Oct. 17 [1903]; EH to JH and Agnes, Jan. 6, 1899; AH, "New Ideals and Old Ideals."

V. Exploring the Dangerous Trades, 1908–1914

1. EDT, pp. 114–115; "Occupational Diseases," *Proceedings, National Conference of Charities and Correction* (1911), p. 197; "Industrial Diseases: With Special Reference to the Trades in Which Women Are Employed," *Charities and The Commons,* 20 (Sept. 5, 1908), 655, 658.

2. The *American Labor Legislation Review,* which started in 1911, includes useful articles, reports of annual meetings, and summaries of early work and legislation on industrial diseases. There is no satisfactory history of the field, but see George Martin Kober, "History of Industrial Hygiene and Its Effects on Public Health," in Mazÿck Ravenel, ed., *A Half Century of Public Health* (New York, 1921), pp. 361–411; George M. Kober and Emery R. Hayhurst, *Industrial Health* (Philadelphia, 1924); Ludwig Teleky, *History of Factory and Mine Hygiene* (New York, 1948), which has an introduction by AH; George Rosen, *A History of Public Health* (New York, 1958); Margaret C. Klem and Margaret F. McKiever, "50-Year Chronology of Occupational Medicine," *Journal of Occupational Medicine,* 8 (April 1966); Daniel M. Berman, *Death on the Job: Occupational Health and Safety Struggles in the United States* (New York, 1978); and Carl Gersuny, *Work Hazards and Industrial Conflict* (Hanover, N.H., 1981). Roy Lubove, *The Struggle for Social Security, 1900–1935* (Cambridge, Mass., 1968) has a useful critique of the workmen's compensation movement, pp. 45–65.

3. J. R. Commons to H. B. Favill, Jan. 21, 1908; Commons to Irene

Osgood, Jan. 29, 1908; Osgood to AH, Dec. 1, 1908; AH to Osgood, Dec. 5 [1908]; all in AALL Papers.

4. AH to Irene Osgood, Feb. 13, 1909, AALL Papers. See also R. Alton Lee, "The Eradication of Phossy Jaw: A Unique Development of Federal Police Power," *The Historian*, 24 (Nov. 1966).

5. [Illinois] Commission on Occupational Diseases, report transmitted by Gov. Charles S. Deneen, April 21, 1909.

6. AH to JH [Feb. 26, 1910]; AH to Agnes, April 16, 1910.

7. [Illinois] Report of Commission on Occupational Diseases; To His Excellency Governor Charles S. Deneen (Chicago, 1911).

8. EDT, pp. 120–121.

9. EDT, p. 128.

10. AH to Webster King Wetherill, May 22, 1911; Wetherill to AH, May 23, 1911; George M. Coates to AH, May 27, 1911; "George D. Wetherill & Bro., May 10, 1911, Notes"; all in AHC, ms. 23. See also *The White-Lead Industry in the United States, with an Appendix on the Lead-Oxide Industry* (Bureau of Labor, Bulletin no. 95, July 1911), esp. pp. 251–253.

11. "George D. Wetherill & Bro., Notes."

12. "Standards of Living and Labor," *Proceedings, National Conference of Charities and Correction* (1912), pp. 376–395.

13. AH interview with Jean Alonzo Curran, Nov. 29, 1963, CLM.

14. *Lead Poisoning in Potteries, Tile Works, and Porcelain Enameled Sanitary Ware Factories* (Bureau of Labor, Bulletin no. 104, 1912); *Hygiene of the Painters' Trade* (BLS, Bulletin no. 120, 1913); *Lead Poisoning in the Smelting and Refining of Lead* (BLS, Bulletin no. 141, 1914); *Lead Poisoning in the Manufacture of Storage Batteries* (BLS, Bulletin no. 165, 1914); *Industrial Poisons Used in the Rubber Industry* (BLS, Bulletin no. 179, 1915); see also *Hygiene of the Printing Trades* (BLS, Bulletin no. 209, 1917, with Charles H. Verrill). Some of the bulletins include summaries of work she had asked others to do, among them laboratory studies on the solubility of lead in human gastric juices and clinical examinations of workers exposed to lead.

15. Quoted in Elizabeth Shepley Sergeant, "Alice Hamilton, M.D.: Crusader for Health in Industry," *Harper's*, 152 (May 1926), 763–764.

16. EDT, pp. 145–146.

17. EDT, p. 138; "Occupational Diseases," p. 206.

18. "Industrial Lead-Poisoning: Presidential Address," *Transactions of the Chicago Pathological Society*, 8 (Aug. 1, 1912). See also "Protection from Lead Poisoning," *American Labor Legislation Review*, 2 (Dec. 1912), 537.

19. EDT, pp. 7, 135–136.

20. AH, "What One Stockholder Did," *The Survey*, 28 (June 1, 1912), 387–389; EDT, pp. 155–159.

21. EDT, pp. 10–11.

22. "Occupational Diseases," p. 197; "Industrial Lead-Poisoning," p. 311; "The Economic Importance of Lead Poisoning," *Bulletin of the American Academy of Medicine,* 15 (Oct. 1914), 304. See also "Lead-Poisoning in Illinois," *Journal of the American Medical Association,* 56 (April 29, 1911), and "The Hygiene of the Lead Industry," typescript, Address at Meeting of Superintendents, National Lead Company, Dec. 7, 1910, AHP, no. 29.

23. John B. Andrews to C. H. Verrill, Feb. 4, 1913, AHP, no. 29, and *Ohio State Journal* (Columbus), Jan. 24, 1913.

24. Theodore Ahrens to AH, Dec. 14, 1914, AHC.

25. EDT, p. 121.

26. Interview with Harriet L. Hardy, May 15, 1974; EDT, p. 129; AH to Agnes, March 8 [1894].

27. EDT, pp. 151–152.

28. "Industrial Diseases," *Charities and The Commons,* 20 (Sept. 5, 1908), 659.

29. EDT, p. 216.

30. NH to JH, Sunday [Feb. 19, 1911]; AH to EH, June 11 [1912].

31. AH to JH, March 1, 1912; EDT, pp. 128–129; AH to Agnes, fragment, Saturday [Dec. 1914 or early 1915].

32. EDT, pp. 91–92; AH to Agnes, Oct. 24 [1913].

VI. The War Years, 1915–1919

1. Marie Louise Degen, *The History of the Woman's Peace Party* (Baltimore, 1939) is still the best secondary account (quotation, p. 70). See also David S. Patterson, "Woodrow Wilson and the Mediation Movement, 1914–17," *The Historian,* 33 (Aug. 1971); John C. Farrell, *Beloved Lady: A History of Jane Addams' Ideas on Reform and Peace* (Baltimore, 1967); Allen F. Davis, *American Heroine: The Life and Legend of Jane Addams* (New York, 1973); JA, *Peace and Bread in Time of War* (New York, 1922); Mercedes Randall, *Improper Bostonian: Emily Greene Balch* (New York, 1964).

2. EDT, p. 167; Emily Greene Balch, "Journal," typescript, p. 6, Emily Greene Balch Papers. See AH, "Is Science for or against Human Welfare?" *The Survey,* 35 (Feb. 5, 1916).

3. International Congress of Women, The Hague, April 28th to May 1st 1914, Report, International Women's Committee of Permanent Peace (Amsterdam, n.d.)

4. AH to her family, May 15, 1915, AHP, no. 3; EDT, pp. 167–169.

5. EDT, pp. 169–180; *Women at The Hague: The International Congress of Women and Its Results* (New York, 1915), which AH coauthored with JA and Emily Greene Balch (AH's section originally appeared in *The Survey,* 34 (Aug. 7, 1915); "The Attitude of Social Workers toward the War," ibid., 36 (June

17, 1916), quotation, p. 308. See also AH's later "Colonel House and Jane Addams," *New Republic,* 47 (May 26, 1926).

6. "As One Woman Sees the Issues," *New Republic,* 8 (Oct. 7, 1916); Davis, *American Heroine,* pp. 212–250, quotation p. 229.

7. EDT, pp. 406–410.

8. EDT, pp. 183–199, quotation p. 184; "Industrial Poisons Encountered in the Manufacture of Explosives," *Journal of the American Medical Association,* 68 (May 19, 1917). See also *Industrial Poisons Used or Produced in the Manufacture of Explosives* (BLS, Bulletin no. 219, 1917).

9. "War Industrial Diseases," *Medical Record,* 95 (June 21, 1919); "Dope Poisoning in the Manufacture of Airplane Wings," *Monthly Review of the United States Bureau of Labor Statistics,* 5 (Oct. 1917); "Effect of the Air Hammer on the Hands of Stonecutters," ibid., 6 (April 1918); "Industrial Poisoning in American Anilin Dye Manufacture," ibid., 8 (Feb. 1919); *Women in the Lead Industries* (BLS, Bulletin no. 253, 1919); Report of the Health Insurance Commission of the State of Illinois (May 1, 1919), esp. pp. 168–173.

10. AH's work with the NRC may be followed in the NRC Papers. See also EDT, pp. 197–199; "War Industrial Diseases." On AH's war work and the conflicts between the Bureau of Labor Statistics and the Public Health Service, see Angela Nugent Young, "Interpreting the Dangerous Trades: Workers' Health in America and the Career of Alice Hamilton, 1910–1935," Ph.D. diss., Brown University, 1982, pp. 68–106.

11. EDT, pp. 192–193.

12. See Melvyn Dubofsky, *We Shall Be All: A History of the Industrial Workers of the World* (Chicago, 1969); Joyce L. Kornbluh, ed., *Rebel Voices: An I.W.W. Anthology* (Ann Arbor, 1964); Robert K. Murray, *Red Scare: A Study in National Hysteria, 1919–1920* (Minneapolis, 1955).

13. AH to EH, Thursday evening [June 6, 1918].

14. "The Bollinger Case," *The Survey,* 35 (Dec. 4, 1915), 266; quoted in FF to CCB, Jan. 25, 1952, FFP.

15. Peyton Rous to Edsall, Oct. 16, 1918, Edsall to Rous, Oct. 18, 1918, NRC Papers; Edsall to AH, Dec. 27, 1918, AHP, no. 5; Edsall to Lowell, Dec. 20, 1918, Lowell Papers, no. 311.

16. Lowell to Henry P. Walcott, Dec. 21, 1918, no. 311, and Lowell to H. H. Moore, April 4, 1919, no. 729, Lowell Papers; AH to EH, Jan. 11 [1919].

17. AH to Agnes, postcard, Jan. 21 [1919]; AH to NH, Jan. 16 [1919]. See also EDT, pp. 208–222; AHP, no. 37.

18. Edsall to AH, Jan. 24, 1919, AHP, no. 5.

19. AH to Family, May 1, 1919, AHP, no. 3; AH to Mary Rozet Smith, May 12, 1919, JA Papers; EDT, pp. 223–233.

20. Report of the International Congress of Women, Zurich, May 12 to 17, 1919 (Geneva, n.d.); Randall, *Improper Bostonian.*

21. "Angels of Victory," *New Republic,* 19 (June 25, 1919). See also "On a German Railway Train," ibid., 20 (Sept. 24, 1919).

22. EDT, pp. 243–251; JA to Mary Rozet Smith, May 23, 1919, JA Papers; AH Diary, 1919.

23. EDT, pp. 236–237, 243–249; JA and AH, "After the Lean Years: Impressions of Food Conditions in Germany When Peace Was Signed," *The Survey,* 42 (Sept. 6, 1919).

24. AH to MH, April 23 [1919].

VII. The Harvard Years, 1919–1927

1. AH to JH, March 6 [1919]; EDT, p. 252; newspaper clippings, April 19, 20, and March 23, 1919, AHP, no. 5.

2. AH to Clara Landsberg, Sunday morning [Feb. 15, 1925]; AH to MH, Feb. 18, 1926.

3. Cecil Drinker, in "Dr. Alice Hamilton, Luncheon, Consumers' League of Massachusetts, Nov. 21, 1935," [p. 9], AHP, no. 69; "Lead Investigation," Dean's Files, HSPH Papers; "The Reminiscences of Joseph Aub," Oral History Research Office, Columbia University, 1956, 1957, vol. 1, pp. 170–178.

4. "The Reminiscences of Joseph Aub," vol. 1, quotation, p. 171. Useful secondary sources on Harvard's industrial hygiene program are: Jean Alonzo Curran, *Founders of the Harvard School of Public Health, with Biographical Notes, 1909–1946* (New York, 1970), esp. pp. 16–19, 35–39, 154–169; Joseph C. Aub and Ruth K. Hapgood, *Pioneer in Modern Medicine: David Linn Edsall of Harvard* (Boston, 1970), esp. pp. 248–262; Angela Nugent Young, "Interpreting the Dangerous Trades: Workers' Health in America and the Career of Alice Hamilton, 1910–1935," Ph.D. diss., Brown University, 1982, pp. 107–155; Louise Joy Short, "Four Pioneers: The Making of the Harvard Program in Industrial Hygiene, 1918–1935," senior thesis, Harvard University, 1982; and George Cheever Shattuck, "Industrial Medicine at Harvard," typescript, 1954, Dean's Correspondence Files, HSPH.

5. "Nineteen Years in the Poisonous Trades," *Harper's,* 159 (Oct. 1929), quotation, p. 587; "Recent Advances in Industrial Toxicology in the United States," in *De Lamar Lectures 1925–1926* (Baltimore, 1927).

6. "Nineteen Years in the Poisonous Trades," pp. 586, 590.

7. *Industrial Poisoning in Making Coal-tar Dyes and Dye Intermediates* (BLS, Bulletin no. 280, 1921); *Carbon-Monoxide Poisoning* (BLS, Bulletin no. 291, 1921); "The Industrial Hygiene of Fur Cutting and Felt Hat Manufacture," *Journal of Industrial Hygiene,* 4 (Aug. 1922).

8. Kober to AH, Sept. 27[?], 1925; Oliver to AH, June 22, 1925; both in AHP, no. 58.

9. AH to Theodore Ahrens, Nov. 1, 1927 and following, AHC; AH to KBC, March 8, 1922, KBC Papers.

10. AH to Katherine Wiley, Dec. 16, 1927, NCL Papers, C-43.

11. AH to KBC, May 17, 1925. See also AH, "What Price Safety? Tetraethyl Lead Reveals a Flaw in Our Defenses," *The Survey*, 54 (June 15, 1925).

12. "Nineteen Years in the Poisonous Trades," pp. 580–581; AH to Florence Kelley, June 17, 1929, NCL Papers, C-43.

13. AH to KBC, April 20 [1921]; AH to Harriet Hardy, June 6, 1947, Hardy Papers. On women reformers, see J. Stanley Lemons, *The Woman Citizen: Social Feminism in the 1920s* (Urbana, 1973); William Henry Chafe, *The American Woman: Her Changing Social, Economic, and Political Roles, 1920–1970* (New York, 1972); Susan Ware, *Beyond Suffrage: Women in the New Deal* (Cambridge, Mass., 1981); and Clarke A. Chambers, *Seedtime of Reform: American Social Service and Social Action, 1918–1933* (Minneapolis, 1963).

14. Elizabeth Glendower Evans, "People I Have Known: Alice Hamilton, M.D., Pioneer in a New Kind of Human Service," *The Progressive*, 2 (Dec. 20, 1930); Dorothy Detzer Denny to author, Aug. 4, 1976. See also AH, "State Pensions or Charity?," *Atlantic Monthly*, 145 (May 1930), and "The Cost of Medical Care," *New Republic*, 59 (June 26, 1929).

15. Untitled, handwritten speech [1931?], AHC, MS. 1.

16. AH to KBC, March 2 [1920], March 12, 1920, April 13, 1926, all KBC Papers; AH interview with Jean Alonzo Curran, Nov. 29, 1963, CLM. See also AH, "Witchcraft in West Polk Street," *American Mercury*, 10 (Jan. 1927), and Elizabeth Dilling, *The Red Network: A 'Who's Who' and Handbook of Radicalism for Patriots* (Kenilworth, Ill., 1934).

17. AH to Agnes, Nov. 3, 1932.

18. AH to Agnes, July 5 [1923]; Agnes to Katherine Hamilton, April 1, 1922; Phoebe Taber Hamilton to Agnes, April 29, 1923.

19. EDT, p. 266; AH Diary, Dec. 1, 3, 4, 16, 1919; AH to KBC, May 17, 1925, HFP.

20. EH to JH, Wednesday [Jan. 4, 1922] and Aug. 18 [1922]; AH to MH, March 3 [1921]; MH to NH, Wednesday [Winter 1920–21]. Information about the Bryn Mawr School crisis appears in the M. Carey Thomas Papers, Bryn Mawr College, microfilm; see esp. Edith Orlady to EH, June 13, 1919, "Report to Entrance Examination Committee," reel 167; EH to M. C. Thomas, July 29, 1919, Thomas to EH, Sept. 8, 1919 (two letters of this date), reel 140; Thomas to EH, Dec. 23, 1920 and Feb. 10, 1921, reel 142; and correspondence of Margaret Thomas Carey (M. C. Thomas' sister and a member of the School's Board of Managers), reel 37.

21. "Protection for Working Women," *The Woman Citizen*, 8 (March 8, 1924), 17; AH to Emma Wold, July 2, 1926, National Woman's Party Papers, reel 33. See also "Protection for Women Workers," part II of "The 'Blanket'

Amendment—A Debate," *Forum,* 72 (Aug. 1924); AHP, no. 11; AH to Mary O'Malley, Sept. 27, 1932, National Woman's Party Papers, reel 17. AH's views on women and industrial poisons are found in *Women in the Lead Industries* (BLS, Bulletin no. 253, 1919) and *Women Workers and Industrial Poisons* (Women's Bureau, Bulletin no. 57, 1926). See also Vilma R. Hunt, "A Brief History of Women Workers and Hazards in the Workplace," which includes a discussion of AH's views, and Ann Corinne Hill, "Protection of Women Workers and the Courts: A Legal Case History," both in *Feminist Studies,* 5 (Summer 1979).

22. AH to JH, March 19 [1922].

23. Typescript dated March 21[?], 1922, and "Statement by President M. Carey Thomas, President of the Board of Managers of the Bryn Mawr School for Girls of Baltimore City," both in M. Carey Thomas Papers, reel 167; *The Sun* (Baltimore), March 21, 22, 24, 25, 1922; *Baltimore American,* March 21, 22, 23, 25, 1922.

24. EH to JH, Sunday [April 23, 1922]; AH to MH, Feb. 9, 1923; AH to NH, Jan. 2, 1926; AH to KBC, June 22, 1959.

25. AH to KBC, April 24 [1922], KBC Papers. See also AH to Agnes, April 13, 1922.

26. Cecil K. Drinker to Roger I. Lee, Nov. 21, 1922; AH to Lee, Dec. 19, 1922; "Conference with Mr. Gerard Swope," Dec. 29 [1922]; [Drinker] to AH, Dec. 30, 1922; all in "General Electric Company" folder, Dean's Correspondence Files, HSPH Papers; AH Diary, Jan. 22, 24, 1923.

27. AH to MH, Feb. 9, 1923; AH to KBC, May 31, 1923, KBC Papers; AH to Clara Landsberg, June 22, 1923, AHC.

28. AH to Florence Kelley, Dec. 15, 1925, NCL Papers, B-13. Material on Workers' Health Bureau in American Fund for Public Service Records.

29. AH to Agnes, Feb. 29, 1924; EDT, pp. 299–317.

30. AH to MH, Sept. 24, 1924, AHP, no. 3.

31. AH to JH, Jan. 4, 1925; EDT, pp. 318–352.

32. EDT, p. 332; AH to JH, Jan. 4, 1925; AH to JA, Nov. 20, 1924, AHP, no. 3. See also AH to "Dear People," Oct. 19, 1924, AHP, no. 3, and AH and Rebecca Edith Hilles, "Industrial Hygiene in Moscow," *Journal of Industrial Hygiene,* 7 (Feb. 1925), 47–61.

33. Clipping, AHP, no. 13; EDT, p. 351.

34. *New York Times,* Jan. 21, 1925, p. 21. See also AH to JA, undated [Jan. 11, 1925?], JA Papers; *Report of the Conference on the Cause and Cure of War, January 18–24, 1925;* and J. M. Jensen, "All Pink Sisters: The War Department and the Feminist Movement in the 1920s," in *Decades of Discontent,* ed. Lois Scharf and J. M. Jensen (Westport, Conn., 1983).

35. The radium case can be followed in two parts of the NCL Papers: C-42 and C-43, "Radium Poisoning" (see esp. AH to Katherine Wiley, Jan. 30, 1925, Feb. 7, 1925, March 16, 1925) and the Raymond H. Berry Papers, a

three-reel microfilm collection by the attorney in the case, and in the Department of Physiology, Correspondence Files, "Radium," HSPH Papers (see esp. AH to Katherine R. Drinker, April 4, 1925). See also Minutes of Annual Meetings, NCL, esp. Nov. 13, 1924, Nov. 20, 1925, Nov. 29, 1926, Nov. 1928, NCL Papers; Sherry Lee Baron, "Watches, Workers, and the Awakening World: A Case Study in the History of Occupational Medicine in America," senior thesis, Harvard University, 1977, and Josephine Goldmark, *Impatient Crusader: Florence Kelley's Life Story* (Urbana, Ill., 1953), pp. 189–204.

36. See also AH to Clara Landsberg, Sunday morning [Feb. 15, 1925].

37. "The Growing Menace of Benzene (Benzol) Poisoning in American Industry," *Journal of the American Medical Association,* 78 (March 4, 1922); quotation from reprint, p. 5. AH's efforts to raise funds to study benzene may be followed in American Fund for Public Service Records, National Research Council Papers, and "Benzol Investigation" File, Dean's Correspondence Files, HSPH Papers.

38. AH to Lovett, Dec. 11, 1925, American Fund for Public Service Records. See also C.-E. A. Winslow, "Summary of the National Safety Council Study of Benzol Poisoning," *Journal of Industrial Hygiene,* 9 (Feb. 1927); AH, "The Lessening Menace of Benzol Poisoning in American Industry," ibid., 10 (Sept. 1928); "Progress Report of Committee on Benzol Poisoning," *Proceedings of the National Safety Council* (1923), esp. remarks by AH, pp. 220–221; AHP, no. 40.

39. AH to Clara Landsberg, Feb. 6, 1926.

40. AH to KBC, Dec. 23, 1924, KBC Papers.

41. AH to Leland E. Cofer, Sept. 21, 1926; Cofer to AH, Oct. 22, 1926; D. H. Kelly to AH, Dec. 30, 1926; all AHC.

42. D. H. Kelly to AH, June 13, 1927, AHC.

43. AH to Philip Drinker, June 16, 1927, AHC.

44. Speech, New York, Aug. 23, 1929, AHP, no. 14; AH to FF, Aug. 10, 1927, FF Papers, Harvard Law School Library.

45. Henry A. Christian to AH, Aug. 23, 1927, AHP, no. 14.

46. Christian to AH, Sept. 2, 1927, AHP, no. 14.

47. AH Speech, Aug. 23, 1929, AHP, no. 14.

VIII. Elder Stateswoman, 1928–1935

1. Clipping, *The Gazette, Montreal,* Sept. 16, 1931, AHP, no. 4; Bradley Dewey to S. P. Miller, Feb. 9, 1933, AHP, no. 40.

2. E.L.B., Review and Editorial on *Industrial Toxicology* [Detroit Medical News], both in AHP, no. 58; Emery R. Hayhurst, Review of *Industrial Toxicology, American Journal of Public Health,* 24 (Sept. 1934), 993.

3. Berry to AH, Jan. 9, 1929, Raymond H. Berry Papers, reel 3 in NCL Papers; see also Chapter 7, note 35. On the tetra-ethyl lead case, see AH, "What Price Safety? Tetra-ethyl Lead Reveals a Flaw in Our Defenses," *The Survey*, 54 (June 15, 1925); AH, Paul Reznikoff, and Grace M. Burnham, "Tetra-Ethyl Lead," *Journal of the American Medical Association*, 84 (May 16, 1925); *Public Health Bulletin*, no. 158 (1925), and no. 163 (1926); Public Health Service Records; and Joseph A. Pratt, "Letting the Grandchildren Do It: Environmental Planning During the Ascent of Oil as a Major Energy Source," *Public Historian*, 2 (Summer 1980).

4. Minutes of the Conference on Radium called by the Surgeon General, Dec. 20, 1928, Public Health Service Records, pp. 4, 35, and passim.

5. Wiley to Kelley, Dec. 26, 1928, NCL Papers, C-43.

6. AH to Kelley, Jan. 15, 1929, NCL Papers, C-43. See also EDT, pp. 415–417, and "Forty Years in the Poisonous Trades," *American Industrial Hygiene Association Quarterly*, 9 (March 1948), 9.

7. "Excessive Child-Bearing as a Factor in Infant Mortality," *Bulletin of the American Academy of Medicine*, 11 (Feb. 1910); "Transactions, Joint Meeting of the Chicago Medical and Chicago Gynecological Societies, Held February 16, 1916," *Surgery, Gynecology and Obstetrics*, 23 (Aug. 1916), 235–236; "Family Limitation Centers in Chicago," *The Survey*, 38 (May 5, 1917); "Poverty and Birth Control," *Birth Control Review*, 9 (Aug. 1925).

8. AH to Agnes, April 3, 1932.

9. See AHP, no. 4; William F. Ogburn Papers, "Woman Member of the Committee," box 15; and Barry D. Karl, "Presidential Planning and Social Science Research: Mr. Hoover's Experts," *Perspectives in American History*, 3 (1969).

10. EDT, p. 292; William F. Ogburn Papers, Recent Social Trends, Transcript of May 1, 1932 session, quotation p. 202, and May 16, 1932 session, pp. 182–212, both box 9. Wolman's reply is in AHP, no. 8.

11. Letter of June 27, 1932 to Macmillan Co., AHP, no. 48.

12. See letter 91 in this volume and AH to Agnes, Thanksgiving Day [Nov. 24, 1932].

13. Cohen to AH, Nov. 7, 1932; AH to Cohen, Nov. 14, 1932; both AHP, no. 48.

14. AH to Cohen, Jan. 4, 1933, AHP, no. 48.

15. Cohen to AH, Jan. 7, 1932, AHP, no. 48.

16. AH Diary, Feb. 1, 1933. On the Lawrence strikes, see EDT, pp. 353–359.

17. AH to Grace Abbott, Nov. 17, 1932, Edith and Grace Abbott Papers.

18. AH to JA, Oct. 18, 1933, JA Papers; AH to Florence Kelley, Jan. 6, 1932, Nicholas Kelley Papers.

19. AH to MH, May 14, 1933. See also EDT, pp. 360–386.

20. AH to JA, April 22, 1933, JA Memorial Collection.

21. The *New York Times* articles were "An Inquiry into the Nazi Mind," Aug. 6, 1933, section 6; "The Youth Who are Hitler's Strength," Oct. 8, 1933, section 6; the editorial was "German Intellectuals," January 7, 1934, section 4. See also AH's "Below the Surface," *Survey Graphic,* 22 (Sept. 1933); "Sound and Fury in Germany," ibid., 22 (Nov. 1933); "The Plight of the German Intellectuals," *Harper's,* 168 (Jan. 1934); "Hitler Speaks," *Atlantic,* 152 (Oct. 1933); AH Diary, Aug. 25, 26, 1933; and AHC, MS. 15–22.

22. AH to JA, Oct. 26, Oct. 18, and Nov. 24, 1933, JA Papers.

23. AH to MH, Sunday afternoon [late Feb. or March 1934].

24. AH to NH, Oct. 26 [1933]; AH to FF, March 12, 1935, FFP; EGE to FF, Sept. 7, 1935, EGE Papers.

25. See AHP, no. 15.

26. Notes on JA's last days by AH, JA Papers.

27. AH to Agnes, Monday [late May or June 1935]; residents' letter to trustees of Hull House, May 25, 1935, AHP, no. 6; AH to Grace Abbott, June 11, 1935, Edith and Grace Abbott Papers, box 55.

IX. Semi-retirement, 1935–1949

1. AH to JH, March 5, 1935; interview with Alberta Pfeiffer, Sept. 3, 1981; Russell Williams to author, ca. Jan. 19, 1977.

2. AH to Frances Perkins, Oct. 18, 1935; Perkins to AH, Oct. 25, 1935; AH to Perkins, April 5, 1933; all in Frances Perkins Papers.

3. AH, "Occupational Diseases and the United States Department of Labor," typescript, AHC, ms. 31; George Martin, *Madam Secretary: Frances Perkins* (Boston, 1976), pp. 420–438. See also Verne A. Zimmer to AH, Aug. 15, 1936, Bureau of Labor Standards Records, box 82.

4. See V. A. Zimmer to Secretary, Nov. 13, 1935, and Perkins to AH, Nov. 16, 1935, for statements of AH's expected duties, in General Subject Files, 1933–1940, Office of the Secretary, Department of Labor; *Recent Changes in the Painters' Trade* (Division of Labor Standards, Bulletin no. 7, 1936); AH to V. A. Zimmer, Feb. 14, 1938, box 29, and Zimmer to AH, March 27, 1940, box 26, both in Bureau of Labor Standards Records; AH, "A Mid-American Tragedy," *Survey Graphic,* 29 (Aug. 1940).

5. EDT, pp. 387–394. See also *Survey of Carbon Disulphide and Hydrogen Sulphide Hazards in the Viscose Rayon Industry* (Occupational Disease Prevention Division, Pennsylvania Department of Labor and Industry, Bulletin no. 46, 1938) and *Occupational Poisoning in the Viscose Rayon Industry* (Division of Labor Standards, Bulletin no. 34, 1940). The story about the cadaver is from AH interview with Jean Alonzo Curran, Nov. 29, 1963, CLM, and Clara M. Beyer to author, Aug. 1, 1982.

6. "Maximum Allowable Concentrations of Dangerous Industrial Dusts

and Gases," speech delivered at 29th National Safety Congress and Exposition, Oct. 7–11, 1940, Bureau of Labor Standards Records, box 87; "Forty Years in the Poisonous Trades," *American Industrial Hygiene Association Quarterly,* 9 (March 1948).

7. AH to KBC, July 8, 1938, KBC Papers; AH and F. H. Lewey, "Carbon Disulphide and Hydrogen Sulphide Hazards in the Viscose Industry in the United States," *Sonderdruck aus Bericht Uber Den VIII Internationalen Kongress für Unfallmedizin und Berufskrankheiten* (Leipzig, 1938).

8. EDT, pp. 395–404.

9. The literature on Frankfurter and the Supreme Court is vast. H. N. Hirsch, *The Enigma of Felix Frankfurter* (New York, 1981); Michael E. Parrish, *Felix Frankfurter and His Times: The Reform Years* (New York, 1982); and Paul L. Murphy, *The Constitution in Crisis Times, 1918–1969* (New York, 1972) were especially helpful.

10. FF to AH, June 13, 1940, AHP, no. 18.

11. EDT, p. 425; AH to KBC [Sept. 1939], fragment, KBC Papers.

12. EDT, pp. 425–427; "Feed the Hungry!" *The Nation,* 151 (Dec. 14, 1940), 597.

13. AH to Agnes, Dec. 18, 1941; AH to JH, Dec. 30 [1941]; interview with Alberta Pfeiffer, Sept. 3, 1981.

14. [Edward A. Weeks] to AH, Jan. 24, 1935, June 26, 1942; AH to Weeks, June 28, 1942; [Weeks] to AH, July 15, 1942; all in *Atlantic Monthly* Papers.

15. FF to AH, April 23, 1964, FFP.

16. AH to Madeleine P. Grant, April 8, 1959, Grant Papers; AH to FF, Feb. 2, 1957, FFP.

17. AH to JH, April 14, 1943; Julian W. Tyler to AH, Dec. 11, 1942; Wilbur Daniel Steele to AH, April 2, 1943; Marion Frankfurter to AH, May 1 [1943]. Tributes on EDT may be found in HFP, nos. 664–668.

18. AH to Elizabeth S. Magee, May 30, 1944, NCL Papers, B-13.

19. AH to Agnes, Jan. 9, 1943; AH to JH, Feb. 26, 1943; AH to Edith Abbott, July 15, 1939, Edith and Grace Abbott Papers.

20. "Diagnosis of Industrial Poisoning," *California and Western Medicine,* 62 (March 1945); "Forty Years in the Poisonous Trades." See also "Industrial Poisons," *American Federationist,* 43 (July 1936); "Some New and Unfamiliar Industrial Poisons," *New England Journal of Medicine,* 215 (Sept. 3, 1936); "Healthy, Wealthy—if Wise—Industry," *American Scholar,* 7 (Winter 1938); "New Problems in the Field of the Industrial Toxicologist," *California and Western Medicine,* 61 (Aug. 1944).

21. AH to Harriet Hardy, Nov. 20, 1946, and Aug. 2, 1949 (quotation); Harriet Hardy interview with author, Nov. 11, 1974; "Interview with Harriet Hardy, M.D.," Oct. 13 and 14, 1977, Oral History Project on Women and Medicine, Medical College of Pennsylvania.

22. Quotation from AH to Clara Haas, March 22, 1948, Haas Papers.

Information about routine from Jarmila and Jane Petrzelka, interviews, May 20, 1977, and Sept. 4, 1981, and Alberta Pfeiffer interview, Sept. 3, 1981.

23. AH to Clara Haas, July 13, 1943, and March 22, 1948, Haas Papers; AH to Elizabeth S. Magee, Dec. 30, 1947, NCL Papers, B-13; AH to Agnes, Jan. 28, 1937.

X. "Old-Old Age," 1950–1970

1. AH to KBC, April 5, 1959; interviews with Jarmila and Jane Petrzelka, May 20, 1977, Sept. 4, 1981.

2. AH to FF, Sept. 28, 1958, FFP.

3. "Dr. Hamilton—Industrial Medicine Pioneer," *Scope Weekly*, 2 (May 8, 1957); AH to Harriet L. Hardy, Sept. 6, 1952, Hardy Papers.

4. AH to Morris L. Ernst, Jan. 3, 1950, FBI file; folders 674–676, HFP, contain copies of appeals that AH signed and preserved.

5. Paul H. Douglas to AH, Oct. 18, 1955, no. 671. See also AH to Douglas, Oct. 8, 1955, no. 657. AH's views on the Cold War and its consequences for civil liberties are documented in AHP, nos. 20 and 21, in HFP, and in letters to the *New York Times*, among them those published on Nov. 12, 1949; Dec. 10, 1950; Jan. 19, 1952. See also *Hartford Courant*, Feb. 4, 1951, and *Washington Post*, Oct. 18, 1954.

6. AH to FF, Nov. 9, 1955, FFP; "The Encroachment of the State upon the Conscience," *Christianity and Crisis*, 11 (May 14, 1951); "Should We Outlaw Anti-Semitism? A Symposium," *New Masses*, 54 (March 20, 1945).

7. FF to S. K. Ratcliff, Aug. 6, 1957, FFP.

8. FF to AH, Feb. 22, 1951, AHP, no. 20.

9. AH to Florence Kitchelt, Aug. 2, 1945, Kitchelt Papers; AH to Elizabeth S. Magee, Aug. 24, 1945, NCL Papers, B-13; AH to Freda Miller, Jan. 18, 1945, Women's Bureau Records, box 810; "Forty Years in the Poisonous Trades," *American Industrial Hygiene Association Quarterly*, 9 (March 1948), 7.

10. Kitchelt to Jane [Norman Smith?] and Alma [Lutz], May 9, 1952, Alma Lutz Papers.

11. AH to Edward A. Weeks, Nov. 21, 1956, *Atlantic Monthly* Papers; FF to Francis Hackett, Sept. 25, FFP. See also AH to CCB, May 1, 1957, CCB Papers.

12. CCB to FF, Sept. 21, 1951; CCB to AH, Oct. 9, 1951; FF to CCB, Feb. 26, 1953; all FFP. See also FF to CCB, Oct. 5, 1951, FFP; obituary, *New York Times*, June 8, 1959; and Philip B. Kurland, ed., *Of Law and Life and Other Things That Matter: Papers and Addresses of Felix Frankfurter, 1956–1963* (Cambridge, Mass., 1965).

13. AH to CCB, March 19, 1953, CCB Papers.

14. AH to FF, Nov. 9, 1955, FFP; letters to *New York Times*, Dec. 10,

1950, and Dec. 3, 1952; Albert Einstein to AH, Dec. 4, 1952, AHP, no. 21; AH to Anna Rochester, Jan. 14, 195[3] (incorrectly dated 1952), Anna Rochester Papers, University of Oregon Library.

15. AH to Clara Haas, Feb. 5, 1946, Haas Papers.

16. AH to Richard Morford, Aug. 9, 1950, AHP, no. 13.

17. Interview with Harriet L. Hardy, Nov. 11, 1974; Manfred Bowditch to Harriet Hardy, Oct. 8, 1958, Hardy Papers; AH to KBC, Jan. 22, 1957; AH to KBC, April 8, 1958.

18. AH to Edward A. Weeks, Nov. 21, 1956, *Atlantic Monthly* Papers; AH to KBC, Nov. 16, Friday [1956]; AH to KBC, June 13, 1958.

19. AH to Clara Haas, March 2, 1954, Haas Papers; AH to KBC, Feb. 28, 1957, and undated [ca. March 9, 1959].

20. Harriet Hardy has a file on arrangements for AH's ninetieth birthday tribute, Hardy Papers. Comments by MH in AH interview with Jean Alonzo Curran, Nov. 29, 1963, CLM. Information also from James L. Whittenberger to author, Aug. 4, 1982.

21. AH to KBC, June 11, 1959.

22. FF to AH, July 8, 1959, HFP, no. 671.

23. FF to AH, July 18, 1959, HFP, no. 671.

24. FF to AH, Aug. 31, 1959, HFP, no. 671, and AH to FF, Sept. 24, 1959, FFP.

25. AH to Agnes, March 8 [1894]; "Dr. Hamilton—Industrial Medicine Pioneer."

26. AH to KBC, Oct. 4, 1957; AH to Madeleine P. Grant, Dec. 4, 1959, Grant Papers; quotation from EH in Doris Fielding Reid, *Edith Hamilton: An Intimate Portrait* (New York, 1967), p. 28. On EH's late career, see Helen H. Bacon, "Edith Hamilton," *Notable American Women: The Modern Period,* ed. Barbara Sicherman and Carol Hurd Green (Cambridge, Mass., 1980).

27. AH to KBC, April 5, 1959; AH to Haas sisters, June 14, 1961, Haas Papers.

28. FF to AH, April 23, 1964, HFP, no. 672.

29. AH to Thomas Dodd, 1961, draft; HFP, no. 676; Abraham Ribicoff to AH, April 7, 1965; AH to Nicholas Katzenbach, July 15, 1965, FBI file.

30. "Words Lost, Strayed, or Stolen," *Atlantic Monthly,* 194 (Sept. 1954); "English is a Queer Language," ibid., 203 (June 1959); "A Woman of Ninety Looks at Her World," ibid., 208 (Sept. 1961); "Edith and Alice Hamilton: Students in Germany," ibid., 215 (March 1965).

31. Interview with Jarmila and Jane Petrzelka, May 20, 1977.

32. *New York Times,* Feb. 28, 1969, pp. 35, 38.

SOURCES

Alice Hamilton's autobiography, *Exploring the Dangerous Trades* (EDT; Boston, 1943) admirably charts the main outlines of her career and public life but is reticent about personal matters. The starting point for serious research on her private life and on the Hamilton family is the voluminous Hamilton Family Papers (HFP), especially the letters of Alice Hamilton (AH), her sisters, Agnes Hamilton, and Phoebe Taber Hamilton and the diaries of Agnes Hamilton, 1883–1897. AH's pocket diaries, available for most years from 1916 to 1940, provide a record of what she did and whom she saw, but include only an occasional personal comment. The Alice Hamilton Papers (AHP) and the Alice Hamilton Collection (AHC) are important collections that concentrate on AH's career in industrial medicine and her public life. They include professional correspondence, drafts of unpublished speeches and writings, and published works. All three collections contain material about her political and reform activities. A taped interview with Jean Alonzo Curran, Nov. 29, 1963, Rare Book Room, CLM, presents AH's recollections late in life. AH's FBI file, which begins in January 1950, was obtained under the Freedom of Information Act.

The only published biography is Madeleine P. Grant, *Alice Hamilton: Pioneer Doctor in Industrial Medicine* (New York, 1967), a work for younger readers, prepared with the cooperation of Alice and Margaret Hamilton. Barbara Sicherman, "Alice Hamilton," in *Notable American Women: The Modern Period,* ed. Barbara Sicherman and Carol Hurd Green (Cambridge, Mass., 1980) is a short overview. Of other published sources, the most discerning are Elizabeth Glendower Evans, "People I Have Known: Alice Hamilton, M.D., Pioneer in a New Kind of Human Service," *The Progressive,* Nov. 29 and Dec. 20, 1930, and Elizabeth Shepley Sergeant, "Alice Hamilton, M.D.: Crusader for Health in Industry," *Harper's,* 152 (May 1926). See also the special issue of the *Journal of Occupational Medicine,* 14 (Feb. 1972), edited by Harriet L. Hardy.

441

There are two dissertations: Wilma Ruth Slaight, "Alice Hamilton: First Lady of Industrial Medicine," Case Western Reserve University, 1974, which has an excellent bibliography of AH's publications, and Angela Nugent Young, "Interpreting the Dangerous Trades; Workers' Health in America and the Career of Alice Hamilton, 1910–1935," Brown University, 1982, the best account to date of AH's career in industrial medicine.

Manuscript Collections

Edith and Grace Abbott Papers, Joseph Regenstein Library, University of Chicago

Jane Addams Memorial Collection, Library, University of Illinois at Chicago

Jane Addams Papers, Swarthmore College Peace Collection

American Association for Labor Legislation Papers, Catherwood Library, Cornell University

American Fund for Public Service Records, Rare Books and Manuscripts Division, The New York Public Library, Astor, Lenox and Tilden Foundations

Atlantic Monthly Papers, Atlantic Monthly Press, Boston

Emily Greene Balch Papers, Swarthmore College Peace Collection

Mary Ritter Beard Papers, Schlesinger Library, Radcliffe College

Bureau of Labor Standards Records [formerly Division of Labor Standards], National Archives and Records Service, Suitland, Maryland

Bureau of Labor Statistics Records, National Archives and Records Service, Washington, D.C.

Charles Culp Burlingham Papers, Harvard Law School Library

Children's Bureau Records, National Archives and Records Service, Washington, D.C.

Katherine Bowditch Codman Papers, Schlesinger Library, Radcliffe College

Connecticut College Archives, Connecticut College, New London

Consumers' League of Massachusetts Papers, Schlesinger Library, Radcliffe College

Consumers' League of New Jersey Papers, Rutgers University

Department of Labor Records, Office of the Secretary, General Subject Files, 1933–1940, National Archives and Records Service, Washington, D.C.

Division of Labor Standards Records [see Bureau of Labor Standards]

Elizabeth Glendower Evans Papers, Schlesinger Library, Radcliffe College

Felix Frankfurter Papers, Harvard Law School Library

Felix Frankfurter Papers, Manuscript Division, Library of Congress

Madeleine P. Grant Papers, Schlesinger Library, Radcliffe College

Clara and Rose Haas Papers, Schlesinger Library, Radcliffe College

Alice Hamilton Collection, Connecticut College Library, New London

Alice Hamilton Papers, Schlesinger Library, Radcliffe College
Allen Hamilton Papers, Indiana Collection, Indiana State Library, Indianapolis
Edith Hamilton Papers, Schlesinger Library, Radcliffe College—on deposit
Hamilton Family Papers, Schlesinger Library, Radcliffe College
Harriet L. Hardy Papers, in possession of Harriet L. Hardy, Concord, Massachusetts
Harvard School of Public Health Records, Francis A. Countway Library of Medicine, Boston: Dean's Office Correspondence Files; Department of Physiology Correspondence Files; Division of Industrial Hygiene Files
Indiana Collection, Allen County Public Library, Fort Wayne
Florence Kelley Papers, Rare Book and Manuscripts Library, Columbia University
Nicholas Kelley Papers, Rare Books and Manuscripts Division, The New York Public Library, Astor, Lenox and Tilden Foundations
Florence L. C. Kitchelt Papers, Schlesinger Library, Radcliffe College
Julia C. Lathrop Papers, Howard Colman Library, Rockford College, Rockford, Illinois
Walter Lippmann Papers, Yale University Library
A. Lawrence Lowell Papers, Harvard University Archives
Alma Lutz Papers, Schlesinger Library, Radcliffe College
Mackinac Island State Park Commission Archives, Mackinac
Michigan Historical Collections, Bentley Historical Library, University of Michigan
Miss Porter's School Archives, Miss Porter's School, Farmington, Connecticut
National Consumers' League Papers, Manuscript Division, Library of Congress
National Research Council Papers, National Academy of Sciences, Washington, D.C.
National Woman's Party Papers, 1913–1972, microfilm edition (Glen Rock, N.J.: Microfilming Corporation of America, 1977)
New England Hospital Papers, Sophia Smith Collection, Smith College
New England Hospital Records, Francis A. Countway Library of Medicine, Boston
William F. Ogburn Papers, Joseph Regenstein Library, University of Chicago
Frances Perkins Papers, Rare Book and Manuscripts Library, Columbia University
Planned Parenthood Federation of Massachusetts Papers, Sophia Smith Collection, Smith College
Public Health Service Records, National Archives and Records Service, Washington, D.C.
Frederick Cheever Shattuck Papers, Francis A. Countway Library of Medicine, Boston

Survey Associates Papers, Social History Welfare Archives Center, University of Minnesota

M. Carey Thomas Papers, Bryn Mawr College (available in microfilm edition: New Haven, Conn.: Research Publications, 1981)

Lillian D. Wald Papers, Rare Books and Manuscripts Division, The New York Public Library, Astor, Lenox and Tilden Foundations

War Labor Policies Board Records, National Archives and Records Service, Washington, D.C.

Edward A. Weeks Papers, Humanities Research Center, University of Texas, Austin

C.-E. A. Winslow Papers, Yale University Library

Women's Bureau Records, National Archives and Records Service, Washington, D.C.

Letters Included in This Volume

1. Jessie Hamilton, February 12, 1888, HFP
2. Agnes Hamilton, July 6, 1890, HFP
3. Agnes Hamilton, July 13, 1890, HFP
4. Agnes Hamilton, March 6, 1892, HFP
5. Agnes Hamilton, March 20, 189[2], HFP
6. Agnes Hamilton, October 2, 1892, HFP
7. Agnes Hamilton, January 22, 1893, HFP
8. Agnes Hamilton, February 19, 189[3], HFP
9. Agnes Hamilton, March 5, 1893, HFP
10. Agnes Hamilton, April 8, 1893, HFP
11. Agnes Hamilton, [July 16, 1893], HFP
12. Agnes Hamilton, July 23, 1893, HFP
13. Agnes Hamilton, September 27 [18]93, HFP
14. Agnes Hamilton, October 15, 1893, HFP
15. Agnes Hamilton, October 26, 1893, HFP
16. Agnes Hamilton, December 5, 1893, HFP
17. Agnes Hamilton, [ca. December 28, 1893], HFP
18. Agnes Hamilton, January 22, 1894, HFP
19. Agnes Hamilton, February 9, 1894, HFP
20. Agnes Hamilton, January 14 [1896], HFP
21. Jessie Hamilton, September 5 [1896], HFP
22. Agnes Hamilton, September 12 [1896], HFP
23. Agnes Hamilton, December 6 [1896], HFP
24. Agnes Hamilton, June 13 [1897], HFP
25. Agnes Hamilton, October 13, 1897, HFP
26. Agnes Hamilton, April 3 [1898], HFP

444

27. Agnes Hamilton, July 3 [1898], HFP
28. Agnes Hamilton, November 26 [18]98, HFP
29. Florence Kelley, May 31 [18]99, Nicholas Kelley Papers
30. Agnes Hamilton, June 23 [18]99, HFP
31. Agnes Hamilton, August 8, 1900, HFP
32. Agnes Hamilton, [mid-June? 1902], HFP
33. Agnes Hamilton, January 28, 1904, HFP
34. Agnes Hamilton, December 1, 1906, HFP
35. Irene Osgood, January 25[?], 1909, AALL Papers
36. Mr. Foster, May 22, 1911, AHC
37. Julia C. Lathrop, June 21, 1911, Julia C. Lathrop Papers
38. Jane Addams, August 14, 1911, JA Papers
39. Charles H. Verrill, February 12, 1913, AHP
40. Agnes Hamilton, March 1, 1914, HFP
41. John B. Andrews, December 4, 1914, AALL Papers
42. Theodore Ahrens, December 4, 1914, AHC
43. Agnes Hamilton, April 5 [1915], HFP
44. Mary Rozet Smith, April 22 [1915], JA Papers
45. Mary Rozet Smith, May 5 [1915], JA Papers
46. Louise deKoven Bowen, May 16 [1915], JA Papers
47. Jane Addams, July 20 [1915], JA Papers
48. Margaret Hamilton, March 19 [1917], HFP
49. Jane Addams, June 13, 1917, JA Papers
50. Edith Hamilton, [May 25, 1918], HFP
51. Margaret Hamilton, June 22, 1918, AHP
52. Edith Hamilton, January 21 [1919], HFP
53. Agnes Hamilton, January 26 [1919], HFP
54. Edith Hamilton, [late January 1919], AHP
55. Mary Rozet Smith, [April 13, 1919], JA Papers
56. Norah Hamilton, May 14 [1919], HFP
57. Jessie Hamilton, May 15 [1919], HFP
58. Mary Rozet Smith, May 19 [1919], JA Papers
59. Paul U. Kellogg, August 27, 1919, Survey Associates Papers
60. Agnes Hamilton, October 13 [1919], HFP
61. Frederick C. Shattuck, December 9, 1919, Frederick C. Shattuck
        Papers
62. Katherine Bowditch Codman, [February 27, 1920], KBC Papers
63. Katherine Bowditch Codman, May 4 [1920], KBC Papers
64. Margaret Hamilton, [May 17, 1921], HFP
65. Edith Houghton Hooker, January 16, 1922, AHP
66. Margaret Hamilton, February 9 [1922], HFP
67. Agnes Hamilton, April 28, 1922, HFP
68. Clara Landsberg, January 24, 1923, HFP

69. Lewis Gannett, February 26, 1923, American Fund for Public Service Papers
70. Margaret Hamilton, March 30, 1923, HFP
71. Julia C. Lathrop, October 3, 1924, Julia C. Lathrop Papers
72. Alice Hamilton's family, November 10, 1924, AHP (excerpted)
73. Jane Addams, January 24, 1925, JA Papers
74. Clara Landsberg, [February 8, 1925], HFP
75. Margaret Hamilton, February 26 [1925], HFP
76. Phoebe Taber Hamilton, [April 17, 1925], HFP
77. Jessie Hamilton, [September 1, 1925], HFP
78. Robert Morss Lovett, November 5, 1925, American Fund for Public Service Papers
79. Margaret Hamilton, October 5 [1926], HFP (excerpted)
80. D. H. Kelly, June 2, 1927, AHC
81. D. H. Kelly, June 16, 1927, AHC
82. Edith Hamilton, August 25, 1927, Edith Hamilton Papers
83. Henry A. Christian, August 30, 1927, AHP
84. Florence Kelley, ca. January 9, 1929, NCL Papers
85. Cornelia James Cannon, January 21, 1932, Planned Parenthood Federation of Massachusetts Papers
86. Agnes Hamilton, [February 5, 1932], HFP
87. Nicholas Kelley, February 27, 1932, Nicholas Kelley Papers
88. Agnes Hamilton, April 27 [1932], HFP
89. Leo Wolman, May 31, 1932, AHP
90. Benjamin V. Cohen, November 5, 1932, AHP
91. Benjamin V. Cohen, January 6, 1933, AHP
92. Edith Hamilton, January 28 [1933], HFP
93. [Agnes Hamilton], [February 1933], HFP
94. Agnes Hamilton, April 4, 1933, HFP
95. Felix Frankfurter, April 24, 1933, FFP
96. Agnes Hamilton, April 28, 1933, HFP
97. Edith Hamilton, May 16, 1933, HFP
98. Jane Addams, July 1, 1933, JA Papers
99. Edith Hamilton, February 23 [1934], Edith Hamilton Papers
100. Agnes Hamilton, February 28 [1934], HFP
101. Margaret Hamilton, January 10, 1935, HFP
102. Gerald J. McMahon, January 11, 1935, AHP
103. Elizabeth Glendower Evans, April 6, 1935, EGE Papers
104. Grace Abbott, May 20, 1935, Edith and Grace Abbott Papers, box 55
105. John B. Andrews, February 17, 1938, AALL Papers
106. Edith Hamilton, October 23 [1938], HFP
107. Felix Frankfurter, January 29, 1939, FFP
108. Felix Frankfurter, [June] 1940, AHP

109. Agnes Hamilton, [late January? 1942], HFP
110. Emily Greene Balch, May 10, 1943, Emily Greene Balch Papers
111. Jessie Hamilton, [1945], HFP
112. Rose Haas, February 28, 1949, Clara and Rose Haas Papers
113. Felix Frankfurter, March 2, 1951, FFP
114. Florence L. C. Kitchelt, May 13, 1952, AHP
115. Charles Culp Burlingham, January 14, 1953, CCB Papers
116. Charles Culp Burlingham, January 23, 195[3], CCB Papers
117. Herbert Brownell, May 16, 1953, FFP
118. Felix Frankfurter, May 16, 1953, FFP
119. Charles Culp Burlingham, March 1, 1955, CCB Papers
120. Katherine Bowditch Codman, January 13, 1959, HFP
121. Felix Frankfurter, July 3, 1959, FFP
122. Felix Frankfurter, July 15, 1959, FFP
123. Felix Frankfurter, August 25, 1959, FFP
124. Katherine Bowditch Codman, January 23 [19]61, HFP
125. Rosemary Park, December 21, 1961, Connecticut College Archives
126. Francesca Molinaro, December 28, 1961, JA Memorial Collection
127. Clara Haas, October 10, 1962, Clara and Rose Haas Papers
128. Arthur Hamilton, March 26 [19]64, HFP
129. Felix Frankfurter, April 27, 1964, FFP
130. Clara Haas, November 25, 1964, Clara and Rose Haas Papers
131. Edward A. Weeks, March 29, 1965, *Atlantic Monthly* Papers
132. Margaret Hamilton to Francesca Molinaro, January 10 [1968], JA
      Memorial Collection

ACKNOWLEDGMENTS

A work as long in the making as this one naturally incurs many debts, and these it is a pleasure to acknowledge. The project would not have been possible without the assistance of W. Rush G. Hamilton, who gave me full access to Alice Hamilton's letters and permission to publish them. He and Frances Hamilton were helpful beyond the necessary at every stage of the project. Other relatives and friends whose reminiscences about Alice and other Hamiltons supplemented the written record were Holman Hamilton, Hildegarde Wagenhals Bowen, Dorothy Detzer Denny, Phoebe Hamilton Soule, Russell Williams, Dorian Reid, Elizabeth Reid Pfeiffer, Edward A. Weeks, Alice Canoune Coates, Harriet L. Hardy, Jarmila Petrzelka, Jane Petrzelka, Alberta Pfeiffer, and Madeleine P. Grant.

I have received generous institutional support over the years for my work on Alice Hamilton. This includes fellowships from the Mary I. Bunting Institute of Radcliffe College and from the National Endowment for the Humanities and a Book Grant from the Commonwealth Fund. Research funds were made available at Trinity College through the William R. Kenan, Jr., Professorship of American Institutions and Values. Both the Bunting Institute and the History of Science Department at Harvard University provided me with institutional affiliations and with intellectual companionship that enhanced the enjoyment of working on the book.

Considerations of space preclude my acknowledging each individual who provided information or every institution that responded to my appeals for help. The notes and the list of manuscript collections included in this volume suggest (but do not exhaust) the number of institutions I consulted; they do not include numerous local libraries or the many college and university archives that provided information about alumnae or faculty. Only a few debts can be acknowledged here. Chief of these is to the staff of the Schlesinger

Library, Radcliffe College, my home base over the years, especially Patricia King, Elizabeth Owen Shenton, Katherine Kraft, Eva Moseley, and Jane Knowles. Many others have answered repeated inquiries or helped beyond any call of duty, notably Mary Lynn McCree, Jane Addams Hull-House, University of Illinois at Chicago; Fred J. Reynolds and other staff members of the Allen County Public Library; Mary Jo Pugh, Bentley Historical Library, University of Michigan; Lucy Fisher West, Bryn Mawr College Archives; Nanette Holben Jones, Bryn Mawr School; W. James MacDonald, Connecticut College Library; Richard J. Wolfe, Rare Books, Countway Library of Medicine; Kathy L. Solloway, First Presbyterian Church of Fort Wayne; John Daly, Illinois State Archives; David Wigdor, Library of Congress; Phil Porter, Mackinac Island State Park Commission; Elizabeth H. Hube and other staff members of Miss Porter's School; Patrick M. Quinn, Northwestern University Library; Daniel Meyer and Albert Tannler, Regenstein Library, University of Chicago; David Klaassen, Social Welfare History Archives, University of Minnesota; Susan L. Boone and Virginia Christenson, Sophia Smith Collection, Smith College; Bernice Nichols, J. Richard Kyle, and Jean R. Soderlund, Swarthmore College Peace Collection; Wilma R. Slaight, Wellesley College Archives; and staff members of Christ Episcopal Church, Tarrytown, N.Y., the Harvard University Archives, and the National Archives and Records Service. I have benefited greatly from access to the libraries of Harvard University and from the assistance of numerous staff members, of whom Veronica Cunningham, Barbara Dames, and Ruth Hoppe deserve special mention.

Those who have generously shared information, research leads, or unpublished materials include Leslie B. Arey, Helen H. Bacon, Clara M. Beyer, Charles Burlingham, Mina Carson, Ronald W. Clark, Allen F. Davis, Marguerite Dorian, Virginia Drachman, Madeleine P. Grant, Harriet L. Hardy, Dolores Hayden, Amy Johnson, Mary Lynn McCree, Midge Mackenzie, George Martin, Barbara Mooney, Regina Markell Morantz, James Reed, Paul Reznikoff, Margaret W. Rossiter, Claudine SchWeber, Louise Short, Kathryn Kish Sklar, Fred Stenn, Louise L. Stevenson, Doris E. Thibodeau, and Imogene Young. I am particularly grateful to Angela Nugent Young for permitting me to read her dissertation on Alice Hamilton; since I read her manuscript only as mine was going to press, I have not incorporated material from it.

I also wish to thank those who helped me with research and proofreading, including Peter Clark, Linda Eisenmann, Sheila Gillooly, Cynthia McLoughlin, Mary Beth Minick, Linda ole-MoiYoi, Catherine Streitwieser, and Louise Short. Special thanks go to Catherine Lord, who was in on the project at the beginning, sorted and made sense of the Hamilton family papers, and was the only other person who could decipher Alice Hamilton's handwriting; to Marilyn Weissman, principal researcher during the middle stage of the project; and to Kate Wittenstein, who helped tie up the loose ends. Grace M. Clark provided her usual incomparable assistance as typist, spotter of infelicities, collector of odd facts, and proofreader par excellence.

Jeannette Bailey Cheek, Alice Kimball Smith, and Mary Kay Risi read the complete text of an earlier version of the manuscript, and Janet Wilson James read critical sections of it. I am especially grateful to Marlene Fisher and Susan Ware for their discerning readings and helpful suggestions for eliminating letters at a time when I had lost perspective. Thanks also go to my fellow members of the Cambridge women's biography group—Joyce Antler, Janet James, Ann J. Lane, and Susan Ware—with whom I not only discussed many of the issues raised by Alice Hamilton's life but also shared numerous good times.

Finally, I want to thank current and former editors of the Harvard University Press for their assistance. Camille Smith brought her considerable skill as a literary editor to bear on this volume and made the last, painstaking stages of publication as pleasant as they can be, and Susan Wallace helped guide the book through to completion. My greatest debt of all is to William Bennett, who believed in the project from the day I first walked into his office with a sheaf of letters. I benefited greatly from his editorial judgment and continued support through the years that followed.

INDEX

Abbott, Edith, 2, 376
Abbott, Grace, 2, 186, 190, 279, 280, 283, 376; letter to, 353–354
Abel, John J., 36, 39, 42
Abel, Mary Hinman, 39
Acheson, Dean, 383–384
Addams, Jane, 125, 126–128, 142, 147, 162, 246, 265, 353–355, 409; as social reformer, 1, 136, 244, 283; and pacifism, 5, 190–194, 198, 201; and book illustrations by Norah Hamilton, 12, 141; and Hull House, 108–115, 120–122, 130–131, 139–140, 148–149, 174; and AH at Hull House, 115–116, 132–134, 182, 346, 349; and immigration, 151; and prostitution, 164–165, 183; health of, 182, 294–296, 315, 319, 346–347; and International Congress of Women, 184–190, 218–222, 229–235; and Women's International League for Peace and Freedom, 219, 278–280, 352; letters to, 165, 194–195, 198–199, 279–280, 343–345
Addison, 18, 20
Ahrens, Theodore: letter to, 177–179
Altgeld, John Peter, 118
American Association for Labor Legislation (AALL), 154, 155, 175–177
American Civil Liberties Union, 205n, 245, 396

American Committee for Protection of Foreign Born, 392–393, 394, 414
American Fund for Public Service, 266–268, 291, 294
American Medical Association, 154, 328; Journal of, 172, 328
American Public Health Association, 183, 311
Amos, Bonté Sheldon, 104, 106, 108
Anderson, Mary, 314
Andrews, John B., 155–156, 158; letters to, 175–177, 359–360
Angell, Katharine Louise (Kate), 47, 48, 109
Anthony, Susan B., 15
Armstrong, Simeon, 119, 121
Ashleigh, Charles, 203
Atlantic Monthly, 372–373, 411, 412, 414
Aub, Joseph C., 238, 264
Augspurg, Anita, 344

Baer, Gertrude, 344
Balch, Emily Greene, 186, 219–220, 280; letter to, 374–375
Baldwin, Roger N., 203, 266
Ball, Frank H., 116, 125
Bartlett, Jessie, 120, 140
Beaver, Gilbert, 246
Benedict, Enella, 126, 207
Benedict XV (pope), 193
Beneš, Eduard, 361, 362

451

Benzene, 1, 241, 291–292, 311, 360
Berry, Raymond H., 312, 315
Biddle, Francis, 403
Bigelow, Albert, 404
Binford, Jessie, 409
Birth control, 6, 316–317, 349–351
Bishop Frances Lewis (Fanny), 38, 43–44, 47, 49–50, 53, 58
Blackwell, Elizabeth, 34
Bleuler, Eugen, 137
Blount, Anna Ellsworth, 116–117
Boardman, Florence Sheffield, 173
Bolshevism, 204–205, 207–208; U.S. re-action against, 244, 245, 283, 286–287; in Soviet Union, 274–277
Borah, William E., 279
Borosini, Victor von, 231
Bowen, Louise deKoven, 168, 187, 199, 229, 315; and Hull House, 162, 209, 347, 353; letter to, 191–193
Bowman, Ella, 140
Bradford, Esther Kelly, see Kelly, Esther
Bradford, Robert, 150
Brandeis, Louis Dembitz, 173
Breckinridge, Sophonisba, 2, 186, 190, 192
Breshkowsky, Catherine, 204, 206
Brockway, Wilfreda (later Deknatel), 117, 120, 125, 128, 129
Bromley, Dorothy Dunbar: Birth Con-trol: Its Use and Misuse, 349
Brooks, Phillips, 70
Brown, Edward Osgood, 27–28
Brown, Helen Gertrude Eagle, 27–28
Brownell, Herbert, 390, 394; letter to, 392–393
Browning, Robert, 8, 400
Bruce, Andrew Alexander, 121, 128
Bryan, William Jennings, 118
Bryn Mawr College: Graduate Depart-ment of Social Economy and Social Research, 249, 259, 375
Buchanan, Sir George, 272
Buchman, Frank, 337
Buchner, Hans, 90
Budenz, Louis, 387
Bulkley, Mary, 259

Bureau of Labor (later Department of Labor), 156, 159; AH's investigations for, 4, 166, 169. See also Bureau of Labor Statistics; Department of Labor
Bureau of Labor Statistics, 166, 209; AH's investigations for, 200–201, 240, 281, 283, 297. See also Bureau of Labor; Department of Labor
Bureau of Mines (U.S.), 154, 325, 327
Burlingham, Charles Culp, 384, 400–401, 403; political views, 382, 394, 404, 405, 406; letters to, 386–391, 396–397
Burnham, Grace, 266, 267

Cabot, Ella Lyman, 250
Cabot, Richard C., 286
Camus, Albert, 2
Cannon, Cornelia James: letter to, 316
Cannon, Walter B., 7
Carbon tetrachloride, 325–331
Carlson, Anton, 384
Carter, Orrin N., 198
Castle, William B., 281
Catt, Carrie Chapman, 278–279
Chamberlain, Neville, 361, 362–363
Charities and The Commons, 147, 153
Chicago Commons, 108, 109–110, 146, 149
Chicago Pathological Society, 183
Childers, Mary Osgood, 286
Children's Bureau, 2, 113, 316, 393
Christian, Henry A., 307; letter to, 308–309
Cocaine, 4, 146, 147, 151
Codman, Ernest Amory, 246, 281–282, 285, 361
Codman, Katherine Bowditch, 282, 286, 349, 401; AH living at home of, 245, 246, 285; and Sacco-Vanzetti case, 303, 306; letters to, 249–251, 399–400, 408
Cohen, Benjamin V.: letters to, 326–331
Collis, Edgar L., 294
Collson, Mary, 139
Commons, John R., 154
Communism, see Bolshevism
Conant, James Bryant, 348–349, 390

Conference on the Cause and Cure of
War, 278–280
Connecticut College, 258–259, 375,
408–409
Coolidge, Calvin, 261
Copper, 172, 210–213, 215–217
Cornish, Edward J., 169
Courtney, Lady Kate, 234
Courtney, Kathleen D., 190
Cox, Kenyon, 117
Croly, Herbert, 199
Croly, Louise Emory, 173, 199
Crouch, Paul, 387
Crowder, Thomas Reid, 285
Crowdy, Dame Rachel, 272
Cumming, Hugh S., 271, 272, 312
Cushing, Harvey, 7

Daladier, Edouard, 361, 362
Dante, 14, 23, 114
Davies, Anna Freeman, 127
Davis, Richard Harding, 195
Day, Dorothy, 404
Debs, Eugene, 202, 367
De Hart, Florence, 65, 67–69, 70, 74
Deknatel, Frederick H., 116, 125, 128,
129, 130, 131, 139–140
Deland, Margaret, 304
Deneen, Charles S., 4, 156
Dennis v. United States, 390
Department of Labor (U.S.), 166, 201,
240, 266, 375, 377
Despard, Charlotte, 232
Detzer, Dorothy (later Denny), 205, 243
Dewey, Alice Chipman, 125
Dewey, Ethel, 353–354
Dewey, John, 125
Dick, George and Gladys, 145
Division of Labor Standards (U.S.),
357–358, 360
Dock, George, 36, 40, 59, 88; AH as-
sistant to, 42, 44, 47, 50, 52
Dodd, Thomas J., 413–414
Donaldson, Henry Herbert, 134, 137
Donnelly, Lucy, 196–197, 258, 269
Doree, E. F., 261
Douglas, Paul, 382
Douglas, William O., 387, 389

Dow, Mary Elizabeth Dunning, 23–24,
53
Draper, Helen Fidelia Hoffman, 220
Drinker, Cecil K., 260n, 262–265, 270,
294; and radium study, 281, 283
Drinker, Katherine Rotan, 258, 259, 281
Drinker, Philip, 263, 265
Dudley, Helena Stuart, 128
Duryee, Susan Rankin (later Fahmy), 96

Edinger, Anna, 90, 233, 343
Edinger, Ludwig, 90, 233, 343
Edinger, Tilly, 343–344
Edsall, David L., 209–210, 217, 349
Ehrlich, Paul, 90
Einstein, Albert, 389
Eisenhower, Dwight D., 400
Eisner, Kurt, 225
Eliot, George, 20, 112
Ellsworth, William W., 283
Equal Rights Amendment, 6, 253–256,
384–385
Erskine, Lillian, 176
Evans, Elizabeth Glendower, 7, 243,
348; and Sacco-Vanzetti case, 303,
306–307; letter to, 352
Evans, Sarah, 70, 80
Everitt, Ella B., 58–60, 61–63

Fifield, Emily W., 63
First Presbyterian Church (of Fort
Wayne), 13, 14, 15, 21, 55
Flexner, Abraham, 396
Flinn, Frederick B., 313, 315
Flynn, Elizabeth Gurley, 6, 266, 380;
and conspiracy conviction, 386, 389,
390
Foreign Policy Association, 244, 275
Fort Wayne College of Medicine, 18, 35
Fosdick, Raymond Blaine, 279
Foster, Mr.: letter to, 160–161
Foxe: Book of Martyrs, 403
Fox, Hugh F., 162
Frankfurter, Felix, 7, 173, 207, 209, 210,
282, 373; and Sacco-Vanzetti case,
303–306; and Benjamin V. Cohen,
326, 329; political and legal views,
382–383, 387–389, 390, 400–401; and

Frankfurter, Felix (*cont.*)
   Charles C. Burlingham, 386, 391; letters to, 338, 364–367, 383–384, 394, 401–407, 412–413
Frankfurter, Marion Denman, 282, 364, 374, 387
Fuchs, Klaus, 404
Fuller, Alvan T., 303–309

Gannett, Lewis S., 221, 266; letter to, 267–268
Garland Fund, *see* American Fund for Public Service
Gates, Allen H., 369–371
Gavit, John Palmer, 198
George, Henry, 162
Gernon, Maud (later Yeomans), 120, 121–122, 140
Gilman, Charlotte Perkins Stetson, 127
Gilman, Elisabeth, 262
Goldburg, Robert, 394
Goncourt, Edmond and Jules de, 106
Goodrich, Helen, 149
Griswold, A. Whitney, 404
Guerard, Albert L., 394
Gyles, Rose Marie, 116, 122

Haas, Clara: letters to, 410, 413
Haas, Rose: letter to, 378–379
Haessler, Gertrude, 277
Haggard, Howard Wilcox, 327
Haines, Dora B., 279
Hamilton, Abigail, 336, 363
Hamilton, Agnes (cousin), 16, 91, 245, 363, 411; as correspondent and friend, 7, 13, 34, 97; youth of, 9, 22, 23–24; religious and social service work, 21, 33, 35, 55–56; and the Lighthouse, 21, 141, 147–148, 150; letters to, 26–31, 37–41, 43–45, 46–55, 57–63, 64–85, 92–97, 101–110, 115–117, 120–129, 132–135, 138–140, 142–144, 148–151, 173–174, 184–185, 214–217, 246–247, 261–262, 318, 319–321, 334–337, 339–340, 347–348, 370–372
Hamilton, Alice, books:
   Exploring the Dangerous Trades (1943), 21, 33, 141, 372–375, 412

Industrial Poisons in the United States (1925), 240, 325
Industrial Toxicology (1934), 311–312
Industrial Toxicology (1949), 7, 377
Hamilton, Allen (grandfather), 14–15, 16, 17
Hamilton, Allen (cousin), 337, 339
Hamilton, Andrew Holman (cousin), 337, 339, 376
Hamilton, Andrew Holman (uncle), 13, 15, 16–17, 21
Hamilton, Arthur (Quint; brother), 12–13, 45, 183, 207, 221, 233, 415; letter to, 411
Hamilton, Edith (sister), 11–12, 20, 57–58, 104, 106, 127, 150, 152, 282, 283–284, 374, 411; youth of, 9, 18, 25–31; as model for AH, 9, 34, 35, 51, 136; and Bryn Mawr School, 12, 97, 252, 253, 257–260; intimate of Jessie Hamilton's, 21, 30, 85; health of, 89, 91, 252, 257, 397; relations with sisters, 109, 110, 124, 135, 142–143, 184, 196–197, 253, 257–261, 265, 269, 336–337, 408; later life in Washington, 376, 381, 387; honorary degree from Yale, 400, 401, 403, 404; acclaim in old age, 407–409; letters to, 202–205, 211–214, 217–218, 304–307, 332–334, 346–347, 340–342, 361–363
Hamilton, Eliza, 16
Hamilton, Emerine Jane Holman (grandmother), 15–16, 17, 19
Hamilton, Gertrude Pond (mother), 55, 98, 131, 137, 139, 182, 185, 196, 199; marriage of, 17–21, 85, 91; relations with AH, 19–21, 71, 73, 136, 183, 184
Hamilton, Helen Knight, 337, 339, 371
Hamilton, Jessie (cousin), 21, 126, 245, 363, 411; and close family ties, 21, 30, 37, 39, 85, 94; letters to, 23–24, 97–100, 224–228, 289–290, 376
Hamilton, Katherine (cousin), 21, 110, 115, 245, 317–318, 411

Hamilton, Margaret (Madge, Peggy; sister), 12, 37, 39, 49, 84, 85, 109, 110, 124, 363, 369, 400, 411, 414; youth of, 22, 23, 25, 29; health of, 87, 88, 91, 144, 150–151, 195, 334, 380, 387, 391, 397, 399, 406, 410; and travel in Europe, 131, 294, 337; and Hadlyme, 223, 224, 336–337, 356, 376, 380, 410; and EH's resignation from Bryn Mawr School, 252–253, 257–260, 262; and Sacco-Vanzetti case, 304, 306–307; death of, 399, 416; letters to, 196–197, 207–209, 253, 257–260, 268–271, 285–287, 295–296, 348–349; letter from, 415

Hamilton, Margaret Vance (aunt), 15, 16, 85

Hamilton, Mary Neal, 207, 411

Hamilton, Montgomery (father), 13, 15, 17–19, 91, 182; relations with daughters, 18–19, 57–58, 85, 136; at Mackinac, 26–27, 29

Hamilton, Norah (sister), 12, 117, 224, 269, 373, 376, 411; and Hull House, 12, 182, 270, 295; youth of, 22, 25, 27, 30; health of, 91, 137–141, 253; and travel in Europe, 131, 294, 361; relations with sisters, 141, 196–197, 337; letter to, 221–224

Hamilton, Phoebe Taber (aunt), 17, 19, 21, 85, 245; letter to, 288–289

Hamilton, Phoebe (cousin), 336

Hamilton, Rush (cousin), 336

Hamilton, Taber (cousin), 39, 260n, 336, 363

Hamilton, Taber, Jr. (cousin), 260, 336

Hammett, Dashiell, 394

Hand, Learned, 390

Hannig, Amalie, 189

Hard, Anne and William, 199

Hardy, Harriet, 7, 377, 397, 400

Harrington, Harriet L., 76

Harrison, Marguerite, 268

Harvard University: AH at, 2, 237–238, 247–249, 294, 307–309; appointment to Div. of Industrial Hygiene, 4–5, 209–210, 211, 217–218; survey of General Electric Company, 262–

266; study of U.S. Radium Corporation, 281–283; AH's resignation requested, 348–349

Hayes, Roland, 282

Hayhurst, Emery Roe, 267

Hektoen, Ludvig, 144–145, 156

Henderson, Charles R., 155, 156

Henderson, Yandell, 327

Herrick, James B., 295–296

Heymann, Lida Gustava, 189, 344

Hill, Mary Dayton (later Swope), 121, 124–125, 129, 131

Hill, William, 121

Hilles, Edith (later Dewees), 285, 289, 348

Hirst, Helena, 396

Hitler, 6, 365, 366, 368; AH's impressions of in 1933, 338, 342, 343, 345; AH's impressions of in 1938, 360–363

Holman, Jesse Lynch, 15

Holman, William Steele, 15

Holmes, Oliver Wendell, Sr., 81

Holmes, Oliver Wendell, Jr., 386, 389

Hood, Mary Gould, 59, 60, 63

Hooker, Edith Houghton: letter to, 254–257

Hooker, George Ellsworth, 116, 128, 131

Hoover, Herbert, 321, 332–333; and American Relief Administration, 226, 228, 231–232, 235

House, E. M., 203

Howe, Gertrude (later Britton), 122, 140

Howe, Mark A. De Wolfe, 389

Howell, Anne Janet, 41–42

Howell, William H., 36, 41n, 89

Huestis, Alexander, 18

Hugenberg, Alfred, 343

Hull, Hannah Clothier, 279–280

Hull House, 201, 213, 247, 261, 283; described, 1, 108–115; AH as resident, 3–4, 5, 115–136, 139–141, 144–152, 182, 244; AH's returning to, 346–347, 349; after Jane Addams, 353–355, 374–375, 409

Ickes, Harold, 352

Illinois Commission on Occupational Diseases, 4, 156–158

Illinois Survey, 4, 156–158, 168, 183
Industrial Workers of the World (IWW):
    and World War I, 202–203, 208–209,
    261, 390; and copper mines, 210,
    217
Ingram, Maria P. de Boiij, 67
International Congress of Women:
    (1915), 184, 188–190, 193–194; (1919),
    218–219, 221–232
International Congress on Occupational
    Accidents and Diseases: (1910), 159;
    (1925), 239, 289; (1938), 360

Jacobs, Aletta H., 192, 233
Jaeckel, Albert F., 325–331
Jay, Louisa, 396
Jennings, Frederick Charles, 103, 207,
    337
Jennings, Louise Hélène Loizeaux, 103,
    207, 269, 337
Johns Hopkins Medical School, 12, 104–
    105
Johnson, Amanda, 116, 119, 122, 128
Johnstone, Rutherford T., 377
Jordan, Edwin Oakes, 259
Journal of Industrial Hygiene, 238, 270,
    281, 314; report on storage battery in-
    dustry, 297–303

Karolyi, Count Michael, 225
Katzenbach, Nicholas, 414
Kelley, Florence, 2, 113, 220, 313, 317–
    319; and AH at Hull House, 115–116,
    133–134; at Hull House, 120, 125,
    127, 128, 140, 182; and child labor,
    280, 283; letters to, 130–131, 314–315
Kelley, Nicholas, 130–131; letter to,
    318–319
Kellogg, Paul U., 233, 304, 307; letter
    to, 234–236
Kellor, Frances, 151
Kelly, Amy, 269–270
Kelly, D. H.: letters to, 297–300, 301–
    303
Kelly, Esther, 141, 142, 148–149, 150
Kelly, Howard Atwood, 164
Kennedy, John F., 407, 408, 414
Keyserling, Mary, 387–388

Kingsbury, John, 394, 396
Kingsbury, Susan M., 335
Kitchelt, Florence L. C., 384; letter to,
    385
Kittredge, Mabel Hyde, 199, 253, 273,
    276–277, 280, 304; visit to Belgium,
    186, 189
Kober, George, 240
Koch, Robert, 90
Kollontai, Aleksandra, 393
Korean War, 381–382

Labor legislation, 153–154, 161–162,
    243, 357; and children, 2, 113, 280,
    282–283, 393; and women, 2, 6, 113,
    253–257, 384–385; American Associa-
    tion for, 154, 175–179. See also Work-
    men's compensation
Ladd, Emily J., 67, 78, 81, 84, 86, 103
Ladd, William H., 67, 81, 82, 84
Lambert, Aloysius A., 121
Landis, Kenesaw Mountain, 261–262, 390
Landsberg, Clara, 163–164, 207, 210,
    221, 294, 370, 415; and Hull House,
    141, 142–143, 174; at Mackinac, 144,
    195; at Hadlyme, 196–197, 199, 349,
    356, 369, 380; in Baltimore, 257, 258,
    262, 280; health of, 280, 380, 397,
    399, 406; trip to Germany, 334, 338,
    340, 345; letters to, 263–265, 281–284
Lathrop, Julia C., 173, 183, 207, 279,
    280, 283, 317; and Children's Bureau,
    2, 113–114, 393; and Hull House,
    113–114, 132, 149, 182, 319; Jane Ad-
    dams's life of, 352–353; letters to,
    162–164, 272–274
Lattimore, Owen, 386
Lead, 1, 153; Illinois survey, 4, 156–158,
    168; AH's investigations for Bureau of
    Labor, 159–161, 163, 166–172;
    "standard bill" of the AALL, 175–
    179; Harvard study, 238–239; and
    storage battery industry, 296–303, 324
League of Nations Health Committee,
    244, 271–274, 279, 289
League of Women Voters, 243
Lee, Roger I., 263–266
Lewey, F. H., 358, 360

Lewis, Louise, 285
Lewis, Lucy Biddle, 232
Libby, Frederick J., 279
Lighthouse, the, 21, 141, 147–148
Lindbergh, Charles A., 368
Linn, James Weber, 353
Linn, John, 117n, 125, 128
Lippmann, Walter, 2, 198, 312
Lodge, Constance Davis, 174
Lombard, Carole, 372
Lombard, Caroline Cook, 48, 53, 80
Lombard, Warren Plimpton, 48
Lomonossoff, George V. and Raissa
  Rosen, 273
Loucheur, Louis, 308
Lovejoy, Esther Pohl, 279
Lovejoy, Owen, 283
Lovett, Ida Mott-Smith, 295–296
Lovett, Robert Morss, 294; letter to,
  291–293
Lowell, A. Lawrence, 5, 210, 365; and
  Sacco-Vanzetti case, 303, 307, 308
Ludington, Katharine, 95, 96, 173, 196,
  285

McCaskey, G. W., 40
McDonald, James G., 273
McDowell, Helen, 127
McKinley, William, 118, 152
McMahon, Gerald J.: letter to, 350–
  351
Macmillan, Chrystal, 190, 232
Malin, Patrick Murphy, 396
Martin, James Nelson, 51–52
Masaryk, Thomas G., 308
Mather, Kirtley, 384, 394
Mathew, Father, 121
Matthews, J. B., 392, 394
Maupassant, Guy de, 30, 106
Mead, George Herbert, 205, 269
Mead, Helen Castle, 269
Mead, Lucia Ames, 220
Meeker, Royal, 174
Meigs, Grace Lynde, 285
Memorial Institute for Infectious Dis-
  eases, 144–146
Merriam, Charles E., 332–333
Mezei, Ignatz, 387–388

Minersville School District v. Gobitis
  (1940), 365–367
Mitchell, Wesley C., 332–333
Molinaro, Francesca: letters to, 409, 415
Mooney, Tom, 364, 365
Moore, Ernest Carroll and Dorothea
  Lummis, 122
Moors, Ethel Paine, 384
Morgenthau, Henry, 224
Morley, John, 391
Morrison, Philip, 394
Moxom, Philip Stafford, 81
Murdock, Kenneth, 390
Murray, George, 121, 146
Mussolini, 308, 361, 362
Muste, A. J., 402, 403

Nancrede, Edith de, 134, 149
Nation, 118, 368, 380
National American Woman Suffrage As-
  sociation, 173, 174, 278
National Conference of Charities and
  Correction, 161–162
National Consumers' League, 242, 243,
  282, 387; and Florence Kelley, 2, 130;
  and AH, 241, 259, 281, 312, 375, 384
National Council of Women, 278–279
National Lead Company, 163, 169, 171,
  172
National Research Council, 200, 209
National Woman's Party, 253–254, 256
Neill, Charles P., 159
New England Hospital for Women and
  Children, 57, 63–80, 86–87
New Republic, 213, 214, 228, 380
Nichols, Rose Standish, 220
Nixon, Richard M., 391, 393, 416
Nocht, Professor, 273
Northwestern Hospital for Women and
  Children, 57–64, 71, 72
Novy, Frederick G., 36, 88

Oberlaender Trust, 319, 334, 338
O'Connell, William Cardinal, 280, 283
Ogburn, William F., 322
O'Hare, Francis Patrick, 261, 262
O'Hare, Kate Richards, 261, 262, 367
Oliver, Sir Thomas, 153, 238, 240

Olsen, J. C., 326–328, 331
Osaki, Yukitaka, 320, 369–371
Osgood, Irene, 154; letter to, 155
Osgood, Margaret Cushing Permain, 286–287
Osler, William, 36, 104, 108
Owen, Philip, 336–337
Owen, Tudor, 336–337

Pacifism, 5, 184; and World War I, 184–185, 188–194, 198–199, 201–206; and World War II, 368
Paderewski, Ignace Jan, 342
Palmer, A. Mitchell, 390
Palthe, Mevrow, 192
Papen, Franz von, 343
Park, Marion Edwards, 258
Park, Rosemary: letter to, 408–409
Parmoor, Lord (Charles Alfred Cripps), 234
Parsons, Edward L., 394
Pasternak, Boris, 2
Pasteur Institute, 145
Paul, Alice, 173, 253
Pauling, Linus, 413
Peck, Gustav, 322
Perkins, Frances, 356–357, 358, 374
Peters, Elizabeth Knight, 371–372
Pethick-Lawrence, Emmeline, 187, 189, 190
Pethick-Lawrence, Frederick William, 187
Petrzelka, Jarmila, Bohus, and Jane, 380, 406, 414
Phosphorus, white, 154–156, 242, 283
Pitkin, May (later Wallace), 116
Pond, Annie Boorman, 100
Pond, Harriet Taylor (grandmother), 17, 20, 91, 411
Pond, Loyal Sylvester (grandfather), 17
Pope, Theodate (later Riddle), 24, 164n
Porter, Ruth Furness, 285
Porter, Sarah, 22–23, 24, 34
Post, Alice Thacher, 186, 220
Pound, Roscoe, 286
Powers, Johnny, 119–120, 121–122
Prescott, Abigail Freeburn, 37, 40–41, 43, 44–45

Prescott, Albert B., 36–37, 40–41, 43, 44, 48
President's Research Committee on Social Trends, 321–322, 332
Prostitution, 164–165, 183
Public Health Service (U.S.), 209, 348, 356; and investigations of industrial hazards, 201, 239, 241, 292, 297, 357; and radium, 241, 312, 314

Radium, 241, 281, 283, 312–315
Rankin, Jeannette, 218, 279
Reid, Dorian, 336
Reid, Doris, 214, 336, 361–362; and Edith Hamilton, 252, 257, 258, 260, 283–284, 411
Reid, Edith Gittings, 213, 265
Ribicoff, Abraham, 414
Rich, Adena Miller, 355
Richardville, John B., 14
Riddle, Joseph B., 149
Ripley, Ida Davis, 335, 344
Ripley, William Z., 344
Roche, Josephine, 356
Roosevelt, Eleanor, 3, 352
Roosevelt, Franklin D., 333, 345, 356, 361, 364, 400
Roosevelt, Theodore, 184
Rosenberg, Ethel and Julius, 382, 386, 396, 404
Rotch, Helen Ludington, 282, 285
Rotch, Josephine, 404
Ruhl, Arthur, 268
Rush Medical College, 111, 142, 144–145
Russell, Alys Pearsall Smith, 104–105, 106, 108
Russell, Bertrand, 104, 106, 108, 373, 412

Sacco-Vanzetti case, 6, 244, 303–310, 364–365, 382
Sanger, Margaret, 316
Savage, Minot Judson, 82
Savage, Philip Henry, 81–82
Schwimmer, Rosika, 189, 190, 192
Scudder, Vida, 94
Sears, Amelia, 163

Sever, Frank, 81
Shapley, Harlow, 384
Shattuck, Frederick C., 247; letter to, 248–249
Shaughnessy v. *U.S.* ex rel *Mezei*, 387–389
Shearman, Margaret Hilles, 106, 348
Sheepshanks, Mary, 194
Shurtleff (Shurcliff), Margaret Homer, 306
Siegmund-Schultze, Friedrich, 344
Silverman, Harriet, 266, 267
Simkhovitch, Mary Kingsbury, 375
Slobodinsky, Rachelle, *see* Yarros, Rachel Slobodinsky
Smith, Alfred E., 244
Smith, Eleanor, 131, 221, 229
Smith, Mary Almira, 75
Smith, Mary Rozet, 207, 251, 265, 346–347; and Jane Addams, 295, 315, 409; letters to, 185–190, 228–232
Spencer, Herbert, 74
Starr, Ellen Gates, 111, 126, 130–131, 140, 148–149, 174
Stelzle, Charles F., 162
Stevens, Alzina Parsons, 121, 122, 140
Stevens, Donald, 284, 286
Stewart, Ethelbert, 283, 313, 314
Stone, Harlan Fiske, 365, 367
Stone, Lucy, 15
Stoner, Cora Lane, 39, 50
Straus, Mary Howe, 284, 285
Stresemann, Gustav, 363
Strong, Anna Louise, 277
Strother, French, 332–333
Suffrage Movement, 15, 105, 173–174, 184, 219, 278–279
Survey, 179, 183, 185, 193, 195, 233
Sutton, H. T., 169–170
Swope, Gerard, 116, 124–125, 129, 262–266

Taft, William Howard, 154, 204
Talbot, Ellen B., 394
Tassin, Algernon, 82
Taylor, Alonzo E., 234
Taylor, Graham, 108, 110
Taylor, Graham Romeyn, 285

Terrell, Mary Church, 220
Thayer, Webster, 308–309, 394
Thomas, Elizabeth H., 122, 125, 126, 127
Thomas, M. Carey, 203, 252, 257, 258, 260
Thomas, Norman, 266, 356
Thompson, Lewis R., 315
Tisza, Count Stephen, 193
Truman, Harry, 382, 387, 400
Tuberculosis, 44, 142–144, 147, 254, 357, 359
Tufts Medical School, 375
Twose, George, 130, 140, 149
Typhoid, 4, 145–146

Underhill, Frank, 270
University of Chicago, 137, 144, 197
University of Michigan, 142; AH as medical student, 34, 35–36, 59; AH as graduate student, 88–89; honorary degrees from, 183, 398
Unmarried mothers, 46–47, 49–52
Uphaus, Willard, 400, 401–402, 403
Urey, Harold C., 404

Valerio, Alessandro Mastro-, 121, 122, 128, 140
Valerio, Amelie Robinson, 116, 122, 128, 140
Van der Vaart, Harriet, 151
Van Hoosen, Bertha, 73
Vaughan, Victor C., 36
Verrill, Charles H., 174; letter to, 170–172
Vietnam war, 6, 414
Vittum, Harriet, 163

Wadsworth, Eliot, 220
Wagenhals, Ellen Hamilton (aunt), 15, 16, 49, 84, 85
Wagenhals, Hildegarde, 32
Wagenhals, Samuel, 16
Wagner, Richard, 53–54
Wald, Lillian D., 127–128, 218
Walker, Marian, *see* Williams, Marian Walker
Wambaugh, Sarah, 273

459

Ware, Harold, 284, 286
Warren, Gretchen Osgood, 286–287
Warthin, Aldred Scott, 44, 48
Watson, Lulu, 116, 122
Weeks, Edward A., 372, 411, 412; letter to, 414
Weigert, Carl, 90
Welch, William H., 104
Wells, H. Gideon, 202
Wetherill, Webster King, 159–160
Wheeler, Everett P., 195
Whinery, Joseph Burgess, 44, 50
White, Grace, 220
Wiley, Katherine G. T., 312–313, 314
Willard, Frances, 15
Williams, Allen Hamilton (cousin), 31, 34, 91, 94, 127; youth of, 13, 22, 28, 30, 47, 52, 55; and women's role, 13, 34, 103, 105–106, 127; and mother, 16, 84, 102–103, 106; marriage of, 26, 32, 100–103, 105–106, 149, 163, 214; in Boston, 67–68, 70, 74, 78, 80–82; in New Mexico, 211, 214; and pacifism, 214, 245
Williams, Creighton, 127, 148
Williams, Henry, 16
Williams, Jesse Lynch, 27, 199
Williams, Marian Walker, 139, 144, 149, 163, 211, 247; marriage of, 100–103, 105–106, 127, 214–215
Williams, Mary Hamilton (aunt), 15, 16, 18, 22, 85; and son Allen, 84, 102–103, 106
Williams, Russell, 215, 336
Williams, Tyrrell, 261
Wilmarth, Mary, 205
Wilson, Woodrow, 194, 219, 222, 230
Winslow, C.-E. A., 1, 294, 311
Winslow, Gertrude, 374
Winslow, Richard, 375
Wolman, Leo: letter to, 322–325
Woman's Medical College of Pennsylvania, 40, 42, 59, 64, 65, 375

Woman's Medical School of Northwestern University: AH as professor of pathology, 4, 108, 111, 117, 118, 133–134; closing of, 137, 141–142
Women, role of, 14, 26, 86, 135–136, 181; and professional opportunities, 1, 7–9, 33–34; and social change, 2, 243; and Harvard, 4–5, 237–238; and marriage and family, 8–9, 31–32, 91–92, 103, 105–106; in Germany, 89–90. *See also* Equal Rights Amendment; International Congress of Women; Suffrage Movement
Women's International League for Peace and Freedom (WILPF), 219, 243, 278–280, 352
Women's Trade Union League, 243, 256
Wood, Carolena, 232, 233
Woods, Robert A., 162
Workers' Health Bureau, 266–268, 292–293
Workmen's compensation, 167, 240, 243, 300–302, 359, 377; primarily for accidents, 154, 239, 324
World War I: and pacifism, 5, 197–199, 201–206; and postwar Europe, 5, 218, 221–235; and International Congress of Women, 184–194; and munitions manufacture, 200–201
World War II, 6, 368–372
Wyatt, Edith Franklin, 165

Yarros, Marie, 277
Yarros, Rachel Slobodinsky, 98, 187, 208; at New England Hospital, 64, 65–66, 74–75, 77; and AH at Hull House, 115, 120, 133
Yarros, Victor S., 64, 120, 133
Yeomans, Charles, 149

Zetkin, Clara, 394
Zimmer, Verne A., 358
Zimmern, Alfred, 273